Microstructure Engineering of Metals and Alloys

Microstructure Engineering of Metals and Alloys

Editor

Konstantin Borodianskiy

 Basel • Beijing • Wuhan • Barcelona • Belgrade • Novi Sad • Cluj • Manchester

Editor
Konstantin Borodianskiy
Chemical Engineering
Ariel University
Ariel
Israel

Editorial Office
MDPI
St. Alban-Anlage 66
4052 Basel, Switzerland

This is a reprint of articles from the Special Issue published online in the open access journal *Materials* (ISSN 1996-1944) (available at: www.mdpi.com/journal/materials/special_issues/micro_metals_alloys).

For citation purposes, cite each article independently as indicated on the article page online and as indicated below:

Lastname, A.A.; Lastname, B.B. Article Title. *Journal Name* **Year**, *Volume Number*, Page Range.

ISBN 978-3-7258-0204-3 (Hbk)
ISBN 978-3-7258-0203-6 (PDF)
doi.org/10.3390/books978-3-7258-0203-6

© 2024 by the authors. Articles in this book are Open Access and distributed under the Creative Commons Attribution (CC BY) license. The book as a whole is distributed by MDPI under the terms and conditions of the Creative Commons Attribution-NonCommercial-NoDerivs (CC BY-NC-ND) license.

Contents

About the Editor vii

Preface ix

Sungsu Jung, Yongho Park and Youngcheol Lee
A Novel Approach to Investigate the Superheating Grain Refinement Process of Aluminum-Bearing Magnesium Alloys Using Rapid Solidification Process
Reprinted from: *Materials* 2023, 16, 4799, doi:10.3390/ma16134799 1

Janik Marius Lück and Joachim Rösler
New Approach in the Determination of a Suitable Directionally Coarsened Microstructure for the Fabrication of Nanoporous Superalloy Membranes Based on CMSX-4
Reprinted from: *Materials* 2023, 16, 3715, doi:10.3390/ma16103715 18

Vladimir Galkin, Andrey Kurkin, Gennady Gavrilov, Ilya Kulikov and Evgeny Bazhenov
Investigation of the Technological Possibility of Manufacturing Volumetric Shaped Ductile Cast Iron Products in Open Dies
Reprinted from: *Materials* 2022, 16, 274, doi:10.3390/ma16010274 35

Łukasz Bohdal, Agnieszka Kułakowska, Radosław Patyk, Marcin Kułakowski, Monika Szada-Borzyszkowska and Kamil Banaszek
Experimental Research on Sheared Edge Formation in the Shear-Slitting of Grain-Oriented Electrical Steel Workpieces
Reprinted from: *Materials* 2022, 15, 8824, doi:10.3390/ma15248824 46

Po-Yu Kung, Wei-Lun Huang, Chin-Li Kao, Yung-Sheng Lin, Yun-Ching Hung and C. R. Kao
Investigation of Low-Pressure Sn-Passivated Cu-to-Cu Direct Bonding in 3D-Integration
Reprinted from: *Materials* 2022, 15, 7783, doi:10.3390/ma15217783 67

B. Hari Prasad, G. Madhusudhan Reddy, Alok Kumar Das and Konda Gokuldoss Prashanth
Fiber Laser Welded Cobalt Super Alloy L605: Optimization of Weldability Characteristics
Reprinted from: *Materials* 2022, 15, 7708, doi:10.3390/ma15217708 79

Pablo Huazano-Estrada, Martín Herrera-Trejo, Manuel de J. Castro-Román and Jorge Ruiz-Mondragón
Characterization of Inclusion Size Distributions in Steel Wire Rods
Reprinted from: *Materials* 2022, 15, 7681, doi:10.3390/ma15217681 98

Saeed Sadeghpour, Vahid Javaheri, Mahesh Somani, Jukka Kömi and Pentti Karjalainen
Heterogeneous Multiphase Microstructure Formation through Partial Recrystallization of a Warm-Deformed Medium Mn Steel during High-Temperature Partitioning
Reprinted from: *Materials* 2022, 15, 7322, doi:10.3390/ma15207322 114

Bernoulli Andilab, Payam Emadi and Comondore Ravindran
Casting and Characterization of A319 Aluminum Alloy Reinforced with Graphene Using Hybrid Semi-Solid Stirring and Ultrasonic Processing
Reprinted from: *Materials* 2022, 15, 7232, doi:10.3390/ma15207232 133

Ling Lin, Lian Zhou, Yu Xie, Weimin Bai, Faguo Li and Ying Xie et al.
Study on the Effect of Pre-Refinement and Heat Treatment on the Microstructure and Properties of Hypoeutectic Al-Si-Mg Alloy
Reprinted from: *Materials* 2022, 15, 6056, doi:10.3390/ma15176056 150

Irina Volokitina, Ekaterina Siziakova, Roman Fediuk and Alexandr Kolesnikov
Development of a Thermomechanical Treatment Mode for Stainless-Steel Rings
Reprinted from: *Materials* **2022**, *15*, 4930, doi:10.3390/ma15144930 . . . . **163**

Liming Xu, Yinsheng He, Yeonkwan Kang, Jine-sung Jung and Keesam Shin
Precipitation Evolution in the Austenitic Heat-Resistant Steel HR3C upon Creep at 700 °C and 750 °C
Reprinted from: *Materials* **2022**, *15*, 4704, doi:10.3390/ma15134704 . . . . **176**

About the Editor

Konstantin Borodianskiy

Prof. Borodianskiy earned a Ph.D. in Materials Chemistry (2011) from Bar Ilan University, Israel. Then, he spent two years as a postdoctoral fellow at the University of Windsor, ON, Canada. In 2015, he was appointed an Assistant Professor at Ariel University, Israel. Prof. Borodianskiy's research interests combine traditional and advanced methods of synthesis and examination in metallurgy and materials science. His works focus on non-ferrous alloys, coatings synthesized by plasma electrolytic oxidation, bioactive surfaces, and microstructure engineering of materials for renewable energy. He has published over 40 peer-reviewed papers and 2 book chapters. His research work has been recognized at the World Innovation Summit in the US, and he received an Outstanding Teaching Award for four years. Prof. Borodianskiy is a member of many international conference organizing committees, a guest editor of numerous peer-reviewed journals, and an active reviewer of several high-quality scientific journals.

Preface

This reprint is a collection of original research papers that cover recent developments in the field of microstructure engineering of metals and alloys. Microstructure and its engineering are of high scientific and industrial interest since it is among the major factors that affect the properties and performance of metals and alloys. This reprint covers studies on microstructure effects on ferrous and non-ferrous alloys. Research studies in the reprint are presented by authors affiliated with different countries worldwide, including the Republic of Korea, Taiwan, India, China, Kazakhstan, Germany, Russia, Poland, Austria, Estonia, Finland, Canada, and Mexico.

I would like to take this opportunity to thank all authors who have contributed to the reprint.

Konstantin Borodianskiy
Editor

Article

A Novel Approach to Investigate the Superheating Grain Refinement Process of Aluminum-Bearing Magnesium Alloys Using Rapid Solidification Process

Sungsu Jung [1,2], Yongho Park [2,*] and Youngcheol Lee [1,*]

1 Energy Component & Material R&BD Group, Korea Institute of Industrial Technology, Busan 46938, Republic of Korea; jungss@kitech.re.kr
2 Department of Materials Science and Engineering, Pusan National University, Busan 46241, Republic of Korea
* Correspondence: yhpark@pusan.ac.kr (Y.P.); yclee87@kitech.re.kr (Y.L.); Tel.: +82-51-510-3967 (Y.P.); +82-51-309-7410 (Y.L.); Fax: +82-51-514-4457 (Y.P.); +82-51-309-7422 (Y.L.)

Citation: Jung, S.; Park, Y.; Lee, Y. A Novel Approach to Investigate the Superheating Grain Refinement Process of Aluminum-Bearing Magnesium Alloys Using Rapid Solidification Process. *Materials* 2023, 16, 4799. https://doi.org/10.3390/ma16134799

Academic Editor: Konstantin Borodianskiy

Received: 26 May 2023
Revised: 20 June 2023
Accepted: 28 June 2023
Published: 3 July 2023

Copyright: © 2023 by the authors. Licensee MDPI, Basel, Switzerland. This article is an open access article distributed under the terms and conditions of the Creative Commons Attribution (CC BY) license (https://creativecommons.org/licenses/by/4.0/).

Abstract: The superheating process is a unique grain refining method found only in aluminum-containing magnesium alloys. It is a relatively simple method of controlling the temperature of the melt without adding a nucleating agent or refining agent for grain refinement. Although previous studies have been conducted on this process, the precise mechanism underlying this phenomenon has yet to be elucidated. In this study, a new approach was used to investigate the grain refinement mechanism of aluminum-containing magnesium alloys by the melting superheating process. AZ91 alloy, a representative Mg-Al alloy, was used in the study, and a rapid solidification process was designed to enable precise temperature control. Temperature control was successfully conducted in a unique way by measuring the temperature of the ceramic tube during the rapid solidification process. The presence of Al_8Mn_5 and $Al_{10}Mn_3$ particles in non-superheated and superheated AZ91 ribbon samples, respectively, manufactured by the rapid solidification process, was revealed. The role of these Al-Mn particles as nucleants in non-superheated and superheated samples was examined by employing STEM equipment. The crystallographic coherence between Al_8Mn_5 particles and magnesium was very poor, while $Al_{10}Mn_3$ particles showed better coherence than Al_8Mn_5. We speculated that $Al_{10}Mn_3$ particles generated by the superheating process may act as nucleants for α-Mg grains; this was the main cause of the superheating grain refinement of the AZ91 alloy.

Keywords: AZ91 alloys; grain refinement; melt superheating; rapid solidification; Al-Mn particles

1. Introduction

Magnesium alloys are among the lightest metals, featuring low density and high specific strength properties [1]. These characteristics are considered as important factors in various industrial fields, particularly in aerospace, automotive, and electronics industries where weight reduction is crucial for specific applications [2]. The utilization of magnesium alloys for creating lightweight and robust structures can reduce fuel consumption, enhance energy efficiency, and decrease environmental impact [3]. However, the strength and ductility of magnesium alloys may be lower compared to other lightweight materials, such as aluminum alloys, which highlights the need to develop methods to enhance the strength and ductility of magnesium alloys in order to make them more competitive in various applications [3].

Grain refinement in magnesium alloys has proven to be an efficient method for enhancing strength and ductility simultaneously [4]. Research on the grain refinement of magnesium alloys is primarily divided into two categories: aluminum-free and aluminum-bearing magnesium alloys [5]. Zirconium is recognized as the most potent grain refiner for aluminum-free magnesium alloys. Qian et al. asserted that the zirconium-rich core structures in Mg-Zr alloys serve as nucleation sites for α-Mg grains, emphasizing zirconium's remarkable grain refining capabilities [6]. However, zirconium forms stable compounds with Al, Mn, Si, Fe, Sn, Ni, Co, and Sb, so it cannot serve as a nucleating agent in magnesium alloys containing aluminum, which are mostly used in commercial applications [7]. Various techniques have been reported for the grain refinement of aluminum-containing magnesium alloys, but, unlike the addition of zirconium in aluminum-free magnesium alloys, there are no commercially available and reliable grain-refining processes for aluminum-containing magnesium alloys. Additionally, the mechanisms of grain refinement in these alloys have not been fully elucidated [8]. Therefore, further research is required to understand the grain refinement mechanisms and develop effective grain refining processes for aluminum-containing magnesium alloys.

One of the grain refinement mechanisms for aluminum-containing magnesium (Mg-Al) alloys is the superheating process, which enables grain refinement through a relatively simple method of controlling only the temperature of the melt without the addition of nucleating or refining agents [9]. In this process, the molten alloy is heated to a temperature well above its liquidus temperature (150–260 °C or higher) for a brief period, after which it is rapidly cooled down to the pouring temperature and held for a short time before being cast [10]. The effectiveness of superheating as a grain refinement method depends on the alloy's composition, particularly its aluminum content. Alloys with higher aluminum content tend to exhibit better grain refinement when subjected to superheating. Other elements, such as Fe, Mn, and Si, can also influence the grain refinement effect, but excessive amounts of these elements can be counterproductive [11]. The key to successful grain refinement through superheating lies in determining the optimal temperature range and holding time. Once these optimal conditions are achieved, further increasing the holding time or repeating the process does not result in additional grain refinement. Rapid cooling and immediate pouring after reaching the superheating temperature are essential to prevent grain coarsening [12].

Although grain refinement by superheating has been known for a long time, the exact mechanism has not yet been identified. Some hypotheses have been proposed to explain the effect, but they all have their limitations. The oxide nucleation theory [13] suggests that grain refinement occurs through the formation of magnesium oxide and other oxide particles produced in the molten metal during the superheating process. However, grain refinement is observed even in a vacuum atmosphere where oxide formation is difficult. Moreover, it cannot explain why grain refinement occurs with compounds composed of Al, C, Fe, and Mn. Temperature-solubility theory [14] posits that particles too large for good nucleants at normal pouring temperatures are dissolved during superheating and re-precipitated as finer particles that act as nucleation sites. However, the theory does not identify specific particle species. Al_4C_3 particle nucleation theory [15] suggests that grain refinement in Mg-Al alloys occurs due to the nucleation of α-Mg grains on Al_4C_3 particles formed during the superheating process. Superheating causes the diffusion of carbon atoms from the steel crucible into the Mg molten metal, leading to the formation of Al_4C_3 particles, which are believed to be nucleation sites for α-Mg grains. Motegi's study [16] on commercial AZ91E alloys showed the presence of numerous Al_4C_3 particles after superheating, supporting this hypothesis. However, there is no direct experimental evidence to confirm the formation of Al_4C_3 particles during the superheating process. Al-Mn intermetallic compound nucleation theory [17] proposes α-Mg grain nucleation by Al-Mn intermetallic compounds precipitated during cooling after superheating. It has been proposed that aluminum contributes to reducing the solubility of manganese in Mg melts, leading to the formation of various Al-Mn intermetallic compounds during the superheating process. Byun et al. [18] suggested that Al_8Mn_5 compounds provided heterogeneous nucleation sites for primary α-Mg grains. However, the edge-to-edge matching model by Zhang et al. [19] indicated that Al_8Mn_5 had low nucleating efficiency as nucleation sites for α-Mg grains. Cao et al. [20] reported ε-AlMn as apotent nucleant for α-Mg grains due to its presence in the Al-60Mn master alloy, which showed high grain refining efficiency. Qin et al. [21] supported these findings, showing that single-phase ε-AlMn refined AZ31 grains, while single-phase Al_8Mn_5 did not. In contrast, Qiu et al. [22] suggested that metastable τ-AlMn, which could be generated during the melt superheating process, had better crystallographic matching with the Mg matrix than ε-AlMn and acted as a nucleant for α-Mg grains. However, no direct evidence currently exists to confirm the presence of τ-AlMn and ε-AlMn in Mg-Al alloys, and methods to control their phase and morphology are still unknown. Duplex nucleation theory [9,10] suggests that Al_4C_3 particles are responsible for grain refinement in Mg-Al alloys, while Mn and Fe may interfere with the nucleation potency of Al_4C_3 particles. During the superheating process, Mn and Fe are dissolved, allowing Al_4C_3 particles to act as nucleants for α-Mg grains. However, prolonged holding at the pouring temperature causes Mn and Fe to re-wrap around the Al_4C_3 particles, coarsening the grains. Han et al. [23] supported these findings, showing that Al_4C_3 particle clusters act as nucleants for α-Mg grains rather than individual particles. In Mn-containing AZ91 alloy, the attachment of Al_8Mn_5 particles to Al_4C_3 reduces nucleation efficiency. Nevertheless, Al_4C_3 particles not attached to Al_8Mn_5 within a nucleating cluster can still play an effective role in refining a-Mg grains. Thus, it is deduced that Al_4C_3 particle clusters are beneficial in overcoming the hindering effect of Mn. However, definitive evidence regarding the formation of duplex structures involving Al_8Mn_5 and Al_4C_3 has not yet been established.

Previous studies have proposed various theories to explain the mechanism of grain refinement by superheating in Mg-Al alloys. However, these studies have certain shortcomings, such as a lack of direct experimental evidence to validate the theories and incomplete understanding due to the inability of any one theory to fully explain all observed phenomena. In this study, the rapid solidification process was employed to identify the nucleants generated by the superheating process. This approach was specifically designed to identify and analyze potential nucleants that may exist as solid phases within the molten Mg-Al alloy during the superheating process. By harnessing the rapid solidification process, these solid particles could be effectively captured and scrutinized, allowing for a deeper and clearer understanding of their structure and composition. Therefore, this study presents new perspectives on the mechanism of superheating grain refinement that have not been reported in any previous studies.

2. Materials and Experiments

2.1. Materials and Casting Process

AZ91D alloy, which is most widely used in industrial fields, was used in this study. Approximately 1000 g of the alloys were melted in an alumina crucible using an electric furnace. Samples were prepared with and without the superheating process. For samples without superheating, the AZ91D alloy was melted at 670 °C under a protective gas of 1.0% SF_6 and 99.0% N_2 and then poured into a mild steel mold preheated to 250 °C. For samples subjected to the superheating process, the alloy was melted at 670 °C under a protective gas, then rapidly heated to 770 °C and held at this temperature for 15 min, followed by rapid cooling to 670 °C, and finally poured into a mild steel mold preheated to 250 °C. The chemical composition of the samples was analyzed by an optical emission spectrometer (Spectro MAXx, SPECTRO, Kleve, Germany).

2.2. Rapid Solidification Process

Rapidly solidified ribbon samples were prepared using the melt spinner equipment (MSE 170, Yein Tech, Seoul, Republic of Korea) shown in Figure 1a. Rapidly solidified ribbon samples were prepared by melting 2 g of AZ91D alloy within a ceramic tube and subsequently injecting the molten metal into a rapidly rotating copper wheel at 1500 rpm. As shown in Figure 1b, each sample was melted using a ceramic crucible that has little reactivity with Mg molten metal. Cracking of the ceramic crucibles due to thermal shock was prevented by wrapping them with SUS304 tube. The thermocouple was attached to a ceramic crucible and was set at a position not affected by the induction heating coil. To enhance the contact between the ceramic crucible and the thermocouple, a copper sheet was used to wrap around the thermocouple. The high thermal conductivity of copper enabled the temperatures of the molten metal and crucible to reach an equilibrium state over time, ensuring proper melting. The process temperature was recorded by a data logger (NI cDAQ-9174, National Instruments, Austin, TX, USA) at the frequency of 20 Hz. Conventional casting samples were produced by melting the molten metal at a temperature of 650 °C, followed by a spraying process onto the rotating wheel. In contrast, superheating samples were prepared by first heating the molten metal to a temperature of 800 °C, then rapidly cooling it down to 650 °C before proceeding with the spraying process.

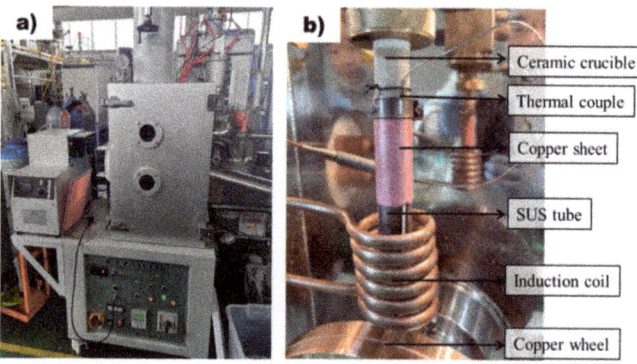

Figure 1. Apparatus for the rapid solidification process: (**a**) Equipment for rapid solidification process. (**b**) Rapid solidification process system with precise temperature control.

2.3. Measurement of Grain Size

Metallographic specimens were polished and then etched using an acetic acid-picral etchant to obtain a clear color contrast that enabled detailed investigation of the microstructure via optical microscopy. Microstructural analysis was performed using a high-resolution optical microscope (Leica MC 170, Leica, Teaneck, NJ, USA) to capture images of the specimen surfaces. The average grain size of the specimens was measured by focusing on the center of the cross-section of each specimen.

2.4. Thermal Analysis

Thermal analysis experiments were carried out using cylindrical graphite crucibles to determine the undercooling during the solidification of the alloys. The graphite crucible was submerged in the melt until its temperature reached the temperature of the molten metal. The crucible, filled with molten metal, was then transferred to a ceramic board. A K-type thermocouple, calibrated using the equilibrium melting temperature of high-purity (99.99%) aluminum, was inserted into the center of the melt to monitor the temperature throughout the solidification process. Cooling curves were recorded using a data logger (NI cDAQ-9174, National Instruments, Austin, TX, USA) at a frequency of 20 Hz.

2.5. Analysis of Microstructure on Rapidly Solidified Sibbon Samples

The microstructure of samples was analyzed using a field emission scanning transmission electron microscope (Talos F200X, Thermo Fisher Scientific, Waltham, MA, USA) after preparing the samples with focused ion beam (Scios2, Thermo Fisher Scientific, Waltham, MA, USA) equipment. The sample was first protected with a thin layer of Pt, then FIB milling was performed to create thin samples. Energy-dispersive x-ray spectroscopy (EDS, Thermo Fisher Scientific, Waltham, MA, USA) was employed to obtain more detailed elemental information, and high-resolution images were acquired to study the microstructure of samples. Crystallographic analysis of the phases observed in each sample was carried out utilizing Gatan software (ver 3.43) and CrysTbox software (ver 1.10).

3. Results

3.1. Microstructure and Chemical Composition

Figure 2 shows the effect of the superheating process on the microstructure and grain size of AZ91D alloy. The average grain size of the non-superheated sample was measured to be 310 μm, while the grain size decreased to 108 μm after the superheating process. Both samples exhibited dendritic microstructures with equiaxed grains, but the superheated sample had grains that were approximately three times smaller than the non-superheated sample. Additionally, the superheated sample showed enhanced uniformity in grain size and shape compared to the non-superheated sample, which had non-uniform grain shapes and sizes.

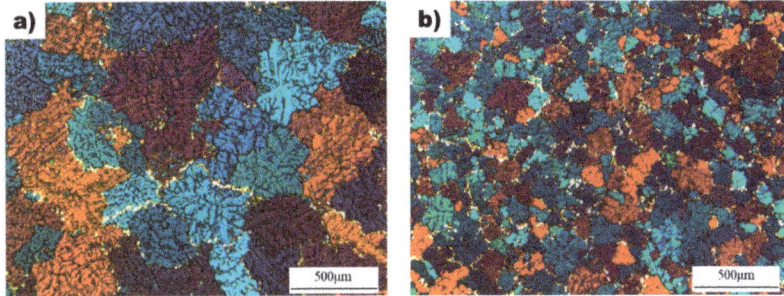

Figure 2. The microstructure of the as-cast samples: (**a**) Non-superheated AZ91D alloy. (**b**) Superheated AZ91D alloy with superheating.

As shown in Table 1, the composition of Al, Zn, and Mn of both alloys is remarkably similar, thus indicating that the possibility of grain refinement effects resulting from constitutional undercooling can be excluded. Furthermore, because both specimens are manufactured under identical casting conditions, a more accurate comparison can be made, eliminating the potential for grain refinement effects due to differences in cooling rates. The consistency in casting conditions and composition ensures that the observed differences in microstructure can be ascribed to the superheating process, rather than variations in the manufacturing process.

Table 1. Chemical compositions on non-superheated and superheated AZ91D alloy.

Alloy	Al	Zn	Mn	Si	Fe	Cu	Ni	Mg
Non-superheated AZ91D	8.93	0.57	0.250	0.015	0.0012	0.0016	0.0012	Bal.
Superheated AZ91D	8.88	0.58	0.251	0.015	0.0012	0.0018	0.0012	Bal.

3.2. Cooling Curve and Undercooling

Figure 3 illustrates the cooling curves of the non-superheated and superheated AZ91 alloys. The non-superheated AZ91 alloy exhibited an undercooling of 2.1 °C, whereas the superheated AZ91 alloy showed negligible undercooling. This suggests that the superheated AZ91 alloy contained effective nucleants that could initiate nucleation with minimal activation energy [24]. Cooling rates were measured at 1.1 °C and 0.9 °C for the non-superheated and superheated samples, respectively, and it was found that the impact of cooling rates on the observed undercooling variations was negligible. Therefore, it can be concluded that the observed differences in microstructure between the two samples can be attributed to the superheating process, rather than variations in cooling rates [25].

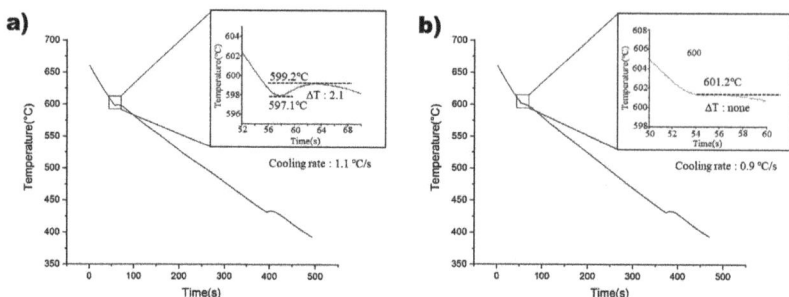

Figure 3. Cooling curves of: (**a**) Non-superheated AZ91 alloy. (**b**) Superheated AZ91 alloy.

3.3. Design of the Precise Temperature Measurement System for the Rapid Solidification Process

In this study, the rapid solidification process was used to detect nucleants within the molten metal that play a significant role in grain refinement. However, measuring the temperature of the molten metal during rapid solidification using melt spinner equipment can be challenging due to the absence of a dedicated temperature measurement system and thermocouple inaccuracies caused by induction coil influence. To overcome this challenge, a method was designed to precisely measure the temperature of the molten metal, and this method was used to manufacture rapidly solidified AZ91 ribbon samples, as shown in Figure 1. By integrating this precise temperature measurement system into conventional melt spinner equipment, it is possible to produce rapidly solidified AZ91 ribbon samples with a precisely controlled superheat process. Figure 4a displays the temperature profile of the non-superheated and superheated AZ91 ribbon samples during the manufacturing process, while Figure 4b illustrates the resulting rapidly solidified AZ91 ribbon samples. The non-superheated AZ91 ribbon sample was heated at a rate of 0.43 °C/s until it reached the target temperature of 650 °C. The molten metal was then immediately sprayed onto the rapidly rotating copper wheel, which led to the formation of rapidly solidified samples. For the superheated AZ91 ribbon sample, a two-stage heating process was employed. First, the sample was heated to 454 °C at a rate of 0.41 °C/s. Then, the heating rate was increased and the sample was rapidly heated to 800 °C at a rate of 1 °C/s. Once the desired superheating temperature was achieved, the molten metal was held at 800 °C for 180 s to ensure uniform temperature distribution. The molten metal was then cooled at a controlled rate of 0.9 °C/s until it reached the target temperature of 650 °C, at which point it was sprayed onto the copper wheel to produce rapidly solidified samples.

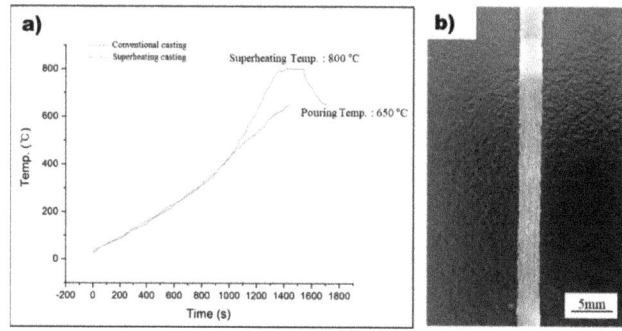

Figure 4. (**a**) The temperature profile during the manufacturing of non-superheated and superheated AZ91 ribbon samples. (**b**) The rapidly solidified AZ91 ribbon samples.

3.4. Cooling Rate Calculation for Rapid Solidification Process

Determining the cooling rate of ribbons fabricated by rapid solidification is essential for the identification and analysis of solid-phase nucleants present in the molten AZ91 magnesium alloy. The cooling rate significantly influences the formation and growth of solid-phase nucleants within molten metals. By establishing and controlling the cooling rate, a systematic investigation of the formation of nucleants and their contribution to grain refinement can be conducted more effectively. Figure 5 shows the cooling rate of AZ91 ribbon samples at different copper wheel speeds. The cooling rate of the rapidly solidified AZ91 magnesium alloy ribbon was determined using an equation derived from a previous study [26]. The thickness of the ribbon samples was measured at different copper wheel speeds, and the data were used to calculate the cooling rate for each. For a specimen with a thickness of 50 µm, the cooling rate reached 2.69×10^6 °C/s, while for a specimen with a thickness of 165 µm, the cooling rate was notably lower at 2.47×10^5 °C/s. It was observed that the cooling rate demonstrated a tendency to increase as the thickness of the specimen decreased. For optimal results in rapid solidification, a higher cooling rate is preferable. However, when the copper wheel speed exceeded 2000 rpm, it became challenging to produce ribbon samples with consistent size and shape, making microstructure analysis difficult due to their small dimensions. Consequently, the rapid solidification process was conducted with a copper wheel speed of 1500 rpm, resulting in a ribbon sample with a thickness of 85 µm.

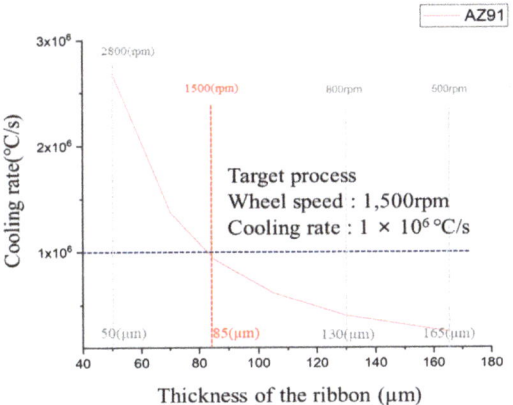

Figure 5. Cooling of AZ91 ribbon samples according Cu wheel to rpm.

3.5. EDS Analysis of Rapidly Solidified AZ91D Ribbon Samples

Figure 6a shows an image of a non-superheated AZ91 alloy ribbon sample obtained using bright-field transmission electron microscopy. The microstructure of the sample contains dispersed particles within the magnesium matrix. One of these particles, indicated by a yellow arrow, was selected for further analysis using energy-dispersive X-ray spectroscopy mapping (Figure 6b). The analysis revealed that the particle did not contain magnesium but had significant amounts of aluminum (Figure 6c) and manganese (Figure 6d). The atomic ratios of aluminum and manganese in the particle were found to be 63.54 and 36.46, respectively, based on EDS point analysis (Figure 6e). This composition ratio is consistent with the Al_8Mn_5 intermetallic compound, which was confirmed through further analysis of other particles in the sample. The results demonstrate that the particles observed in the non-superheated AZ91 alloy ribbon sample were Al_8Mn_5 intermetallic compounds [27]. Upon conducting further analysis of the other particles using the same method, results consistent with the previous observation were obtained.

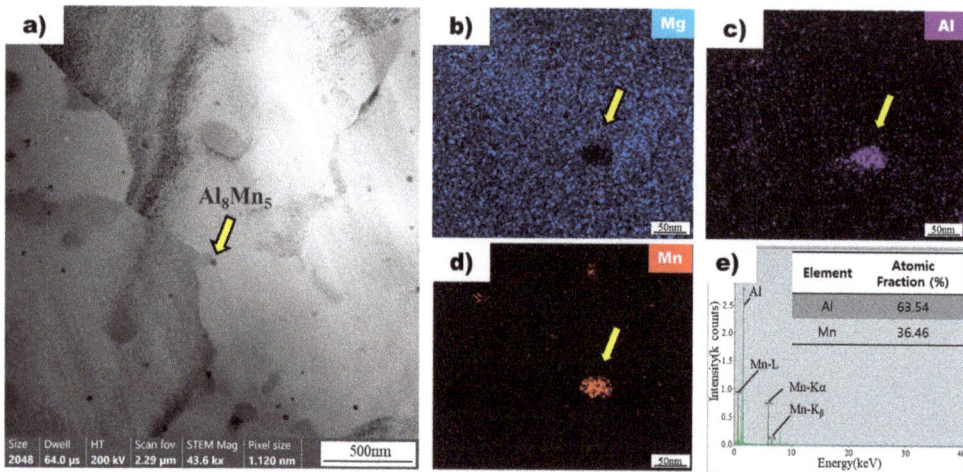

Figure 6. (**a**) Bright field TEM image of non-superheated AZ91 ribbon samples. The yellow arrow points to the Al$_8$Mn$_5$ particle. EDS mapping profile pointed by the yellow arrow of Mg (**b**), Al (**c**), and Mn (**d**) for the particle observed in non-superheated AZ91 ribbon samples. (**e**) EDS point analysis of the particle observed in non-superheated AZ91 ribbon samples.

The image in Figure 7a shows a bright-field transmission electron microscopy image of a superheated AZ91 alloy ribbon sample, revealing the presence of particles dispersed throughout the magnesium matrix. One particle, indicated by a yellow arrow, was chosen for detailed analysis using energy-dispersive X-ray spectroscopy mapping. The analysis confirmed the absence of magnesium in the particle (Figure 7b) but the presence of aluminum (Figure 7c) and manganese (Figure 7d), consistent with the particles observed in the non-superheated sample. However, the atomic ratios of aluminum and manganese in the particle of the superheated sample (Figure 7e) were different from those in the non-superheated sample (Figure 6e). Specifically, the particle in the superheated sample exhibited an atomic ratio of 78.34 for aluminum and 21.66 for manganese, whereas the particle in the non-superheated sample had a ratio of 63.54 for aluminum and 36.46 for manganese. Based on these ratios, the particles were identified as Al$_{10}$Mn$_3$ in the superheated sample and Al$_8$Mn$_5$ intermetallic compounds in the non-superheated sample [27].

3.6. HR-TEM Images of Rapidly Solidified AZ91D Ribbon Samples

Table 2 shows the crystal structures of Mg, Al$_8$Mn$_5$, and Al$_{10}$Mn$_3$ [27]. It was observed that the crystal structure and lattice parameters of Al$_8$Mn$_5$ significantly diverge from those of Mg. Conversely, Al$_{10}$Mn$_3$ and Mg share the same crystal structure and even belong to the same space group; however, a distinction in their lattice parameters was observed. Considering the crystal structures of each phase, Al$_{10}$Mn$_3$ appears to present a higher propensity to function as a nucleat for α-Mg grains compared to Al$_8$Mn$_5$. However, to elucidate this with higher precision, it is essential to evaluate the interplanar spacings and investigate potential crystallographic mismatches between the planes interfacing with each phase.

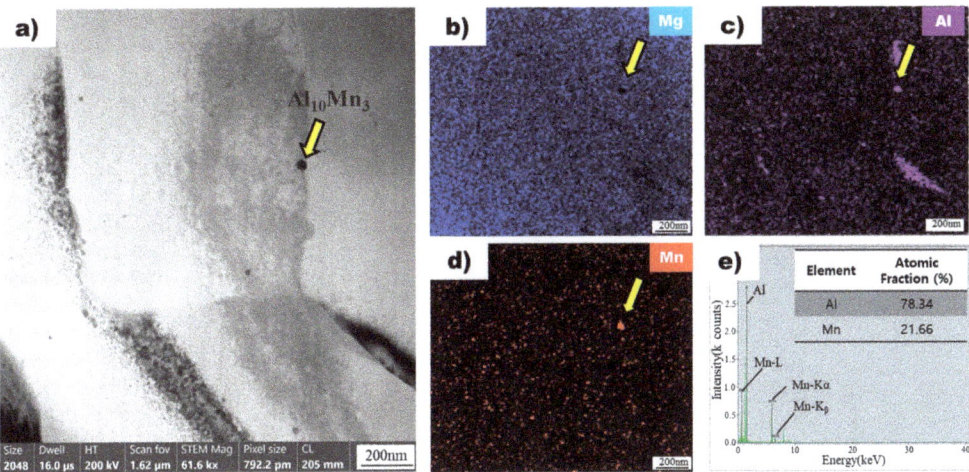

Figure 7. (a) Bright field TEM image of superheated AZ91 ribbon samples. The yellow arrow points to the Al$_8$Mn$_5$ particle. EDS mapping profile pointed by the yellow arrow of (b) Mg, (c) Al, and (d) Mn for particle observed in superheated AZ91 ribbon samples. (e) EDS point analysis of particle observed in superheated AZ91 ribbon samples.

Table 2. The crystal structures for Mg, Al$_8$Mn$_5$, and Al$_{10}$Mn$_3$ [27].

Phase	Crystal Structure	Space Group	Space Group Number	Lattice Parameter (nm)	
Mg	Hexagonal	P6$_3$/mmc	194	a and b = 0.320	c = 0.521
Al$_8$Mn$_5$	Rhombohedral	R3m	160	a and b = 1.264	c = 1.585
Al$_{10}$Mn$_3$	Hexagonal	P6$_3$/mmc	194	a and b = 0.750	c = 0.783

Figure 8a shows a high-resolution transmission electron microscopy (HR-TEM) image of the magnesium matrix and Al$_8$Mn$_5$ particles in the non-superheated AZ91 alloy ribbon sample. A local inverse fast Fourier transform (IFFT) image of white circle A (Mg) is shown in Figure 8b. The spacing of white line in the IFFT image denotes the interplanar spacing of the region being measured. Using the IFFT image as a reference, a profile image was derived, as shown in Figure 8c, enabling the measurement of the white line distance. As there is a variance between the peak intervals, an average value was acquired by measuring over an extended distance, which resulted in a value of 0.2479 nm. Table 3 presents the planes and corresponding d-spacing (in nm) for Mg, Al$_8$Mn$_5$, and Al$_{10}$Mn$_3$, derived using the CrysTBox software (ver 1.10). The $(0\ 1\ \bar{1}\ 1)$ plane of Mg, with a D-spacing of 0.2453 nm, displayed the closest proximity to the 0.2479 nm value obtained through the IFFT profile image. The IFFT image of the white circle B, representing Al$_8$Mn$_5$, is shown in Figure 8d, and Figure 8e displays the profile image derived from this IFFT image. The Al$_8$Mn$_5$ particles were identified on the (1 4 1) plane and the interplanar spacing was calculated to be 0.2275 nm, using the same method employed for the determination of the plane and interplanar spacing of Mg. Through the analysis of the HR-TEM image, IFFT images were acquired for the contact area between magnesium and Al$_8$Mn$_5$ particles, leading to the identification of the magnesium plane as $(0\ 1\ \bar{1}\ 1)$ and the Al$_8$Mn$_5$ plane as (1 4 1). Furthermore, the interplanar spacing was determined for both planes: the magnesium $(0\ 1\ \bar{1}\ 1)$ plane had an interplanar spacing of 0.2453 nm, while the Al$_8$Mn$_5$ particle (1 4 1) plane had a slightly smaller interplanar spacing of 0.2275 nm.

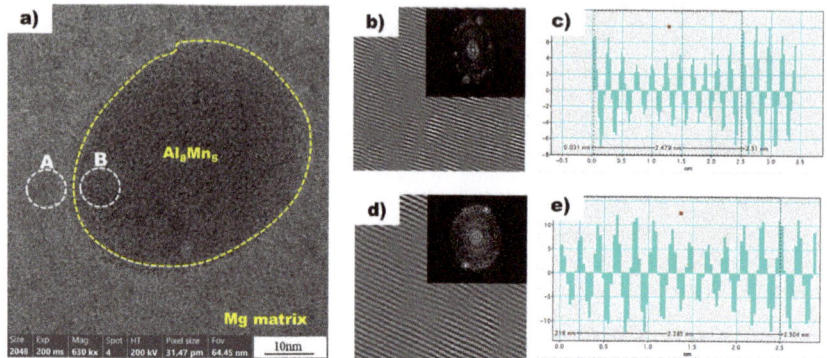

Figure 8. (a) HRTEM image of Mg (background) and Al$_8$Mn$_5$ (yellow dot circle) in the non-superheated AZ91 ribbon sample. (b) The local IFFT image of the white circle A (Mg) in (a). (c) The profiles of the local IFFT image (b). (d) The local IFFT image of the white circle B (Al$_8$Mn$_5$) in (a). (e) The profiles of the local IFFT image (d).

Table 3. List of Plane and D-Spacing (nm) for Mg, Al$_8$Mn$_5$, and Al$_{10}$Mn$_3$ (used CrysTbox ver1.10).

Magnesium		Al$_8$Mn$_5$		Al$_{10}$Mn$_3$	
Plane	D-Spacing (nm)	Plane	D-Spacing (nm)	Plane	D-Spacing (nm)
0 0 0 1	0.5210	0 0 1	0.7690	0 0 1	0.7789
⋮	⋮	⋮	⋮	⋮	⋮
0 0 0 2	0.2605	0 2 3	0.2320	0 1 3	0.2412
0 1 $\bar{1}$ 1	0.2453	1 4 1	0.2275	1 2 1	0.2347
0 1 $\bar{1}$ 2	0.1901	0 5 0	0.2183	0 3 0	0.2147
⋮	⋮	⋮	⋮	⋮	⋮
5 5 $\overline{10}$ 3	0.0316	5 5 3	0.1131	5 5 3	0.0772

Figure 9a shows a high-resolution transmission electron microscopy (HRTEM) image of the magnesium matrix and Al$_{10}$Mn$_3$ particles in the superheated AZ91 alloy ribbon sample. Local inverse fast Fourier transform (IFFT) images of two white circles, A and B, are also shown in Figure 9b,d, respectively. In addition, Figure 9c,d are profile images derived through IFFT images of two white circles, A and B, respectively. In the profile images of Mg and Al$_{10}$Mn$_3$, the average distance between the peaks was calculated to be 0.2406 nm and 0.2328 nm, respectively. The values obtained from the profile images were compared with the D-spacing values listed in Table 3, allowing for the confirmation of the Mg and Al$_{10}$Mn$_3$ planes. In the superheated sample, the magnesium plane was identified as $(0\ 1\ \bar{1}\ 1)$, which is consistent with the non-superheated sample. However, the Al$_{10}$Mn$_3$ plane was observed as (1 2 1), which is distinct from the Al$_8$Mn$_5$ plane found in the non-superheated sample. Furthermore, a detailed examination of the interplanar spacing was conducted, revealing an interplanar spacing of 0.2347 nm for the Al$_{10}$Mn$_3$ particle (1 2 1) plane in the superheated sample.

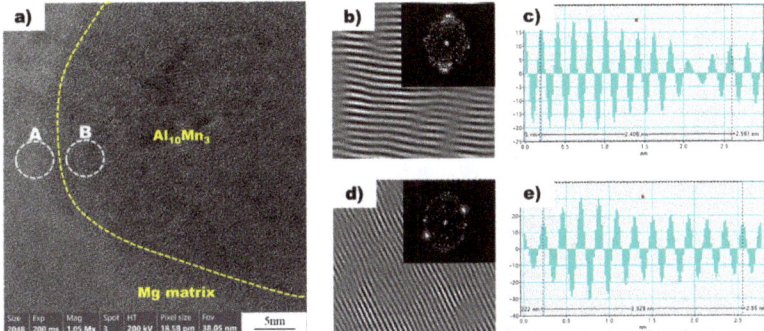

Figure 9. (a) HRTEM image of Mg (background) and $Al_{10}Mn_3$ (yellow dot circle) in the non-superheated AZ91 ribbon sample. (b) The local IFFT image of the white circle A (Mg) in (a). (c) The profiles of the local IFFT image (b). (d) The local IFFT image of the white circle B ($Al_{10}Mn_3$) in (a). (e) The profiles of the local IFFT image (d).

4. Discussions

The microstructure of as-cast AZ91D samples was investigated to determine the effect of superheating, and the results are presented in Figure 2. The non-superheated alloy exhibited an average grain size of 310 μm, while superheating reduced it to 108 μm. This indicates that the superheating process has a significant impact on the grain size of the alloy. Furthermore, the superheated AZ91D alloy had a more uniform and finer grain size than the non-superheated sample. The average grain size of the superheated sample was approximately three times smaller than that of the non-superheated sample. The chemical composition of the cast AZ91D samples was analyzed and is shown in Table 1. The primary additives, Al, Zn, and Mn, were found to have similar compositions in both alloys, thereby eliminating constitutional undercooling as a possible cause of grain refinement. In addition, because the samples were manufactured under the same casting conditions, any grain refinement effect due to the difference in cooling rate can be excluded. Thus, the observed difference in grain size between the non-superheated and superheated samples can be attributed to the superheating process. The cooling curves of non-superheated and superheated AZ91 alloys are shown in Figure 3. The extent of undercooling during solidification can reveal the presence of nucleants in the melt. When potent nucleants are present in the melts, the thermodynamic driving force required for the transition from the liquid phase to the solid phase is decreased, facilitating the formation of solid-phase nuclei with minimal undercooling [24]. This suggests that the cooling curves of the non-superheated and superheated AZ91 alloys provide information about the extent of undercooling during solidification, which can reveal the presence of nucleants in the melt. When potent nucleants are present, the thermodynamic driving force required for the transition from liquid to solid phase is decreased, facilitating the formation of solid-phase nuclei with minimal undercooling. Although a quantitative analysis of the nucleants was not conducted in this study, the degree of undercooling measured during solidification provides a basis for qualitatively deducing that the superheated AZ91 alloy possessed more nucleants promoting grain refinement than the non-superheated AZ91 alloy.

The rapid solidification process was used in this study to investigate the effect of nucleants on grain refinement in molten metal, as it can detect nucleants present in the molten metal. Ribbon samples were fabricated using non-superheated and superheated AZ91 alloys to study the impact of the superheating process on grain refinement and the contribution of nucleants to the formation of α-Mg grains. Rapid solidification also allows for the preservation of metastable phases within the molten metal, which is important for understanding the underlying mechanisms of grain refinement during the superheating process. Accurately measuring the temperature of the molten metal during the production of rapidly solidified ribbon samples using conventional melt spinner equipment is a significant challenge in this study. This is because precise temperature measurement is crucial to understanding the mechanism of grain refinement induced by the superheating process. However, conventional melt spinner equipment lacks an integrated temperature measurement system, and obtaining precise temperature control using a thermocouple is difficult due to the interference of the induction coil responsible for heating the metal material. In this study, a system capable of precise temperature measurement was designed and used to produce these rapidly solidified ribbon samples by modifying the existing melt spinner equipment. The temperature profiles during the fabrication of non-superheated and superheated AZ91 ribbon samples are presented in Figure 4a. It is evident that both samples were manufactured using a rapid solidification process at a precisely controlled temperature. The cooling rate is an important factor in identifying and analyzing solid-phase nucleants in AZ91 magnesium alloy produced by rapid solidification. It significantly affects the formation and growth of solid-phase nucleants, and controlling it allows for a more systematic investigation of their formation and contribution to grain refinement. The study by Wang et al. [26] investigated the heat transfer during rapid solidification of a ribbon prepared by the melt spinning process. They used a one-dimensional heat conduction equation to model the heat transfer, which allowed them to determine the temperature distribution and cooling rate within the ribbon. The cooling rate was found to be inversely proportional to the square of the thickness of the ribbon [26], and the same principle was applied for estimating the cooling rate of those rapidly solidified AZ91 alloy ribbon samples in this study. Various factors such as melting temperature, thermal conductivity, specific heat capacity, density, and latent heat were adjusted for the AZ91 magnesium alloy. The molten metal was rapidly solidified by spraying it onto a copper wheel rotating at a speed of 1500 rpm, resulting in the manufacturing of a ribbon sample with a thickness of 85 μm. This approach builds upon the findings of previous research and allowed for a systematic investigation of the cooling rate and its effect on the formation of solid-phase nucleants and subsequent grain refinement in the AZ91 magnesium alloy.

Bright-field transmission electron microscopy images of ribbon samples from non-superheated and superheated AZ91 alloys are shown in Figures 6a and 7a, respectively. In both samples, particles are dispersed throughout the magnesium matrix. The EDS results showed that the particle in the non-superheated sample was an intermetallic compound with a composition of Al_8Mn_5, while the particle in the superheated sample was identified as $Al_{10}Mn_3$. Although manganese is an element added to enhance the corrosion resistance of the AZ91 alloy [28], it reacts with aluminum to form Al-Mn intermetallic compounds. Among these, Al_8Mn_5 is known to form prior to the primary magnesium phase [29], and its role as a nucleant for α-Mg grains has been debated in numerous studies. However, no definitive mechanism has been established [17,30]. One widely accepted theory is the edge-to-edge matching model proposed by Zhang et al. [19], which suggests that, crystallographically, Al_8Mn_5 cannot act as a nucleant for α-Mg grains. There are various theories regarding the mechanism of the superheating process, and one of them is related to the Al-Mn intermetallic compound nucleation [17]. Cao et al. [20] found that ε-AlMn, which is present in Al-60Mn master alloy, can act as a potent nucleant for α-Mg grains, showing excellent grain refinement efficiency. This was further supported by Qin et al. [21], who reported that single-phase ε-AlMn could refine AZ31 grains, whereas single-phase Al_8Mn_5 could not. However, Qiu et al. [22] suggested that metastable τ-AlMn, potentially formed during the melt superheating process, could exhibit better crystallographic matching with the Mg matrix than ε-AlMn, making it a more effective nucleant for α-Mg grains. Nonetheless, there is currently no clear evidence to verify the presence of both τ-AlMn and ε-AlMn in Mg-Al alloys, and there are no established methods to regulate their phase and morphology. The finding of a new particle, $Al_{10}Mn_3$, during the superheating process of the AZ91 alloy is significant as it has not been previously reported in the literature. This new particle may provide novel insights into the mechanism of grain refinement during the superheating process. However, further research is needed to fully understand the thermodynamic calculations and physical and chemical processes involved in the formation of $Al_{10}Mn_3$ during the superheating process.

The coherent relationship between two phases occurs when their interatomic distances and atomic arrangements on their crystal faces are similar, enabling one phase to act as a highly efficient heterogeneous nucleation site for the other phase [31]. It is important to note that for a favorable coherent relationship to exist at the interface of the two phases, comparable interplanar distances are essential. Classical nucleation theory suggests that a substrate's ability to promote nucleation depends on the interfacial energy between the substrate and the nuclei. This interfacial energy is influenced by the crystallographic structures of both the substrate and the nuclei [32]. Turnbull and Vonnegut [33] proposed a one-dimensional misfit model to describe the relationship between substrate particle effectiveness and crystallographic mismatch for nuclei. However, this model has limitations when the crystallographic structures of the nuclei and substrate particles differ significantly. Bramfitt [34,35] improved upon Turnbull's model by creating a two-dimensional model that takes into account the angle between the crystal orientations of both phases. This two-dimensional model enables the calculation of two-dimensional lattice misfit for matching planes.

In this modified Turnbull's model, the two-dimensional lattice misfit of two matching planes was calculated based on Equation (1),

$$\delta_{(hkl)_n}^{(hkl)_s} = \sum_{i=1}^{3} \frac{\left|\left(d_{[uvw]_s}^i \cos\theta\right) - d_{[uvw]_n}^i\right|/d_{[uvw]_n}^i}{3} \times 100\% \quad (1)$$

where $(h\ k\ l)_s$ is a plane of the substrate, $[uvw]_s$ a direction in $(h\ k\ l)_s$, $(h\ k\ l)_n$ a plane of the nucleated solid, $[uvw]_n$ a direction in $(h\ k\ l)_n$, $d[uvw]_s$ the interatomic distance along $[uvw]_s$, $d[uvw]_n$ the interatomic distance along $[uvw]_n$, and θ is the angle between the $[uvw]_s$ and $[uvw]_n$.

This study found that there is crystallographic coherence between the interfaces of Al_8Mn_5 and $Al_{10}Mn_3$ particles with the magnesium matrix. The high-resolution transmission electron microscopy (HR-TEM) and inverse fast Fourier transform (IFFT) images of the non-superheated AZ91 alloy ribbon samples are shown in Figure 8, which revealed that the contact planes for magnesium and Al_8Mn_5 particles were the magnesium plane $(0\ 1\ \bar{1}\ 1)$ and the Al_8Mn_5 plane $(1\ 4\ 1)$. Similarly, the HR-TEM and IFFT images of the superheated AZ91 alloy ribbon samples are shown in Figure 9, which identified the contact planes for magnesium and $Al_{10}Mn_3$ particles as the magnesium plane $(0\ 1\ \bar{1}\ 1)$ and the $Al_{10}Mn_3$ plane $(1\ 2\ 1)$. The modified Turnbull's model is effective in analyzing the coherence of two planes when their crystallographic structures are not significantly different. However, in the case of the $(1\ 4\ 1)$ plane of Al_8Mn_5 and the $(0\ 1\ \bar{1}\ 1)$ plane of Mg, their crystallographic coherence was poor, and thus the application of the modified Turnbull's model was not feasible. On the other hand, the $(1\ 2\ 1)$ plane of $Al_{10}Mn_3$ and the $(0\ 1\ \bar{1}\ 1)$ plane of Mg showed good crystallographic coherence, and the modified Turnbull's model could be used to analyze their coherence. Figure 10 shows the typical planar atomic arrangements in the $(1\ 2\ 1)$ plane of $Al_{10}Mn_3$ and in the $(0\ 1\ \bar{1}\ 1)$ plane of Mg. The brown dash circles represent the $Al_{10}Mn_3$ atoms in the $(1\ 2\ 1)$ plane and the blue circles represent the Mg atoms in the $(0\ 1\ \bar{1}\ 1)$ plane. The $Al_{10}Mn_3$ and Mg atoms applied to the lattice misfit analysis are indicated by filling the center with the same color as the circle, and the place where the $Al_{10}Mn_3$ and Mg atoms exactly match is marked in red. The i_1 indicate a line connecting the arrangement of $Al_{10}Mn_3$ atoms in the [1 0 1] direction with an interatomic distance of 1.224 nm and a line connecting the arrangement of Mg atoms in the [3 3 −3] direction with an interatomic distance of 1.084 nm; the angle (θ) between both lines is 4°. The i_2 and the i_3 are marked in the same way as the i_1; the required parameters for Equation (1) are listed in Table 4. The disregistry, δ, for Mg $(0\ 1\ \bar{1}\ 1)\ ||\ Al_{10}Mn_3\ (1\ 2\ 1)$ is as follows:

$$\delta^{(0\,1\,\bar{1}\,1)\text{Mg}}_{(1\,2\,1)_{\text{Al}_{10}\text{Mn}_3}} = \frac{(|1.081-1.244|/1.244)+(|2.956-3.21|/3.21)+(|1.734-1.524|/1.524)}{3} \times 100\% \approx 11\%$$

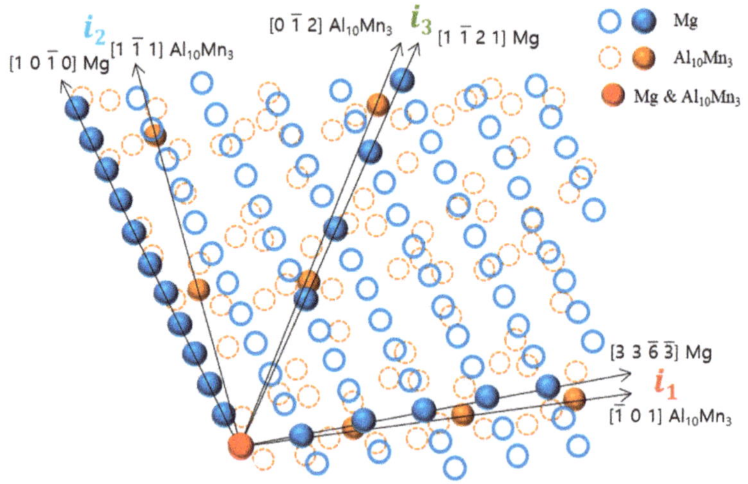

Figure 10. The crystallographic relationship at the interface between the (1 2 1) plane of $Al_{10}Mn_3$ and the $(0\,1\,\bar{1}\,1)$ plane of Mg.

Table 4. Parameters for Equation (1).

Case	i	$d[u\,v\,w]_s$	$d[u\,v\,w]_n$	θ (°)
Mg $(0\,1\,\bar{1}\,1)$ ‖ $Al_{10}Mn_3$ (1 2 1)	1	1.084	1.224	4
	2	3.012	3.210	11
	3	1.736	1.524	2

This indicates that the $Al_{10}Mn_3$ particle is a better nucleating site for α-Mg grains than the Al_8Mn_5 particle, and α-Mg can nucleate on the $Al_{10}Mn_3$ nucleation substrates. Previous research has shown that a substrate can effectively act as a nucleant for a solid if the disregistry between the two is less than 5% [36]. However, this is not always the case. Despite an 11% disregistry between $Al_{10}Mn_3$ particles and Mg, it is not conclusive that $Al_{10}Mn_3$ particles cannot act as nucleants for α-Mg grains. This study found that Al_8Mn_5 particles, which have low crystallographic coherence with Mg, can transform into $Al_{10}Mn_3$ through superheating, allowing them to act as nucleants for α-Mg grains. This finding provides a new perspective on investigating the mechanism of superheating grain refinement in the AZ91 alloy.

This study investigated the mechanism of superheated grain refinement in the AZ91 magnesium alloy through rapid solidification with precise temperature control. The results showed that Al_8Mn_5 may transform into $Al_{10}Mn_3$ during the superheating process. At a higher temperature, the solubility of Mn in the Mg melt will be increased and that would also affect the phase transformation of Mn intermetallics. From the HR-TEM and IFFT images of the samples, the $Al_{10}Mn_3$ can effectively act as a nucleant for grain refinement in AZ91. The formation of $Al_{10}Mn_3$ particles was found to play a crucial role in the superheating grain refinement process of the alloy. However, the study did not provide objective thermodynamic evidence to support the phase transformation of Al_8Mn_5 to $Al_{10}Mn_3$ or the generation of $Al_{10}Mn_3$, which will be explored in future research. Overall, this study offers a new perspective on the superheating grain refinement mechanism in AZ91 and provides insights for future research.

5. Conclusions

In this work, the mechanism on superheating grain refinement of the AZ91 alloy was investigated using a rapid solidification process. The main conclusions can be summarized as follows:

(1) The average grain size of the non-superheated sample manufactured by the mold casting methods was measured to be 310 μm, while the grain size decreased to 108 μm after the superheating process.

(2) The non-superheated AZ91 alloy exhibited an undercooling of 2.1 °C, whereas the superheated AZ91 alloy showed negligible undercooling. This suggests that nucleants, which influence the refinement of α-Mg grains, were generated by the superheating process.

(3) Through the utilization of a rapid solidification process with precise temperature control, Al_8Mn_5 particles were observed in the non-superheated AZ91 ribbon samples; the $(1\ 4\ 1)$ plane of these particles and the $(0\ 1\ \bar{1}\ 1)$ plane of magnesium was found to be in contact. However, it was confirmed that the crystallographic coherence between the two planes was so inconsistent that the modified Turnbull–Vonnegut equation, which is used for quantitative crystallographic coherence analysis, could not be applied.

(4) $Al_{10}Mn_3$ particles were observed in the superheated AZ91 ribbon samples; the $(1\ 2\ 1)$ plane of these particles and the $(0\ 1\ \bar{1}\ 1)$ plane of magnesium was found to be in contact. The 11% mismatch between the two planes was calculated using the modified Turnbull–Vonnegut equation.

(5) It is thought that the superheating process contributes to grain refinement of AZ91 alloy by generating $Al_{10}Mn_3$, which exhibits more good crystallographic matching with magnesium compared to Al_8Mn_5. However, the study did not provide objective thermodynamic evidence to support the phase transformation of Al_8Mn_5 to $Al_{10}Mn_3$ or the generation of $Al_{10}Mn_3$, and additional thermodynamic studies are planned to clarify our results.

Author Contributions: Planning and designing of experiments were performed by Y.L., and experiments were carried out by S.J., with the supervision of Y.L. Writing was carried out by S.J. and reviewed by Y.L.; Writing—review and editing, Y.P. and Y.L. All authors have read and agreed to the published version of the manuscript.

Funding: This work was financially supported by the Korea Institute of Industrial Technology ("Development of Root Technology for multi-product flexible production (Grant number: KITECH EO–23–0008)").

Institutional Review Board Statement: Not applicable.

Informed Consent Statement: Not applicable.

Data Availability Statement: The data used to support the findings of this study are available from the corresponding author upon request.

Acknowledgments: We would like to thank all ASTL (Advanced Solidification Technology Lab) members and very special thanks to Gwang-seok Son in Donga University for his valuable help in TEM analysis.

Conflicts of Interest: The authors declare no conflict of interest.

References

1. Kaya, A.A. A Review on Developments in Magnesium Alloys. *Front. Mater.* **2020**, *7*, 1–26. [CrossRef]
2. Mordike, B.L.; Ebert, T. Magnesium Properties-applications-potential. *Mater. Sci. Eng. A* **2001**, *302*, 37–45. [CrossRef]
3. Song, J.; She, J.; Chen, D.; Pan, F. Latest research advances on magnesium and magnesium alloys worldwide. *J. Magnes. Alloy.* **2020**, *8*, 1–41. [CrossRef]
4. Lee, Y.C.; Dahle, A.K.; Stjohn, D.H. The role of solute in grain refinement of magnesium. *Metall. Mater. Trans. A Phys. Metall. Mater. Sci.* **2000**, *31*, 2895–2906. [CrossRef]

5. Song, J.; Chen, J.; Xiong, X.; Peng, X.; Chen, D.; Pan, F. Research advances of magnesium and magnesium alloys worldwide in 2021. *J. Magnes. Alloy.* **2022**, *10*, 863–898. [CrossRef]
6. Qian, M.; Stjohn, D.H.; Frost, M.T. Characteristic zirconium-rich coring structures in Mg-Zr alloys. *Scr. Mater.* **2002**, *46*, 649–654. [CrossRef]
7. Stjohn, D.H.; Qian, M.; Easton, M.A.; Cao, P. Grain Refinement of Magnesium Alloys. *Metall. Mater. Trans. A Phys. Metall. Mater. Sci.* **2005**, *36*, 1669–1679. [CrossRef]
8. Ali, Y.; Qiu, D.; Jiang, B.; Pan, F.; Zhang, M.X. Current research progress in grain refinement of cast magnesium alloys: A review article. *J. Alloys Compd.* **2015**, *619*, 639–651. [CrossRef]
9. Cao, P.; Qian, M.; StJohn, D.H. Mechanism for grain refinement of magnesium alloys by superheating. *Scr. Mater.* **2007**, *56*, 633–636. [CrossRef]
10. Cao, P.; Qian, M.; StJohn, D. Grain refinement of commercial purity Mg-9%Al alloys by superheating. *Magnes. Technol.* **2005**, *6*, 297–302.
11. Jung, S.S.; Son, Y.G.; Park, Y.H.; Lee, Y.C. A Study on the Grain Refining Mechanisms and Melt Superheat Treatment of Aluminum-Bearing Mg Alloys. *Metals* **2022**, *12*, 464. [CrossRef]
12. Zhao, P.; Geng, H.; Wang, Q. Effect of melting technique on the microstructure and mechanical properties of AZ91 commercial magnesium alloys. *Mater. Sci. Eng. A* **2006**, *429*, 320–323. [CrossRef]
13. Karakulak, E. A review: Past, present and future of grain refining of magnesium castings. *J. Magnes. Alloy.* **2019**, *7*, 355–369. [CrossRef]
14. Vinotha, D.; Raghukandan, K.; Pillai, U.T.S.; Pai, B.C. Grain refining mechanisms in magnesium alloys—An overview. *Trans. Indian Inst. Met.* **2009**, *62*, 521–532. [CrossRef]
15. Stjohn, D.H.; Easton, M.A.; Qian, M.; Taylor, J.A. Grain refinement of magnesium alloys: A review of recent research, theoretical developments, and their application. *Metall. Mater. Trans. A Phys. Metall. Mater. Sci.* **2013**, *44*, 2935–2949. [CrossRef]
16. Motegi, T. Grain-refining mechanisms of superheat-treatment of and carbon addition to Mg-Al-Zn alloys. *Mater. Sci. Eng. A* **2005**, *413–414*, 408–411. [CrossRef]
17. Han, G.; Liu, X. Phase control and formation mechanism of Al–Mn(–Fe) intermetallic particles in Mg–Al-based alloys with FeCl3 addition or melt superheating. *Acta Mater.* **2016**, *114*, 54–66. [CrossRef]
18. Byun, J.; Kwon, S.I.; Ha, H.P.; Yoon, J. A Manufacturing Technology of AZ91-Alloy Slurry for Semi Solid Forming. *Magnesium* **2003**, *27*, 713–718. [CrossRef]
19. Zhang, M.X.; Kelly, P.M.; Qian, M.; Taylor, J.A. Crystallography of grain refinement in Mg-Al based alloys. *Acta Mater.* **2005**, *53*, 3261–3270. [CrossRef]
20. Cao, P.; Qian, M.; Stjohn, D.H. Effect of manganese on grain refinement of Mg-Al based alloys. *Scr. Mater.* **2006**, *54*, 1853–1858. [CrossRef]
21. Qin, G.W.; Ren, Y.; Huang, W.; Li, S.; Pei, W. Grain refining mechanism of Al-containing Mg alloys with the addition of Mn-Al alloys. *J. Alloys Compd.* **2010**, *507*, 410–413. [CrossRef]
22. Qiu, D.; Zhang, M.X.; Taylor, J.A.; Fu, H.M.; Kelly, P.M. A novel approach to the mechanism for the grain refining effect of melt superheating of Mg-Al alloys. *Acta Mater.* **2007**, *55*, 1863–1871. [CrossRef]
23. Han, M.; Zhu, X.; Gao, T.; Liu, X. Revealing the roles of Al4C3 and Al8Mn5 during α-Mg nucleation in Mg-Al based alloys. *J. Alloys Compd.* **2017**, *705*, 14–21. [CrossRef]
24. Men, H.; Jiang, B.; Fan, Z. Mechanisms of grain refinement by intensive shearing of AZ91 alloy melt. *Acta Mater.* **2010**, *58*, 6526–6534. [CrossRef]
25. Xu, J.; Yang, T.; Li, Z.; Wang, X.; Xiao, Y.; Jian, Z. The recalescence rate of cooling curve for undercooled solidification. *Sci. Rep.* **2020**, *10*, 1380. [CrossRef] [PubMed]
26. Wang, X.J.; Chen, X.D.; Xia, T.D.; Yu, W.Y.; Wang, X.L. Influencing factors and estimation of the cooling rate within an amorphous ribbon. *Intermetallics* **2004**, *12*, 1233–1237. [CrossRef]
27. Smith, J.F. (Ed.) *Phase Diagrams of Binary Vanadium Alloys*; Series: Monograph series on alloy phase diagrams; ASM International: Metals Park, OH, USA, 1989.
28. Esmaily, M.; Svensson, J.E.; Fajardo, S.; Birbilis, N.; Frankel, G.S.; Virtanen, S.; Arrabal, R.; Thomas, S.; Johansson, L.G. Fundamentals and advances in magnesium alloy corrosion. *Prog. Mater. Sci.* **2017**, *89*, 92–193. [CrossRef]
29. Zeng, G.; Xian, J.W.; Gourlay, C.M. Nucleation and growth crystallography of Al8Mn5 on B2-Al(Mn, Fe) in AZ91 magnesium alloys. *Acta Mater.* **2018**, *153*, 364–376. [CrossRef]
30. Wang, Y.; Xia, M.; Fan, Z.; Zhou, X.; Thompson, G.E. The effect of Al8Mn5 intermetallic particles on grain size of as-cast Mg-Al-Zn AZ91D alloy. *Intermetallics* **2010**, *18*, 1683–1689. [CrossRef]
31. Rigsbee, J.M.; Aaronson, H.I. A computer modeling study of partially coherent f.c.c.:b.c.c. boundaries. *Acta Metall.* **1979**, *27*, 351–363. [CrossRef]
32. Glicksman, M.E.; Childs, W.J. Nucleation catalysis in supercooled liquid tin. *Acta Metall.* **1962**, *10*, 925–933. [CrossRef]
33. Turnbull, D.; Vonnegut, B. Nucleation Catalysis. *Ind. Eng. Chem.* **1952**, *44*, 1292–1298. [CrossRef]
34. Bruce, L. BRAMFITT The Effect of Carbide and Nitride Additions on the Heterogeneous Nucleation Behavior of Liquid Iron. *Metall. Trans.* **1970**, *1*, 1987–1995.

35. Wang, D.; Chang, W.; Shen, Y.; Sun, J.; Sheng, C.; Zhang, Y.; Zhai, Q. The role of lattice mismatch in heterogeneous nucleation of pure Al on Al2O3 single-crystal substrates with different termination planes. *J. Therm. Anal. Calorim.* **2019**, *137*, 791–797. [CrossRef]
36. Kim, B.; Hwang, J.; Park, Y.; Lee, Y. Microstructural improvement of eutectic al + mg2si phases on al–zn–si–mg cast alloy with tib2 particles additions. *Materials* **2021**, *14*, 2902. [CrossRef] [PubMed]

Disclaimer/Publisher's Note: The statements, opinions and data contained in all publications are solely those of the individual author(s) and contributor(s) and not of MDPI and/or the editor(s). MDPI and/or the editor(s) disclaim responsibility for any injury to people or property resulting from any ideas, methods, instructions or products referred to in the content.

Article

New Approach in the Determination of a Suitable Directionally Coarsened Microstructure for the Fabrication of Nanoporous Superalloy Membranes Based on CMSX-4

Janik Marius Lück *[ ] and Joachim Rösler

Institute for Materials Science, Technische Universität Braunschweig, Langer Kamp 8, 38106 Braunschweig, Germany
* Correspondence: janik.lueck@tu-braunschweig.de

Citation: Lück, J.M.; Rösler, J. New Approach in the Determination of a Suitable Directionally Coarsened Microstructure for the Fabrication of Nanoporous Superalloy Membranes Based on CMSX-4. *Materials* 2023, 16, 3715. https://doi.org/10.3390/ma16103715

Academic Editor: Konstantin Borodianskiy

Received: 18 April 2023
Revised: 10 May 2023
Accepted: 11 May 2023
Published: 13 May 2023

Copyright: © 2023 by the authors. Licensee MDPI, Basel, Switzerland. This article is an open access article distributed under the terms and conditions of the Creative Commons Attribution (CC BY) license (https://creativecommons.org/licenses/by/4.0/).

Abstract: The pore size of nanoporous superalloy membranes produced by directional coarsening is directly related to the γ-channel width after creep deformation, since the γ-phase is removed subsequently by selective phase extraction. The continuous network of the γ'-phase thus remaining is based on complete crosslinking of the γ'-phase in the directionally coarsened state forming the subsequent membrane. In order to be able to achieve the smallest possible droplet size in the later application in premix membrane emulsification, a central aspect of this investigation is to minimize the γ-channel width. For this purpose, we use the $3w_0$-criterion as a starting point and gradually increase the creep duration at constant stress and temperature. Stepped specimens with three different stress levels are used as creep specimens. Subsequently, the relevant characteristic values of the directionally coarsened microstructure are determined and evaluated using the line intersection method. We show that the approximation of an optimal creep duration via the $3w_0$-criterion is reasonable and that coarsening occurs at different rates in dendritic and interdendritic regions. The use of staged creep specimens shows significant material and time savings in determining the optimal microstructure. Optimization of the creep parameters results in a γ-channel width of 119 ± 43 nm in dendritic and 150 ± 66 nm in interdendritic regions while maintaining complete crosslinking. Furthermore, our investigations show that unfavorable stress and temperature combinations favor unidirectional coarsening before the rafting process is completed.

Keywords: CMSX-4; superalloy membranes; creep test; directional coarsening; premix membrane emulsification

1. Introduction

Previous studies by Kohnke et al. [1] have shown that there is a promising application opportunity for nanoporous superalloy membranes in the field of premix membrane emulsification, a process that can be used to produce nanoemulsions, e.g., for pharmaceutical applications. Particularly promising results were achieved for directionally coarsened γ'-membranes based on the single crystalline superalloy CMSX-4. Key processing steps to produce this kind of superalloy membrane are (i) solution and precipitation heat treatment of CMSX-4 resulting in cubic γ'-particles embedded in the γ-matrix; (ii) directional coarsening, also referred to as rafting, of the γ'-particles by creep deformation to a bicontinuous γ/γ'-microstructure; and (iii) selective phase extraction of the γ-phase so that a nanoporous material consisting of the γ'-phase and channel-like porosity at the location of the dissolved γ-phase results. In premix membrane emulsification, a coarse emulsion is then pushed through the membrane, resulting in droplet size reduction. To achieve the finest possible droplet size and a narrow droplet size distribution during this process, it is of great importance to reduce the γ-channel width, i.e., the subsequent pore size, by adjusting the process parameters in the creep test [1–5]. Therefore, for the production of nanoporous superalloy membranes with a directionally coarsened microstructure, the process of rafting is of central importance. This depends on several factors such as the direction of loading, the γ'-volume fraction, and the misfit, which is defined as $\delta = 2(a_{\gamma'} - a_\gamma)/(a_{\gamma'} + a_\gamma)$ with a_γ and $a_{\gamma'}$ being the lattice parameters of the γ- and γ'-phase, respectively [6–9]. Single crystalline Ni-based superalloys such as CMSX-4 have a negative misfit of the magnitude of $\delta \approx -0.001$ at room temperature. Due to the different thermal expansion coefficients of γ and γ', the misfit δ becomes increasingly negative with increasing temperature [6,10,11]. If the starting material, which contains cube-shaped γ'-particles (L_{12} crystal structure) coherently embedded in the fcc γ matrix, is loaded in the [001]-direction at elevated temperatures, a change in microstructure occurs and a raft structure is formed [4,12]. This is due to the introduction of dislocations in the γ-channels between the γ'-precipitates. In case of a negative misfit δ, the superposition of applied tensile force and internal misfit stresses leads to shear stresses, which are highest in the horizontal γ-channels so that plastic deformation starts there. Dislocations of type $a/2 <0\overline{1}1> \{111\}$ preferentially bulge through the γ-channels leaving interfacial dislocations of screw or mixed character in the γ/γ'-interfaces [13–16]. The dislocations introduced in this way shift the misfit δ to more positive or negative values in the horizontal and vertical channels, respectively [17]. For alloys with negative misfit, this means that the associated coherency stresses are reduced in the horizontal channels while they are elevated in the vertical ones. Consequently, it becomes energetically favorable to widen the horizontal channels at the expense of the vertical ones, leading eventually to the coalescence of the γ'-precipitates at the vertical γ-channels. This process is called rafting [1,4,6,12,14–16]. Thus, a negative misfit δ results in rafts oriented perpendicular to the tensile loading direction, referred to as type-N. If the misfit δ is positive, rafts form parallel to the load direction and are referred to as type-P [7]. Altogether, a bicontinuous γ/γ'-network is formed, where both phases interpenetrate each other while each phase is connected in itself [3]. This understanding led to the invention and development of nanoporous superalloy membranes [3,5].

During past research on the fabrication of nanoporous superalloy membranes based on CMSX-4, mainly directional coarsening and crosslinking at a temperature of 1000 °C and a tensile stress of 170 MPa was investigated. The γ-channel widths obtained were 250–400 nm [18]. An important component of this investigation is the γ-channel widening rate experimentally determined by Epishin et al. [19,20] for the horizontal γ-channels of dendritic regions. Two different rates $\dot{w}_1$ and $\dot{w}_2$ were found for the process of raft formation and subsequent coarsening, respectively. Furthermore, it was found that at the time the raft formation is completed, the channel width of the horizontal channels is about three times the initial channel width w_0 in the precipitation heat-treated condition. This was attributed to the fact that there are twice as many vertical channels as horizontal ones and that the entire γ-phase of the closing vertical channels is transferred to the horizontal ones. This is called the $3w_0$-criterion [19]. Taking these conditions into account, an optimal heat treatment was therefore worked out in a previous study, combining the most regular arrangement and angularity of the γ'-particles with the minimum size w_0 of the γ-channels. The optimized precipitation heat treatment at 1140 °C for 30 min results in a γ'-particle size of 224 ± 52 nm and a γ-channel width of 35 ± 19 nm [21].

Using the above-mentioned heat path development and the resulting initial microstructure, we investigate in this work the influence of stress and time on the microstructure development during creep deformation with the aim of identifying optimal creep parameters for the production of superalloy membranes with the smallest possible γ-channel width. Since the subsequent membranes are subjected to mechanical stress in premix membrane emulsification, the selection of the process parameters should include an additional increase in strength via an increased γ'-content. As the γ'-volume fraction depends on the temperature and previous investigations were carried out at 1000 °C, the temperature in the creep tests will be reduced to 950 °C. According to Epishin et al. [22], this increases the γ'-volume content from 70% to 72%. To compensate for the reduced channel widening rate due to the reduced temperature, the stress in the creep tests is increased up to 250 MPa [20]. In order to be able to investigate the influence of different creep stresses at a constant temperature, stepped creep specimens are used. These have three different diameters in three sections so that stresses of 250 MPa, 183 MPa, and 140 MPa can be set in one sample. A similar variant with different dimensions has already been used by Cheng et al. [23] in tensile and compressive creep tests. Epishin et al. [19] used a so-called flat wedge-shaped sample. The designs differ in that we used clearly separated sections with transition radii rather than a rectangular, continuously converging cross-section. However, in this investigation, the combination of 950 °C and 250 MPa is decisive for the subsequent membrane production. In summary, this investigation pursues two goals in the context of superalloy membrane development:

1. To identify suitable creep parameters for the fabrication of nanoporous superalloy membranes with the application target of premix membrane emulsification with minimized pore size.
2. To investigate if stepped creep specimens are suitable for the fast and cost-effective investigation of multiple stress and temperature combinations in membrane development.

2. Methodology

2.1. Materials and Processing

To investigate the influence of creep duration and stress on the microstructure evolution during directional coarsening, the single-crystalline Ni-based superalloy CMSX-4 was used. For this purpose, three out of four creep specimens were manufactured from rods with a length of 220 mm and a diameter of 21 mm produced by Access Aachen e.V. using the Bridgeman process (Specimen ID starting with "A-"). The deviation of the [001]-orientation from the rod centerline was determined by Röntgenlabor Eigenmann using the Laue diffraction method. It was less than 10° for the used rods. One of the four samples used was made from a single crystal plate. The deviation of the plate center axis from the [001]-orientation was measured using electron backscatter diffraction (EBSD) in a scanning electron microscope (FEI Helios NanoLab 650) and was 6–10° (Specimen ID starting with "I-"). All samples were heat treated using the following homogenization (HT) and precipitation heat treatment: 1277 °C/2 h + 1288 °C/3 h + 1296 °C/3 h + 1304 °C/2 h + 1313 °C/2 h + 1316 °C/2 h + 1318 °C/2 h + 1321 °C/2 h + AC to RT + 1140 °C/30 min + AC to RT (AC: Air cooling). The precipitation heat treatment at 1140 °C for 30 min is the result of previous studies on γ'-precipitate size and morphology as mentioned before [21]. For completeness, Figure 1a shows the γ/γ'-microstructure in the precipitation-hardened state. The average edge length of the cubic γ'-particles is 224 ± 52 nm, and the channel width between the γ'-particles 35 ± 19 nm. Stepped creep specimens were then fabricated with a total length of 72 mm. The three sections along the gauge length have diameters of 6, 7, and 8 mm, and the specimens have a metric M12 × 1.75 thread at each end. Figure 1b shows the technical drawing of the creep test sample and a manufactured sample from CMSX-4. Table 1 summarizes the test conditions of the investigated stepped creep specimens. The minimal creep duration of 140 h at a stress of 250 MPa was estimated using the channel widening rate $\dot{w}_1 = 5.25 \times 10^{-1} \frac{nm}{h}$ determined experimentally by Epishin et al. [19,20] in dendritic regions for that test condition. Creep tests were carried out with a constant weight applied to the creep frame. Following the creep test, the samples were cleaned and the three sections were separated. For further investigations of the microstructure, each section was cut in half parallel to the [001] direction.

In order to be able to check the crosslinking in dendritic and interdendritic areas, sheets for the production of membranes were taken from the creep specimens. The sheets were ground to a thickness of approximately 0.3 mm after cutting (including P2500). Electrochemical etching was used to dissolve the γ-phase, leaving the γ'-phase as a solid membrane and a channel-like porosity in place of the γ-phase. A solution of 800 mL H_2O, 8 g $(NH_4)_2SO_4$, and 8 g $C_6H_8O_7$ and an extraction voltage of 1.3 V were used. For a detailed description as well as a detailed experimental setup, see [5]. The specimens with the specimen ID A-CMSX-4/140h/250, A-CMSX-4/170h/250, A-CMSX-4/310h/250, and I-CMSX-4/506h/250 were investigated with this method.

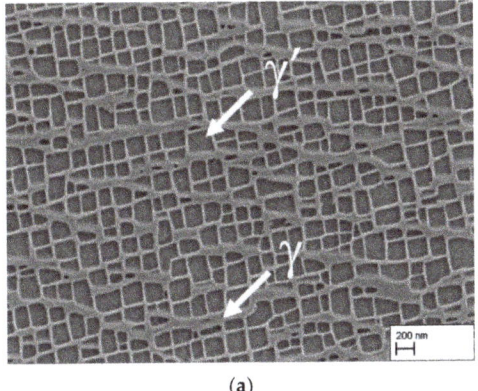 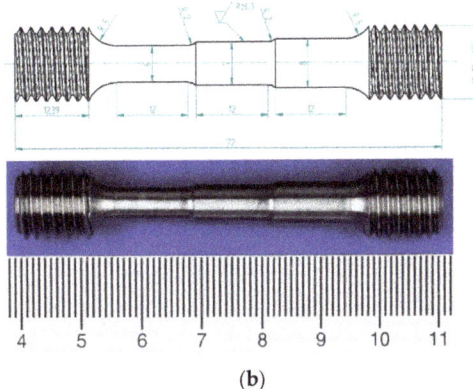

Figure 1. (a) γ/γ'-microstructure of the CMSX-4 creep samples used after homogenization and precipitation heat treatment. The γ'-particle size is 224 ± 52 nm, the γ-channel width 35 ± 19 nm, (b) technical drawing of the stepped creep specimens used in this study (top) and exemplary stepped creep specimen made of CMSX-4 before the creep test (bottom).

Table 1. Loading conditions and geometrical details of the stepped creep test samples; θ is the misorientation angle between the crystallographic [001]-orientation and the long axis of the single crystal rods.

Specimen ID	Cross Section (mm²)	θ (°)	Stress (MPa)	Duration (h)
A-CMSX-4/140h/250		9.9		140
A-CMSX-4/170h/250		9.8		170
A-CMSX-4/310h/250	28.27	9.6	250	310
I-CMSX-4/506h/250		6–10		506
A-CMSX-4/140h/183		9.9		140
A-CMSX-4/170h/183		9.8		170
A-CMSX-4/310h/183	38.48	9.6	183	310
I-CMSX-4/506h/183		6–10		506
A-CMSX-4/140h/140		9.9		140
A-CMSX-4/170h/140		9.8		170
A-CMSX-4/310h/140	50.26	9.6	140	310
I-CMSX-4/506h/140		6–10		506

2.2. Microstructure Analysis

For the investigation and analysis of the microstructure, the samples were first ground and polished up to and including oxide polishing suspension (OPS). Then the γ'-phase was etched with the help of molybdic acid (1 part H_2O, 1 part HCl (37%), 1 part HNO_3, 1 weight-% MoO_3) for 3 s to obtain sufficient contrast in the scanning electron microscope (SEM). For each condition, five SEM images were taken with a ZEISS LEO 1550 GEMINI at a magnification of 8 k and then used for the measurement with the line intersection method. To be able to carry out the measurement of the γ/γ'-microstructure, first, a binarization was necessary. For this purpose, after cropping the image to 688 × 688 pixels, the software ImageJ [24] and the plugin Trainable Weka Segmentation [25] were used. The plugin is trained by the user's input; in the broadest sense, this training is conducted by determining which pixels with which grey values belong to the γ-phase or γ'-phase. The software can use various settings, such as taking the neighboring pixels into account, to make the distinction between the adjacent pixels. Since the recorded images were usually taken with identical brightness and contrast settings for each creep state, the software only had to be trained once for each state to the corresponding settings. Afterward, the image was imported into a program for the line intersection method.

In the line intersection method, a defined set of parallel lines is placed in the binarized image in an angular range of 0° to 180° degrees. The software then detects where the line is interrupted by black or white areas and measures the length of these areas according to the previously defined distance between adjacent pixels. This method was carried out for every creep state on five images each in dendritic and interdendritic areas, after which the values were summarized. The mean value and the associated standard deviation are then calculated for each angle. This procedure is equivalent to the one described by Näth et al. [18]. The only difference is that the analysis here was not carried out on membranes, but on polished and etched samples.

The results of the line intersection method can be plotted polar so that the results are displayed as an ellipse. The vertical axis a refers to the γ-channel and γ'-ligament width. Since the γ-phase is removed for membrane production in the selective phase extraction, the γ-channel width is associated with the pore width in the nanoporous superalloy membrane. The horizontal axis c refers to the length of the γ- and γ'-ligaments, respectively. In the case of the line intersection method, it must be noted that due to a non-perfect alignment or the vertical connection of two γ-channels, an overestimation of the γ-channel width along the axis a may occur. In the case of the γ'-ligament length along the horizontal axis c, this can correspondingly lead to an underestimation.

3. Results

3.1. Creep Stress: 250 MPa

The results of the creep tests at a stress of 250 MPa are shown in Figure 2a–d. Exemplary images from the dendritic areas of the specimens are shown. After a creep time of 140 h, the microstructure in (a) shows an irregular raft structure. In many cases, vertical struts are present between the γ-channels. The individual channels are mostly not parallel to each other and have a curved shape. This contrasts with the microstructure in (b) after a creep period of 170 h. Compared to (a), the raft structure appears to be finer and the channels themselves are more regular and parallel in many places. Nevertheless, irregular areas and vertical connecting struts are also present here.

A further increase in the creep duration to 310 h results in a comparatively homogeneous raft structure with predominantly parallel γ-channels. Although the γ-channels have a slightly curved shape, they differ in appearance from the creep durations considered so far. There are also only a few vertical struts connecting the rafts. Overall, the structure appears very homogeneous and regular. After 506 h, the structure of the γ-channels changed in comparison to 310 h. The shape of the rafts is now more curved or bent and resembles a wavy shape. The structure is still homogeneous and regular, but the dimensions, especially the width of the channels, have increased.

The development of the dimensions of the rafting structure can also be measured and plotted using the line intersection method. Figure 3 shows the structural ellipses of the creep tests performed at a stress of 250 MPa. The results have been plotted as polar plots. Accordingly, axis a refers to width, and axis c to length. The left half of the plot shows the results from dendritic regions, and the right half from interdendritic ones. The values are given in units of nanometers. Furthermore, the distinction between γ-channels and γ'-ligaments is made, where γ-channels are indicated as dashed lines and the solid lines refer to the γ'-ligaments. For further presentation of the results from the line intersection method, the values from Table 2 will be considered in more detail. These summarize the most important parameters for the characterization of the rafting structure. Values marked a refer to the vertical axis a and correspond to the results of the line intersection procedure in which the parallel lines were placed at an angle of 90° in the image. They are used to determine the width of the pores and ligaments. The label c corresponds to the angle 0°, i.e., along the axis c, and describes the length. The subscripts d and id indicate dendritic and interdendritic areas, respectively.

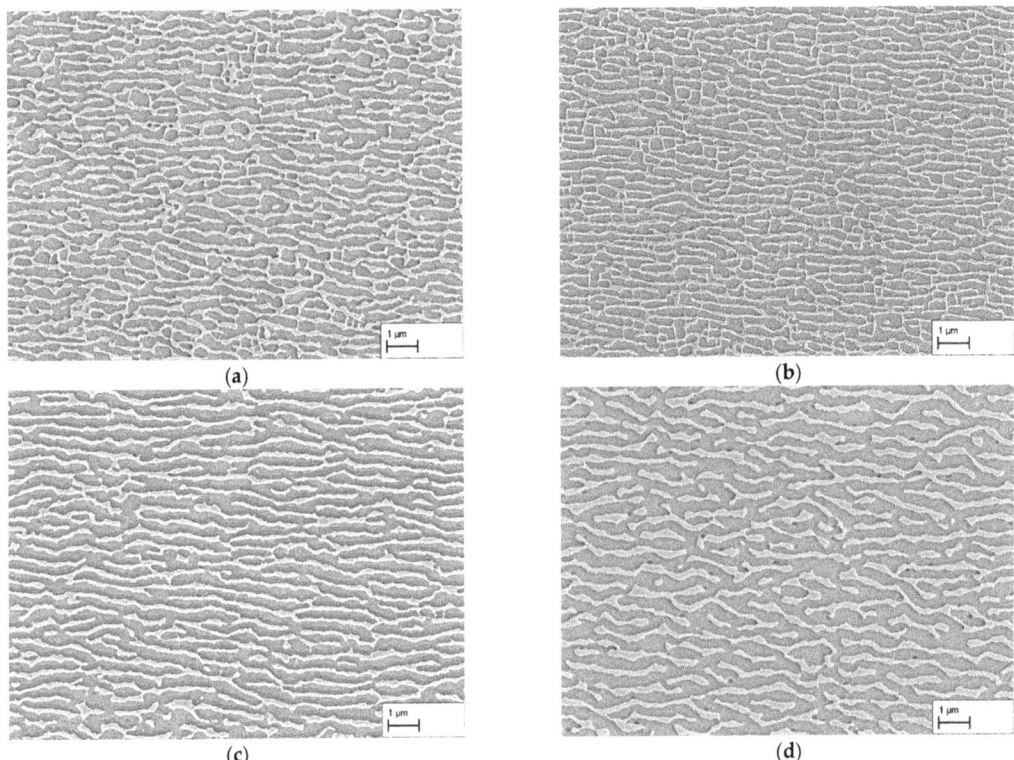

Figure 2. SEM images of dendritic areas after creep testing at 250 MPa/950 °C for durations of (**a**) 140 h, (**b**) 170 h, (**c**) 310 h, and (**d**) 506 h. Note, due to the etching of the γ'-phase, it appears darker than the γ-phase.

In both the dendritic and interdendritic regions, the progression of the γ-channel width is identical. The minimum in both cases is after a creep time of 170 h and is 108 ± 65 nm for γ-channels in the dendritic and 147 ± 123 nm in the interdendritic region. For the homogeneous structures in Figure 2c,d, 119 ± 43 nm and 171 ± 60 nm could be measured in the dendritic and 150 ± 66 nm and 164 ± 69 nm in the interdendritic regions. For the lowest creep duration of 140 h, γ-channel widths of 132 ± 70 nm in dendritic and 166 ± 137 nm in interdendritic regions are obtained. The γ-channel length shows a similar trend; the smallest channel length is found after 170 h and is 280 ± 273 nm in the dendritic and 198 ± 197 nm in the interdendritic region. It increases for the creep tests in the order 140 h, 310 h, and 506 h up to 403 ± 343 nm and 423 ± 354 nm for dendritic and interdendritic areas, respectively.

With regard to the width of the γ'-rafts, there are strong similarities to the γ-channels in the dendritic areas. The smallest width was also measured here after a creep time of 170 h and amounts to 219 ± 92 nm. This increases in the sequence 140 h, 310 h, and 506 h up to 337 ± 163 nm. In the interdendritic areas, the results are more constant. The minimum is 325 ± 163 nm after 140 h, after 170 h 338 ± 147 nm, after 310 h 334 ± 147 nm, and after 506 h 348 ± 170 nm.

Considering the interdendritic regions, a pronounced shape difference between the two ellipses for the shorter test durations (140 h, 170 h) and those for the longer ones (310 h, 506 h) is noticeable. While the ellipses show c/a-ratios of 2.50 and 2.58 for the γ-channels as well as 2.40 and 2.43 for the γ'-ligaments after 310 h and 506 h, respectively (see Table 2), the c/a-ratios after 140 h and 170 h are 1.43 and 1.35 for the γ-channels as well as 1.36 and

1.18 for the γ′-ligaments. This indicates that the interdendritic regions of the specimens with creep durations of 140 h and 170 h have a different microstructure compared to the dendritic regions after 140 h and 170 h as well as to the dendritic and interdendritic regions after 310 h and 506 h, respectively.

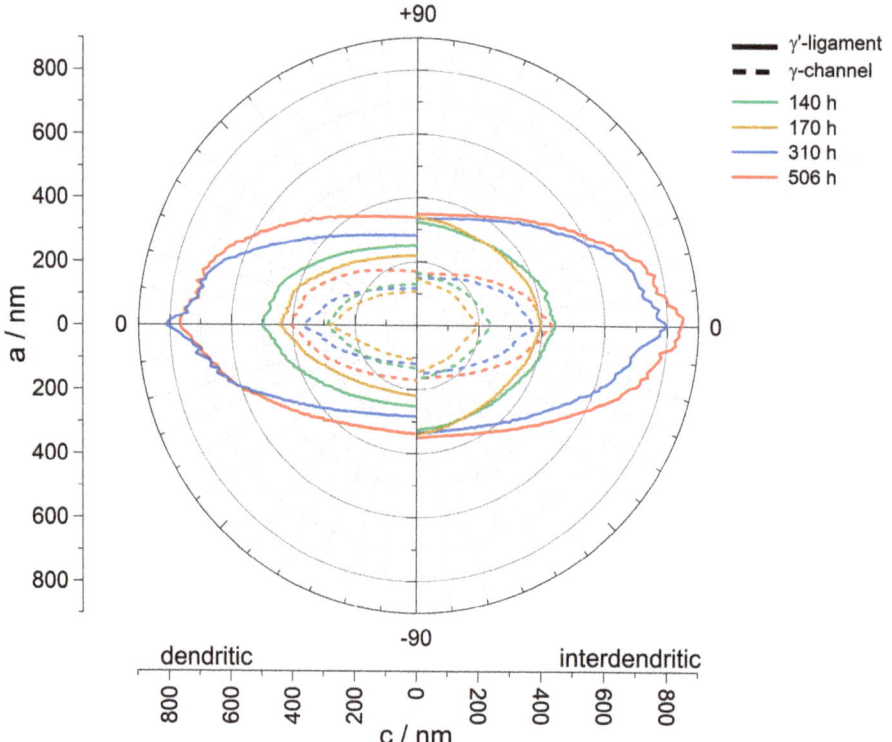

Figure 3. Structural ellipse of creep specimens A−CMSX−4/140h/250, A−CMSX−4/170h/250, A−CMSX-4/310h/250, and I−CMSX−4/506h/250. The values for γ-channels (dotted line) and γ′-ligaments (solid line) are shown. Furthermore, the left side shows the results of the dendritic, and the right side the results of the interdendritic areas. Each of these ellipses illustrates the associated length of the lines as a function of their orientation to the microstructure.

In this context, Figure 4a–d show an example comparison of the dendritic and interdendritic regions after creep durations of 140 h and 310 h, respectively. In (a), the microstructure from a dendritic region can be seen. As described previously, the microstructure shows irregular γ′-rafts and γ-channels. In contrast, the microstructure imaged in (b) is from an interdendritic region and shows mostly isolated, coarsened γ′-cubes and elongated γ-channels. Overall, the structure shown is very inhomogeneous. If we now compare Figure 4c,d from dendritic and interdendritic regions after 310 h, we see that the microstructure in both regions has a directional appearance. Although the γ′-rafts and γ-channels in (c) are more uniform and finer than in (d), a fully developed raft structure is present in both figures. This comparison shows that dendritic and interdendritic regions display different stages of directional coarsening depending on the creep duration. This applies not only to the state after 140 h creep shown here, but also to the state after 170 h creep.

Table 2. Results of the line intersection measurements of the longitudinal sections of the creep tests with a stress of 250 MPa at a test temperature of 950 °C. The values for the vertical axis a and horizontal axis c of the structural ellipse are given. The additions d and id refer to the analyzed dendritic and interdendritic areas; the additions γ- and γ′-channels po denote the respective structure.

	A-CMSX-4/140h/250	A-CMSX-4/170h/250	A-CMSX-4/310h/250	I-CMSX-4/506h/250
$c_{d,\,\gamma\text{-channel}}$	290 (261)	280 (273)	363 (329)	403 (343)
$c_{id,\,\gamma\text{-channel}}$	238 (232)	198 (197)	375 (326)	423 (545)
$a_{d,\,\gamma\text{-channel}}$	132 (70)	108 (65)	119 (43)	171 (60)
$a_{id,\,\gamma\text{-channel}}$	166 (137)	147 (123)	150 (66)	164 (59)
$c_{d,\,\gamma'\text{-channel}}$	500 (391)	443 (347)	812 (786)	766 (582)
$c_{id,\,\gamma'\text{-channel}}$	442 (334)	398 (236)	801 (679)	845 (691)
$a_{d,\,\gamma'\text{-channel}}$	251 (106)	219 (92)	284 (101)	337 (163)
$a_{id,\,\gamma'\text{-channel}}$	325 (163)	338 (169)	334 (147)	348 (170)
$(c/a)_{d,\,\gamma\text{-channel}}$	2.20	2.59	3.05	2.36
$(c/a)_{id,\,\gamma\text{-channel}}$	1.43	1.35	2.50	2.58
$(c/a)_{d,\,\gamma'\text{-ligament}}$	1.99	2.02	2.86	2.27
$(c/a)_{id,\,\gamma'\text{-ligament}}$	1.36	1.18	2.40	2.43

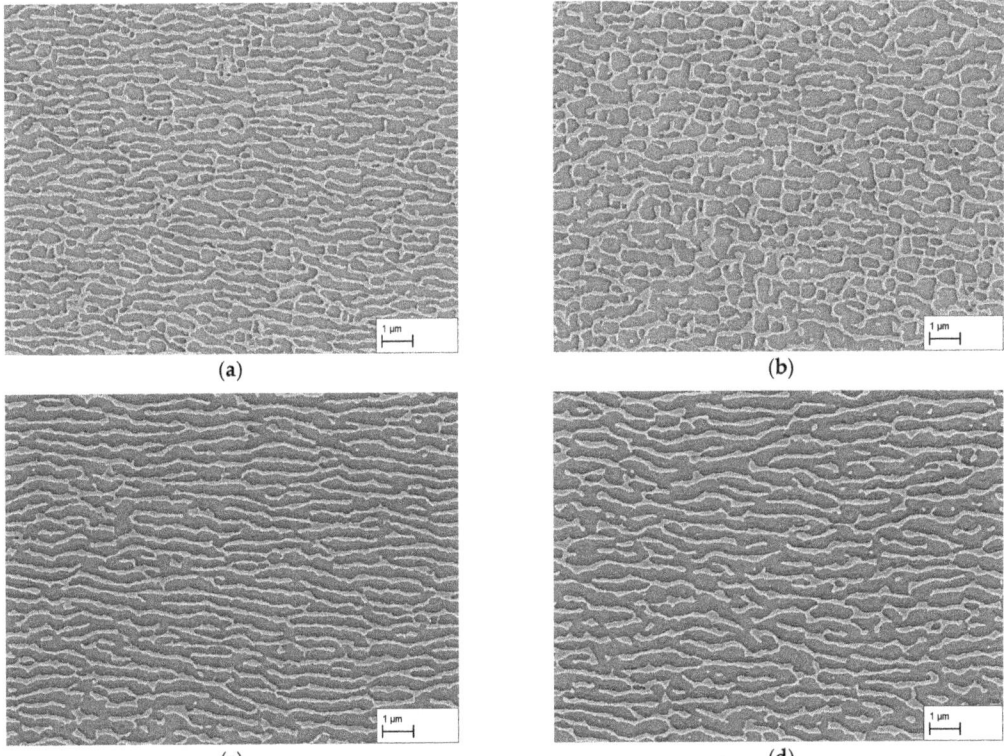

Figure 4. SEM images after creep testing at 250 MPa/950 °C for (a) dendritic and (b) interdendritic areas after 140 h; (c) dendritic and (d) interdendritic areas after 310 h. Note, due to the etching of the γ′-phase, it appears darker than the γ-phase.

The obtained results for the dendritic and interdendritic areas after 140 h and 310 h creep duration show that, depending on the location, the raft formation process is at different stages. Since complete crosslinking is required for membrane production and a connected structure must be present for the subsequent application purpose, the surfaces of the membranes manufactured from the creep samples are examined in the following. Figure 5a–d show the membrane surfaces at dendritic areas after 140 h, 170 h, 310 h, and 506 h creep deformation. In both (a) and (b), holes or breakouts of the γ'-phase can be seen on the surface demonstrating that the γ'-phase was not completely connected at those locations. In contrast, in (c) and (d), the surfaces do not show any breakouts or irregularities. Thus, crosslinking of the γ'-phase is now complete everywhere. As previously observed in Figure 2c,d, a regular microstructure is present.

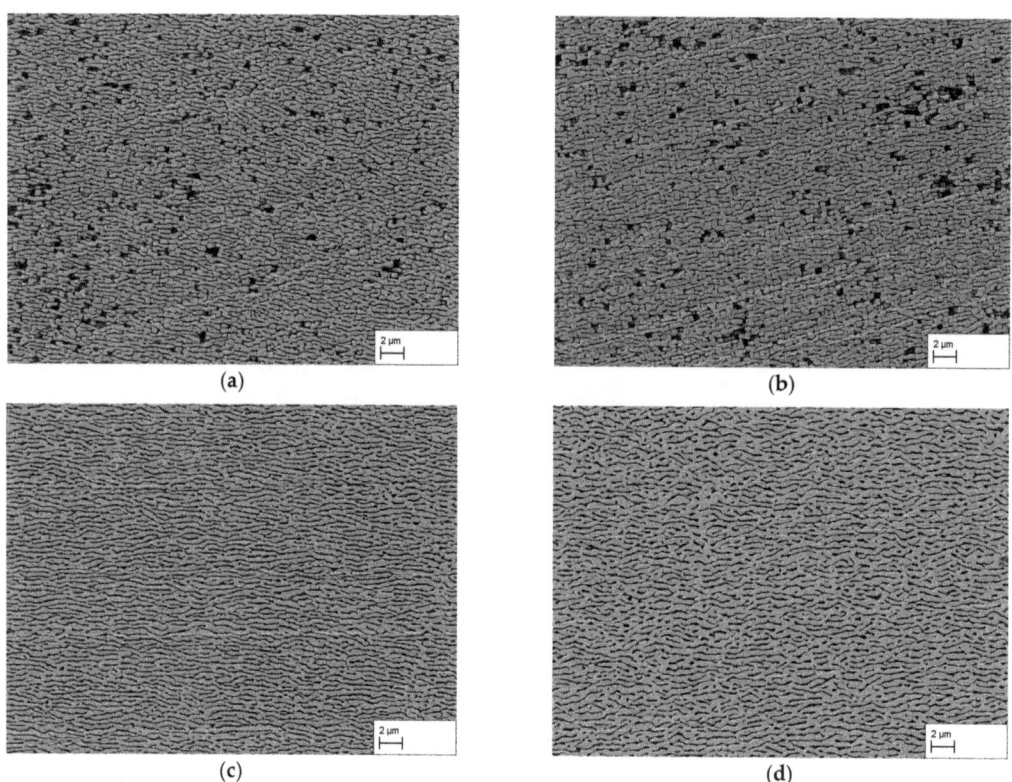

Figure 5. Membrane SEM images of dendritic areas after creep testing at 250 MPa/950 °C for durations of (**a**) 140 h, (**b**) 170 h, (**c**) 310 h, and (**d**) 506 h. Note, due to the dissolving of the γ-phase, pores appear black.

Figure 6a–d show the membrane surfaces at interdendritic regions after 140 h, 170 h, 310 h, and 506 h. After a creep duration of 140 h, a fissured surface can be seen in (a). In several areas of the image, the γ'-phase is missing over a large area, and a kind of skeleton of the remaining γ'-phase forms a fissured structure. In addition, small breakouts can be seen in the areas of the remaining γ'-phase. In (b), a similar situation after 170 h can be seen. The large-scale breakouts of the γ'-phase are still present, but are smaller than in (a). Furthermore, in (b), there are also small breakouts of the γ'-phase between the ligaments. Overall, the structures on the surface of (a) and (b) look very similar. After a creep time of 310 h, the surface in (c) shows no breakouts of the γ'-phase except near larger casting pores. This can be attributed to the altered stress state around pores during creep deformation, hampering directional coarsening at certain locations. After 506 h of creep deformation, breakouts are absent everywhere (see Figure 6d).

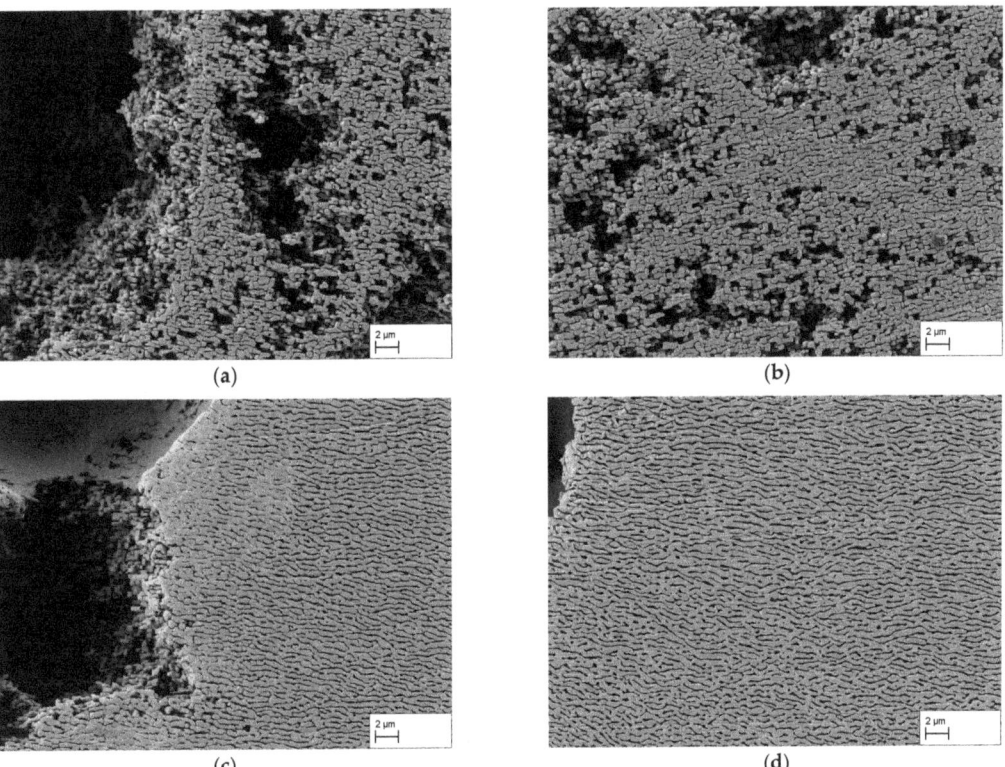

Figure 6. Membrane SEM images of interdendritic areas after creep testing at 250 MPa/950 °C for durations of (**a**) 140 h, (**b**) 170 h, (**c**) 310 h, and (**d**) 506 h. Note, due to the dissolving of the γ-phase, pores appear black.

Finally, it is possible to plot the results of this investigation for a stress of 250 MPa in a graph. Figure 7 shows the γ-channel width in nanometers as a function of the creep time in hours for the dendritic regions. Marked as a red circular dot is the initial channel width of 35 nm in the precipitation heat-treated condition. The blue point marks the result of the $3w_0$-criterion and represents the minimum channel width of 105 nm when raft formation is just completed. According to Epishin et al. [19], this occurs in the dendritic regions after 133 h for the stress of 250 MPa at 950 °C. In green, the γ-channel widths measured in the dendritic regions are indicated with the corresponding standard deviations. The diagram visualizes the results from Table 2. The obtained γ-channel width for the creep time of 140 h is 132 ± 70 nm, which is higher than the value of 105 nm given by the $3w_0$-criterion. After 170 h, the value for the γ-channel width decreases to 108 ± 65 nm and is thus very close to the value of the $3w_0$-criterion. For the creep duration of 310 h, the value increases slightly to 119 ± 43 nm and for 506 h to 171 ± 60 nm. Furthermore, it becomes clear that the γ-channel width in general increases with increasing creep duration. Nevertheless, the results for the creep duration of 140 h show a larger γ-channel width than after 170 h and 310 h. On the one hand, this can be attributed to the measurement uncertainty (see large standard deviations); on the other hand, the cross-links in the vertical direction and the non-parallel alignment of the γ-channels and γ'-ligaments after 140 h have an influence on the measurements. This can lead to an overall overestimation of the γ-channel width.

With increasing creep time, the γ-channel width increases, the cross-links decrease and the microstructure aligns better. If the decrease in the cross-links and the alignment of the γ'-ligaments and γ-channels outweigh the real channel widening, the used measurement method may result in a minimum as observed after 170 h creep duration.

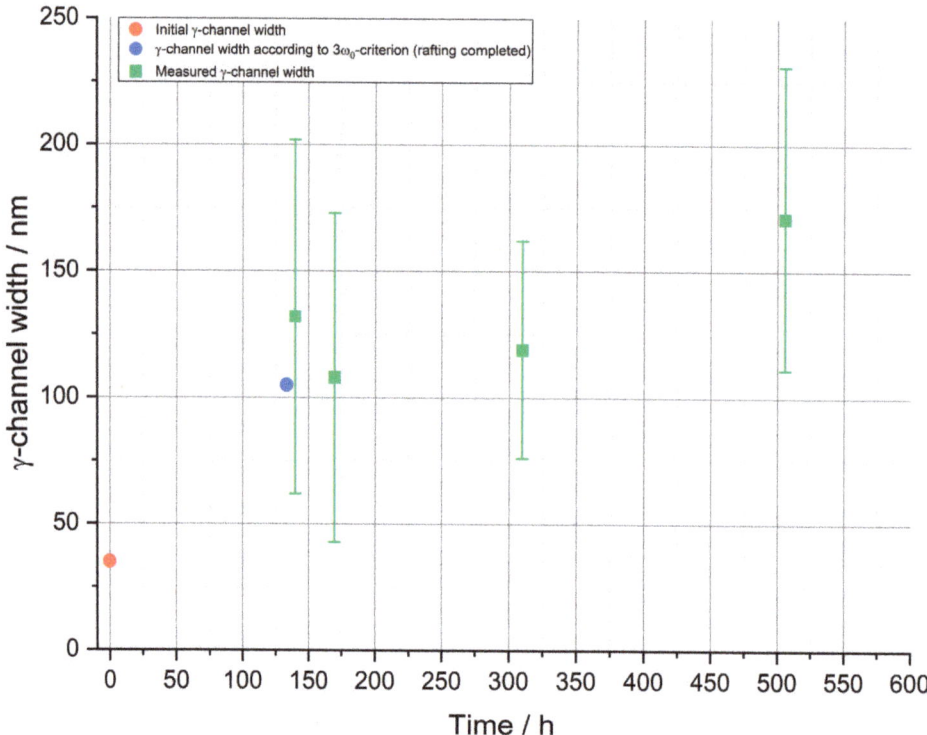

Figure 7. Summary of the γ-channel widths for the creep tests at a temperature of 950 °C and a stress of 250 MPa. The initial channel width in the precipitation heat-treated state, the minimum channel width after completion of the rafting process according to the $3w_0$-criterion and the measured channel widths (dendritic) of the creep tests are plotted.

In summary, the results of the investigation at a stress of 250 MPa show that dendritic and interdendritic regions coarsen at different rates. With the measurement method used in this study, the γ-channel width of 105 nm predicted via the $3w_0$-criterion could not be achieved in dendritic regions after a creep period of 140 h. Firstly, the still existing crosslinking between the γ-channels can be stated as a reason here, and secondly, the non-parallel alignment influences the results of the line intersection method. On the basis of the measurements and taking into account the influencing factors, however, it can be concluded that the actual γ-channel width is overestimated and that the real channel width must be smaller than the measured 132 ± 80 nm. Nevertheless, the results show that an approximation to the γ-channel width established by the $3w_0$-criterion is possible via variation in creep duration. Using two-dimensional SEM images, the smallest γ-channel width with a complete directionally coarsened microstructure in dendritic and interdendritic regions could be observed after 310 h, whereas the creep durations of 140 h and 170 h show directionally coarsened structures only in the dendrites and after 506 h larger structures have already emerged. Considering the later application in premix membrane emulsification, membranes were prepared from the tested creep samples. The results from the observation of the surfaces show that the membranes from the creep samples with a creep time of 140 h and 170 h do not have complete crosslinking in dendritic and interdendritic areas. While in dendritic areas, isolated parts of the γ'-phase fell out during the dissolution of the γ-phase, the interdendritic areas of both membranes show more extensive breakouts of the γ'-phase. Complete crosslinking can be observed in dendritic areas after a creep time of 310 h. Contrary to the results from the longitudinal sections, breakouts can also be seen in interdendritic areas in the region around the casting pores. Note that this does not hinder membrane application so that a creep duration of 310 h is considered optimal when the finest possible pores are the objective. Finally, after a creep time of 506 h, dendritic and interdendritic areas show complete crosslinking everywhere.

3.2. Creep Stress: 183 MPa and 140 MPa

Figure 8a–d show the dendritic microstructures at a stress of 183 MPa as a function of time. In (a), after 140 h, as well as in (b), after 170 h, a cubic precipitation structure can be seen. In contrast to (a), the structure in (b) appears more inhomogeneous, and the γ'-precipitates are in many places no longer cube-shaped but rectangular. In (c), a raft structure is already present after 310 h, and a large number of vertical γ-channels continues to be present, connecting the horizontal and parallel γ-channels. The structure in (c) is comparatively homogeneous. Measurements after 310 h result in a γ-channel width of 141 ± 130 nm. In (d), the dimensions of the γ-channels have increased after 506 h. Moreover, the γ'-rafts have increased in size. Vertical γ-channels are still present but they are more irregular with respect to their shape or run increasingly in a curved form. Measurements after 506 h result in a γ-channel width of 147 ± 85 nm.

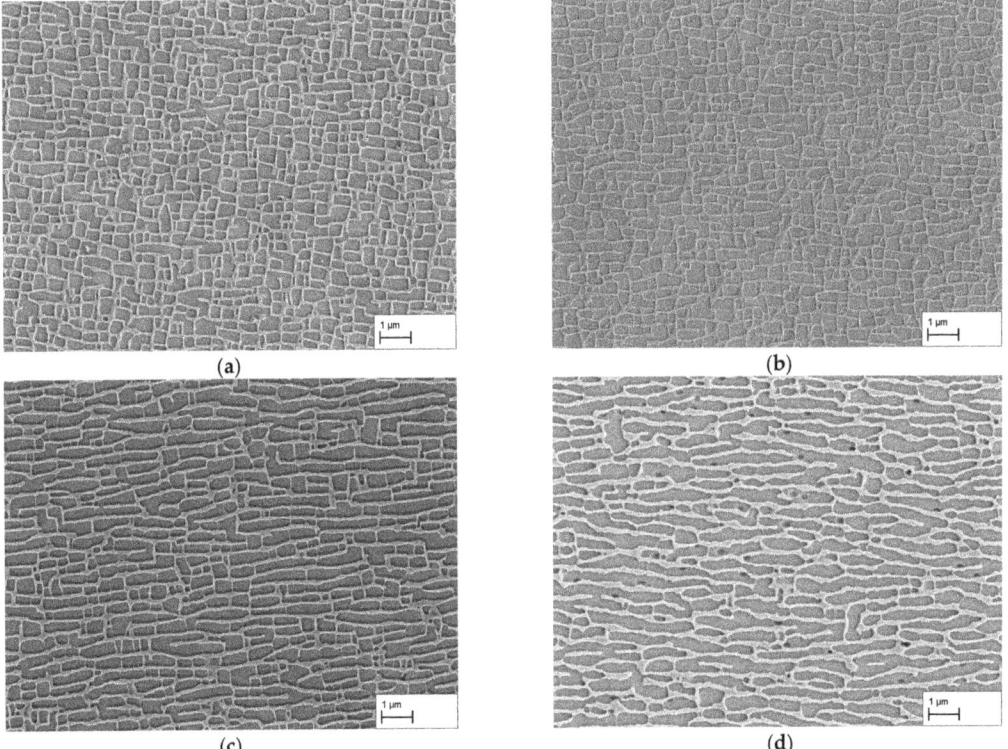

Figure 8. SEM images of dendritic areas after creep testing at 183 MPa/950 °C for durations of (**a**) 140 h, (**b**) 170 h, (**c**) 310 h, and (**d**) 506 h. Note, due to the etching of the γ'-phase, it appears darker than the γ-phase.

Finally, Figure 9a–d show the microstructures of the creep specimens with a stress of 140 MPa. In (a), after 140 h, and (b), after 170 h, a cubic γ'-precipitate structure can be seen; although, the γ'-particles are less regularly arranged in (b) than in (a), so that the microstructure appears more inhomogeneous. In (c), after a creep time of 310 h, the first rectangular γ'- particles are visible in addition to cubic γ'-precipitates. In comparison to (d), it is noticeable that the γ'-particles are very angular, whereas in (d), they have lost their angularity after 510 h. At the same time, several elongated γ'-precipitates are visible. Nevertheless, it is not possible to speak of a raft structure here.

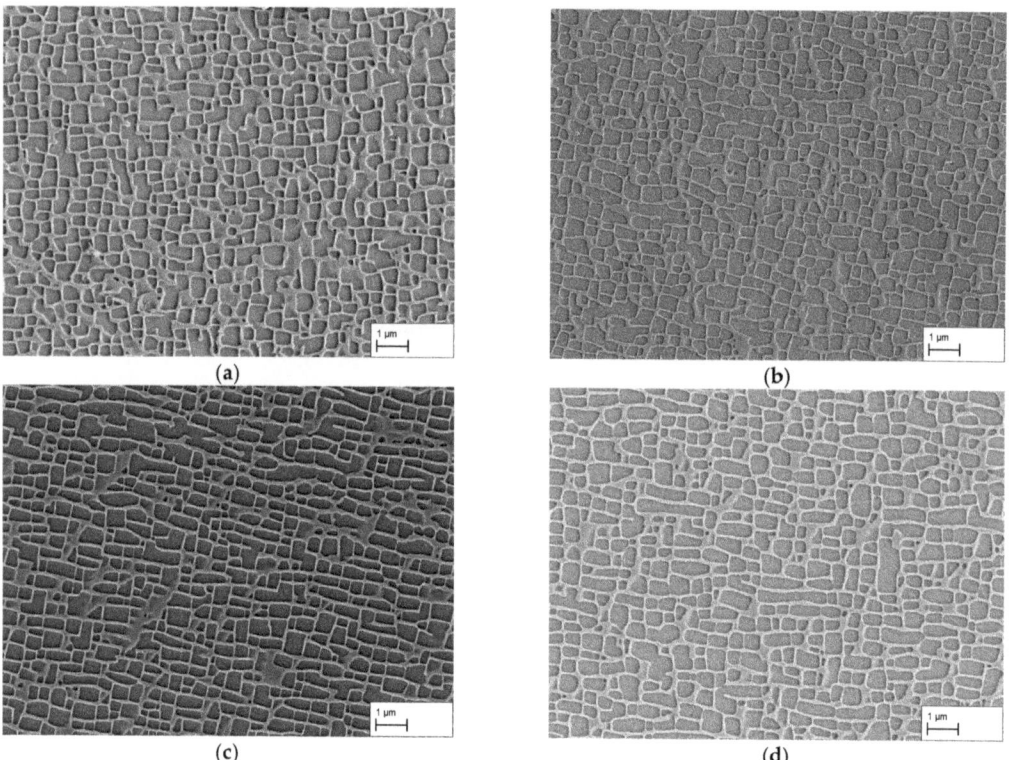

Figure 9. SEM images of dendritic areas after creep testing at 140 MPa/950 °C for durations of (a) 140 h, (b) 170 h, (c) 310 h, and (d) 506 h. Note, due to the etching of the γ'-phase, it appears darker than the γ-phase.

4. Discussion

By loading the creep specimens along the [001]-orientation at constant temperature and stress, a directionally coarsened γ/γ'-microstructure was observed after a sufficiently long creep period. This is an N-type raft structure that occurs at negative misfit δ of the alloy. The alloy CMSX-4 has a negative misfit δ at room temperature, which becomes more negative with increasing temperature. The formation of type-N rafts is thus consistent with results from the literature, for example [6–8,14,19,20,26–29]. For a stress of 250 MPa, a directionally coarsened microstructure in the dendritic and interdendritic regions was observed at a creep time of 310 h. Moreover, after 140 h and 170 h, rafting of the γ'-precipitates is essentially complete in the dendritic domains. However, the interdendritic areas show only partial areas in which directional coarsening, i.e., raft formation, has taken place (see Figure 4). According to Reed et al. [14] the reason for these different rates of raft formation is the chemical imbalance between dendritic and interdendritic areas, which leads to different γ'-volume contents, different misfits, and chemical compositions of the phases. Aluminum and tantalum, in particular, preferentially accumulate in interdendritic regions, so that the γ'-volume content is greater here than in dendritic regions. This leads to the γ-channels being narrower, which results in greater creep resistance. Additionally, Völkl et al. have shown that the misfit is more positive in the interdendritic regions than in the dendritic ones in the case of CMSX-4 [30]. These aspects explain the observation made here that raft formation in the interdendritic regions is delayed compared to the dendritic ones, which is why the creep durations of 140 h and 170 h are not suitable to obtain a completely directionally coarsened microstructure. Nevertheless, the results show that approximating the optimal creep duration for a directionally coarsened microstructure via the channel widening rate $\dot{w}_1$ determined by Epishin et al. [20] is reasonable, since the results can be used as a starting point for further experiments. Since the channel widening rate was determined in dendritic regions and the results of our investigation show raft structures in the same already after 140 h, a consistency is given. However, because the raft structure after 140 h and 170 h contains vertical γ-channels to a significant extent, there is a considerable overestimation of the γ-channel width using the line intersection method, especially after 140 h, which results in a γ-channel width of 132 ± 70 nm. Taking this into account, the actual channel width appears to be close to three times its initial value $3w_0 = 35$ nm, in agreement with the $3w_0$-criterion for just-completed raft formation proposed by Epishin et al. [19]. Note that the same holds true for the width of 108 ± 65 nm measured after 170 h.

The results from Figure 7 show that with the completion of raft formation in the dendritic regions at about 140 h to 170 h, the widening rate of the γ-channels decreases considerably. This is also consistent with the results of Epishin et al. [20] and has been explained by the higher stability of the plate-like γ'-structure compared to the cube-like one as well as different coarsening mechanisms associated with these different γ'-morphologies. Considering the fabrication of nanoporous superalloy membranes where a bicontinuous γ/γ'-microstructure is required prior to selective phase extraction, the slowed coarsening kinetics is of great benefit. It allows creep durations to be selected where rafting is completed not only in the dendritic but also in the interdendritic regions with limited further coarsening of the already-rafted dendritic areas. In this respect, a creep duration of 310 h at 250 MPa appears to be favorable. Then, the interdendritic regions are fully rafted (see Figure 4d) while the γ-channel width in the dendritic ones is still close to the minimum possible value (see Figure 7). Even though isolated breakouts still occur at casting pores in the interdendritic regions (see Figure 6c), they are not expected to affect the mechanical strength or functionality of the membrane in any appreciable way because they neither significantly reduce the membrane's cross-section nor do they cause a short-circuit for the flow of a medium through the membrane.

The other results for the stresses of 183 MPa and 140 MPa show that the raft formation process becomes less complete with decreasing stress at constant creep duration (see Figures 8 and 9). According to Kamaraj et al. [6] and Pollock et al. [7], the driving force for the raft formation process is the increase in dislocation density in the γ-channels when the creep specimen is loaded. The introduced dislocations reduce or increase the misfit δ in the horizontal and vertical channels, respectively, and thus the associated elastically stored energy. Due to the resulting change in thermodynamic driving forces and chemical potentials, raft formation occurs, eliminating the energetically less favorable vertical γ'-channels. Thus, the greater the stress in the creep test, the more dislocations are introduced into the γ-channels within a certain time period and the greater the thermodynamic driving force for the raft formation process. This is the reason for the increasingly incomplete raft formation process with decreasing stress at constant temperature and creep duration. It can also be inferred from the results that the channel widening rate during the process of raft formation becomes smaller with decreasing stress, which is in agreement with the results of Epishin et al. [20]. This is simply due to the fact that the dislocations in the γ'-channels, and with them the thermodynamic driving force for rafting, evolve more slowly with decreasing stress.

Another observation was made during the creep test at 183 MPa and 506 h. According to Epishin's $3w_0$-criterion and the stress-dependent widening rate [20], the calculated minimum channel width of 105 nm would have to be reached after a longer creep time at 183 MPa than at a50 MPa. In certain respects, this is consistent with the experimental results. They show that raft formation is complete after 506 h/183 MPa in the dendritic areas in a similar way as after 140 h and 170 h at 250 MPa (compare Figure 8d with Figure 2a,b) with numerous vertical γ-channels still present in all these cases. However, as the width of the horizontal channels is already 141 ± 130 nm in the former case, i.e., larger than $3w_0$ and the widths after 140–170 h/250 MPa, there is an additional factor influencing raft formation at this point. It is reasonable to assume that, as already observed by Epishin et al. [20], undirectional coarsening of the γ/γ'-microstructure by Ostwald ripening takes place in parallel to the ongoing raft formation process, becoming the more important the slower the rafting is, i.e., the lower the applied stress. For this reason, the minimum possible raft width increases with decreasing stress and the $3w_0$-criterion has to be seen as a lower limit in the absence of undirectional coarsening. This is also suggested by the observations of Chen et al. [31] who reasoned that the increase in thickness of the γ'-rafts is related to Ostwald ripening as a function of the initial size of the γ'-particles and the coarsening rate. This results not only in the coarsening of the γ'-rafts but also of the γ-channels. It can be concluded that the channel widening rate at 183 MPa is so low that undirectional coarsening already takes place to a significant extent during the raft formation process. Considering the application of interest here, it means that a creep stress of 250 MPa is favored over the lower ones inspected here. As the achieved channel width is already close the lower limit of $3w_n$, even higher stresses are expected to be of limited benefit.

5. Conclusions

In this study, the γ-channel width was investigated as a function of creep duration and stress using staged creep specimens. Stresses of 250 MPa, 183 MPa, and 140 MPa and creep durations of 140 h, 170 h, 310 h, and 506 h were considered. The test temperature was 950 °C in all cases. The following conclusions can be drawn from this investigation:

1. The use of stepped creep specimens with three stress ranges allows the simultaneous investigation of different creep states within one creep test. This can reduce material usage and make it easier to find the optimum combination of temperature, stress, and creep duration.
2. The $3w_0$-criterion is suitable for approximating the minimum γ-channel width in the dendritic regions of the single-crystalline Ni-based superalloy CMSX-4 achieved at the point in time when rafting is just completed. The calculated minimum γ-channel width of 105 nm could be essentially reached after 170 h/250 MPa with a measured γ-channel width of 108 ± 65 nm in the dendritic regions. Due to the chemical imbalances between dendritic and interdendritic regions, no raft structure was present in the interdendritic regions at this point in time, which does not allow for the use of this microstructure for membrane fabrication.
3. The combination of a temperature of 950 °C, a stress of 250 MPa, and a creep time of 310 h achieves the requirements we set for a membrane structure, leading to a γ-channel width of 119 ± 43 nm in dendritic and 150 ± 66 in interdendritic regions and complete crosslinking. Further characterizations such as mechanical strength or emulsification behavior of nanoporous superalloy membranes produced by these parameters have to be performed.
4. In combination with the test temperature of 950 °C, the stresses 183 MPa and 140 MPa do not lead to suitable microstructures for the fabrication of membranes for premix membrane emulsification as the lower rafting speed in combination with undirectional coarsening results in larger γ-channels than required.

5. The use of previous research results to investigate the microstructure development of CMSX-4 in the high-temperature range could be successfully transferred to the field of membrane development and contribute to further optimization of Ni-based nanoporous membranes. For future application in the field of premix membrane emulsification and the production of colloidal lipid-based excipients, further experiments have to be carried out, for example, on the mechanical properties or resulting droplet size.

Author Contributions: Conceptualization, J.R. and J.M.L.; methodology, J.M.L.; validation, J.M.L.; investigation, J.M.L.; resources, J.R.; writing—original draft preparation, J.M.L. and J.R.; writing—review and editing, J.R.; visualization, J.M.L.; supervision, J.R.; project administration, J.R.; funding acquisition, J.R. All authors have read and agreed to the published version of the manuscript.

Funding: This research was funded by Deutsche Forschungsgemeinschaft (DFG, German Research Foundation), grant number 451891484. We also acknowledge support by the Open Access Publication Funds of Technische Universität Braunschweig.

Institutional Review Board Statement: Not applicable.

Informed Consent Statement: Not applicable.

Data Availability Statement: Not applicable.

Acknowledgments: We are grateful for Heiko Meißner's support in setting up and conducting the experiments.

Conflicts of Interest: The authors declare no conflict of interest.

References

1. Kohnke, M.; Finke, J.H.; Kwade, A.; Rösler, J. Investigation of Nanoporous Superalloy Membranes for the Production of Nanoemulsions. *Metals* **2018**, *8*, 361. [CrossRef]
2. Rösler, J.; Näth, O.; Jäger, S.; Schmitz, F.; Mukherjiis, D. Nanoporous Ni-based superalloy membranes by selective phase dissolution. *J. Miner. Met. Mater. Soc.* **2005**, *57*, 52–55. [CrossRef]
3. Rösler, J.; Mukherji, D. Design of Nanoporous Superalloy Membranes for Functional Applications. *Adv. Eng. Mater.* **2003**, *5*, 916–918. [CrossRef]
4. Rösler, J.; Voelter, C. Nanoporous Superalloy Membranes: A Review. *Adv. Eng. Mater.* **2018**, *20*, 1–13. [CrossRef]
5. Rösler, J.; Näth, O.; Jäger, S.; Schmitz, F.; Mukherji, D. Fabrication of nanoporous Ni-based superalloy membranes. *Acta Mater.* **2005**, *53*, 1397–1406. [CrossRef]
6. Kamaraj, M. Rafting in single crystal nickel-base superalloys—An overview. *Sadhana* **2003**, *28*, 115–128. [CrossRef]
7. Pollock, T.; Argon, A. Directional coarsening in nickel-base single crystals with high volume fractions of coherent precipitates. *Acta Met. Mater.* **1994**, *42*, 1859–1874. [CrossRef]
8. Nabarro, F.R.N. Rafting in Superalloys. *Metall. Mater. Trans. A Phys. Metall. Mater. Sci.* **1996**, *27*, 513–530. [CrossRef]
9. Baldan, A. Review Progress in Ostwald ripening theories and their applications to the γ'-precipitates in nickel-base superalloys Part II Nickel-base superalloys. *J. Mater. Sci.* **2002**, *37*, 2379–2405. [CrossRef]
10. Siebörger, D.; Knake, H.; Glatzel, U. Temperature Dependence of the Elastic Moduli of the Nickel-Base Superalloy CMSX-4 and Its Isolated Phases. 2001. Available online: www.elsevier.com/locate/msea (accessed on 29 November 2021).
11. Link, T.; Epishin, A.; Brückner, U.; Portella, P. Increase of misfit during creep of superalloys and its correlation with deformation. *Acta Mater.* **2000**, *48*, 1981–1994. [CrossRef]
12. Touratier, F.; Andrieu, E.; Poquillon, D.; Viguier, B. Rafting microstructure during creep of the MC2 nickel-based superalloy at very high temperature. *Mater. Sci. Eng. A* **2009**, *510–511*, 244–249. [CrossRef]
13. Buffiere, J.; Ignat, M. A dislocation based criterion for the raft formation in nickel-based superalloys single crystals. *Acta Met. Mater.* **1995**, *43*, 1791–1797. [CrossRef]
14. Matan, N.; Cox, D.; Rae, C.; Reed, R. On the kinetics of rafting in CMSX-4 superalloy single crystals. *Acta Mater.* **1999**, *47*, 2031–2045. [CrossRef]
15. Matan, N.; Cox, D.C.; Carter, P.; Rist, M.A.; Rae, C.M.F.; Reed, R.C. Vreep Of CMSX-4 Superalloy Single Crystals: Effects Of Misorientation And Temperature. *Acta Mater.* **1999**, *47*, 1549–1563. [CrossRef]
16. Kamaraj, M.; Serin, K.; Kolbe, M.; Eggeler, G. Influence of stress state on the kinetics of γ-channel wideningduring high temperature and low stress creep of the single crystalsuperalloy CMSX-4. *Mater. Sci. Eng.* **2001**, *319–321*, 796–799. [CrossRef]
17. Biermann, H.; Kuhn, H.-A.; Ungár, T.; Hammer, J.; Mughrabi, H. Internal Stresses, Coherency Strains and Local Lattice Parameter Changes in a Creep-Deformed Monocrystalline Nickel-Base Superalloy. *High Temp. Mater. Process.* **1993**, *12*, 21–30. [CrossRef]
18. Rösler, J.; Näth, O. Mechanical behaviour of nanoporous superalloy membranes. *Acta Mater.* **2010**, *58*, 1815–1828. [CrossRef]

19. Epishin, A.; Link, T.; Klingelhöffer, H.; Fedelich, B.; Brückner, U.; Portella, P. New technique for characterization of microstructural degradation under creep: Application to the nickel-base superalloy CMSX-4. *Mater. Sci. Eng. A* **2009**, *510–511*, 262–265. [CrossRef]
20. Epishin, A.; Link, T.; Nazmy, M.; Staubli, M.; Nolze, G. Microstructural Degradation of CMSX-4: Kinetics and Effect on Mechanical Properties. *Superalloys* **2008**, 725–731. [CrossRef]
21. Lück, J.M.; Rösler, J. Reducing the γ'-Particle Size in CMSX-4 for Membrane Development. *Materials* **2022**, *15*, 1320. [CrossRef] [PubMed]
22. Epishin, A.; Fedelich, B.; Finn, M.; Künecke, G.; Rehmer, B.; Leistner, C.; Nolze, G.; Petrushin, N.; Svetlov, I. Investigation of Elastic Properties of the Single-Crystal Nick-el-Base Superalloy CMSX-4 in the Temperature Interval be-tween Room Temperature and 1300 °C. *Crystals* **2021**, *11*, 152. [CrossRef]
23. Cheng, K.; Jo, C.; Jin, T.; Hu, Z. Influence of applied stress on the γ' directional coarsening in a single crystal superalloy. *Mater. Des.* **2009**, *31*, 968–971. [CrossRef]
24. ImageJ. ImageJ—Image Processing and Analysis in Java. Available online: https://imagej.net/ij/index.html (accessed on 18 June 2021).
25. Trainable Weka Segmentation. Available online: https://imagej.net/plugins/tws/ (accessed on 18 June 2021).
26. Durand-Charre, M. *The Microstructure of Superalloys the Microstructure of SUPERALLOYS*; Routledge: London, UK, 2017. [CrossRef]
27. Epishin, A.; Link, T. Mechanisms of high-temperature creep of nickel-based superalloys under low applied stresses. *Philos. Mag.* **2004**, *84*, 1979–2000. [CrossRef]
28. Reed, R.C. *The Superalloys: Fundamentals and Applications*; Cambridge University Press: Cambridge, UK, 2006.
29. Miura, N.; Kondo, Y.; Ohi, N. The Influence of Dislocation Substructure on Creep Rate during Accelerating Creep Stage of Single Crystal Nickel-based Superalloy CMSX-4. *Superalloys* **2000**, 377–385. [CrossRef]
30. Völkl, R.; Glatzel, U.; Feller-Kniepmeier, M. Measurement of the lattice misfit in the single crystal nickel based superalloys cmsx-4, srr99 and sc16 by convergent beam electron diffraction. *Acta Mater.* **1998**, *46*, 4395–4404. [CrossRef]
31. Chen, W.; Immarigeon, J.-P. Thickening behaviour of γ' precipitates in nickel base superalloys during rafting. *Scr. Mater.* **1998**, *39*, 167–174. [CrossRef]

Disclaimer/Publisher's Note: The statements, opinions and data contained in all publications are solely those of the individual author(s) and contributor(s) and not of MDPI and/or the editor(s). MDPI and/or the editor(s) disclaim responsibility for any injury to people or property resulting from any ideas, methods, instructions or products referred to in the content.

Article

Investigation of the Technological Possibility of Manufacturing Volumetric Shaped Ductile Cast Iron Products in Open Dies

Vladimir Galkin [1], Andrey Kurkin [2,*], Gennady Gavrilov [3], Ilya Kulikov [1] and Evgeny Bazhenov [3]

1 Department of Mechanical Engineering Technological Complexes, Nizhny Novgorod State Technical University n.a. R.E. Alekseev, 603950 Nizhny Novgorod, Russia
2 Department of Applied Mathematics, Nizhny Novgorod State Technical University n.a. R.E. Alekseev, 603950 Nizhny Novgorod, Russia
3 Department of Materials Science, Materials Technology and Heat Treatment of Metals, Nizhny Novgorod State Technical University n.a. R.E. Alekseev, 603950 Nizhny Novgorod, Russia
* Correspondence: aakurkin@nntu.ru

Citation: Galkin, V.; Kurkin, A.; Gavrilov, G.; Kulikov, I.; Bazhenov, E. Investigation of the Technological Possibility of Manufacturing Volumetric Shaped Ductile Cast Iron Products in Open Dies. *Materials* 2023, *16*, 274. https://doi.org/10.3390/ma16010274

Academic Editor: Konstantin Borodianskiy

Received: 2 December 2022
Revised: 22 December 2022
Accepted: 25 December 2022
Published: 28 December 2022

Copyright: © 2022 by the authors. Licensee MDPI, Basel, Switzerland. This article is an open access article distributed under the terms and conditions of the Creative Commons Attribution (CC BY) license (https:// creativecommons.org/licenses/by/ 4.0/).

Abstract: Information about the technological possibility of stamping in open dies, round-shaped forgings made of ductile cast iron, is outlined herein. Cast iron's propensity for plastic deformation under complex loading conditions is analyzed from the standpoint of the morphology of graphite inclusions, depending on the degree and mechanical scheme of deformation. The research methodology included: the choice of a brand of cast iron with spherical graphite and the technological process of its deformation; mathematical modeling of the deformation process using the DEFORM–3D software package; stamping of an experimental batch of forgings; and microstructural studies of the forging material, together with the data of its stress–strain state and the direction of flow of the material. Under the conditions of comprehensive compression of the workpiece material at the beginning of the deformation process, lateral pressure was created from the tool walls. It was carried out by selecting the size and shape of the initial blank by mathematical modeling. When analyzing the morphology of graphite inclusions, the stress state scheme was determined as the main influencing factor. The greatest change in the shape and size of graphite inclusions of cast iron, including their crushing, corresponds to the condition of the disappearance of all-round compression. When stamping in open dies, this occurs in the area where the material exits from the stamp engraving into the flash gutter. In the process of forming, cast iron showed the possibility of deformation in the deformation intensity index $\varepsilon_i = 2.5$.

Keywords: forging in open die; ductile cast iron; ultimate plasticity; analysis of the stress–strain state of the material; microstructural studies

1. Introduction

In mechanical engineering, one of the important practical tasks is the manufacture of products from hard-to-deform metal materials, one of which is cast iron. As a structural material, grey cast iron contains, in the microstructure, graphite, which is widely used for the manufacture of various parts. Compared with steels, it has a number of advantages: it works well on friction, is wear-resistant, and dampens vibration. However, cast iron is inferior in strength due to the low level of viscosity and fracture resistance, which is due, on the one hand, to the presence of graphite particles, and on the other, to defects in the structure of the cast metal [1–5]. The first studies on plastic deformation of cast iron containing graphite in its structure date back to the end of the 1940s. The starting point of these studies was the experiments of T. Karman, Becker [4,5] and the fundamental works of P.V. Bridgman (1947) [6–8], the results of which showed that even brittle materials, such as marble and red sandstone, can be plastically deformed under conditions of uneven all-round compression. In practice, the scheme of uneven triaxial compression was implemented in the 1950s in the technological processes of extrusion, pressing, and volumetric stamping.

As concrete examples of the manufacture of cast iron blanks, one can cite: hot rolling of a sheet to a thickness of 1.5–2.1 mm from a rectangular-shaped slouch in several passes and the production of pipes by pressing, stitching, and rolling, obtaining grinding bodies (balls) by the process of cross rolling [7]. The mechanical properties of cast iron depend on the shape of graphite [1]. Analysis of the deformability of cast irons with different graphite shapes has shown that as graphite inclusions approach the spherical shape, the ductility of cast iron increases significantly. The spheroidal shape of graphite does not have a strong incising effect on the metal base, as a result of which a stress concentration occurs to a lesser extent around graphite spheroids. For comparison, the value of elongation in tensile tests for cast iron with lamellar graphite is 0.2–0.5%, and with spherical graphite, it is 10–12%. Ductile cast iron has a high ratio of conditional yield strength $\sigma_{0.2}$ to the temporary tensile resistance σ_b, which is 0.6–0.7, which allows it to be used as a structural material along with steel (steel: 0.5–0.6) [9]. The second decade of our century has been marked by a large number of studies in the field of hot plastic deformation of ductile cast iron, in particular, those carried out by scientists of the Belarusian school of deformation [10–12]. The research direction concerned changes in the structure, in particular, the shape of graphite particles, under various deformation schemes and their influence on mechanical properties. The main results of the research were formulated into the following conclusions:

- The structural, mechanical, and physical properties of cast iron are determined by the shape of graphite inclusions, the changes in which are determined by the deformation of the metal base (matrix) of cast iron [13,14].
- Depending on the mechanical deformation scheme implemented, graphite inclusions of cast iron acquire a different shape: a disc-shaped shape under precipitation conditions, an elongated shape during extrusion [11].
- Studies of plastic deformation of cast iron under conditions of complex loading are rationally carried out using technological samples in which deformation schemes are implemented in relation to the manufacture of a certain class of products [15].
- The plastic flow of brittle graphite inclusions inside a metal matrix must be performed under conditions of uneven all-round compression; however, it is possible to use deformation schemes with the presence of tensile stresses [16,17]. As a confirmation of the use of deformation schemes that do not fully correspond to comprehensive compression, examples of the developed processes are given: backward extrusion in one pass of the "electric drill chuck body" part, hot transverse three-roll rolling of parts such as rods and shafts, and cold surface rolling with rollers [18,19].

When analyzing the developed processes of plastic deformation of cast iron to date, the practical interest in the manufacture of volumetric shaped products using deformation technological schemes that differ from the schemes of the implemented processes should be noted. In particular, this applies to the process of hot volumetric stamping in open dies, which makes it possible to produce products of various shapes. The shaping of the material in open dies is characterized by uneven all-round compression at an average total pressure and incomplete lateral pressure of the metal on the rigid walls of the tool. During deformation, part of the metal is displaced from the engraving in the form of a technological burr, which causes local free broadening of the metal in the area of the exit into the burr groove, creating additional tensile stresses that contribute to destruction. The structural features of the deformation focus and the periods of its change during stamping in open dies are described in detail in the works of M.V. Storozhev, S.I. Gubkin, E. I. Semenov, and other scientists [20,21]. To date, hot stamping of volumetric shaped ductile cast iron products in open dies has no practical implementation, which allows us to conclude that this research topic is relevant to determine its technological capabilities for the manufacture of specific classes of products. The mechanical properties of the material depend on its metallurgical and structural condition. In the process of plastic deformation of the material, the defect and structure parameters change: grain size, morphology of the structure, residual micro, and macro stresses. Therefore, the problem of determining the technological feasibility of manufacturing volumetric shaped ductile cast iron products in open dies requires solving theoretical issues to establish patterns of changes in its structural state depending on the deformation conditions, which was the purpose of this work.

2. Materials and Methods

The solution to this problem involves the application of a comprehensive research methodology, which includes: mathematical modeling of the process, with the determination of the optimal dimensions of the workpiece and the assessment of the stress–strain state and the nature of the flow of the material in its volume during deformation; stamping of the experimental batch of forgings; and structural studies of the deformed material.

Ductile cast iron in a ferrite–perlite matrix with the following characteristics was selected as the material under study:

- The strength characteristics are hardness 170–180 HB, which corresponds to the ultimate strength 490–530 MPa.
- The permissible degree of deformation in hot precipitation is ε_h = 12%.
- The chemical composition of cast iron is shown in Table 1.

Table 1. Chemical composition of cast iron.

Content of the Elements, %							
C	Si	Mn	Ni	Cr	Mg	S	P
2.60	2.15	0.60	0.38	0.14	0.07	0.019	0.09

In order to clarify the ductility parameters of cast iron used for deformation, tests were carried out on the hot sediment of turned cylindrical samples with a diameter of 30 mm and a height of 60 mm from the batch of material used. The samples were heated in the furnace to a temperature of 1025–1050 °C and deformed by precipitation on the press with degrees of deformation of 10, 15, and 20%. In order to avoid contact with cold strikers, the deflection was carried out between heated plates. The results obtained showed that the cast iron used for testing has a ductility with a degree of deformation at a hot draft of ε_h = 20%. It was also noted that the cooling of the deformed material must be carried out in the furnace at a speed of 100 °C per hour [22]. As a technological test, the process of hot volumetric forging of a round-shaped "flange" part in an open die on a crank hot-stamping press (JSC "Tjazhmekhpress", Voronezh, Russia) was chosen. In the manufacture of forgings, a technological scheme was chosen for deforming a blank piece of a round shape into an end face without the use of a precipitation operation.

Mathematical modeling was carried out using the DEFORM–3D software package (SFTC, Columbus, OH, USA, Ver. 6.0).

The optimal dimensions of the initial workpiece were determined based on the possibility of creating conditions of all-round uneven compression in it during deformation. According to the modeling results, the dimensions of the workpiece were selected: height 30 mm and diameter 56 mm, which is 4 mm less than the internal size of the flange section of the working cavity of the stamp.

Evaluation of the stress–strain state in the volume of the product was carried out in areas corresponding to microstructural studies. The schemes of the deformed state of the workpiece material are shown in Figure 1. The diagram of the stressed state of the workpiece material during the stamping process is shown in Figure 2.

Stamping was carried out on a crank hot-stamping press with a force of 10 kN. The ground workpieces were heated in an electric furnace at a temperature of 1025–1050 °C for one hour. A sketch (a) and a photograph (b) of the stamped forging of the "flange" part are shown in Figure 3. There are no surface defects in the form of cracks on the forging surface, which indicates that the dimensions of the initial workpiece are correctly determined. After stamping, the forgings were cooled in a furnace of 900 °C and a reduction of 100 °C per hour.

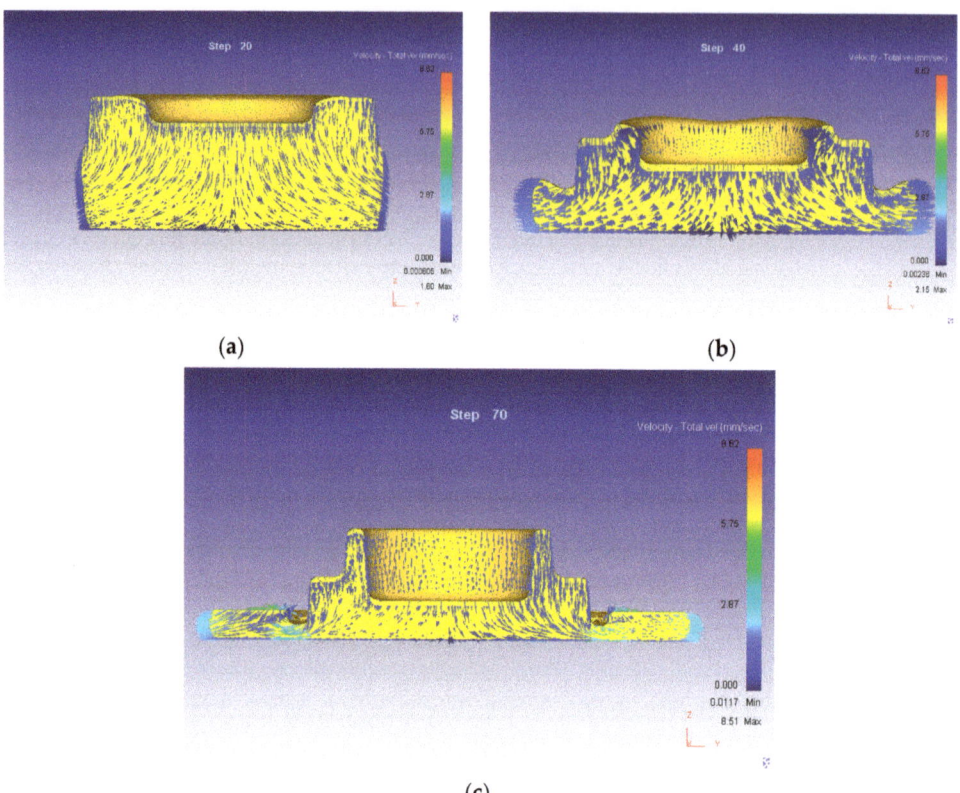

Figure 1. Schemes of the deformed state of the workpiece material: beginning of the process (**a**), intermediate state (**b**), end of the process (**c**).

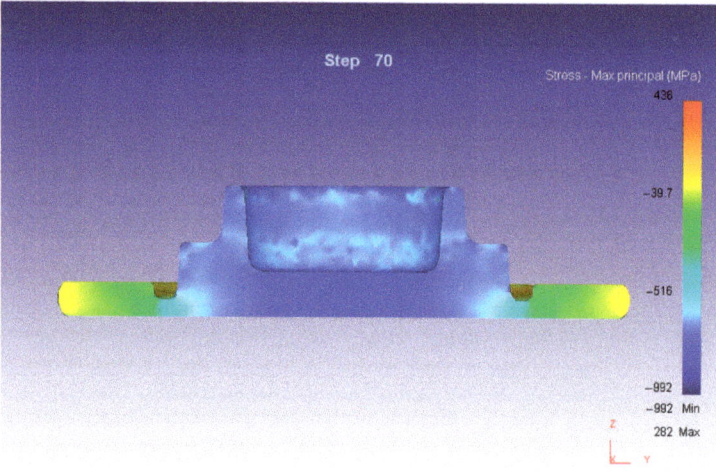

Figure 2. Diagram of the stressed state of the workpiece material at the end of the stamping process.

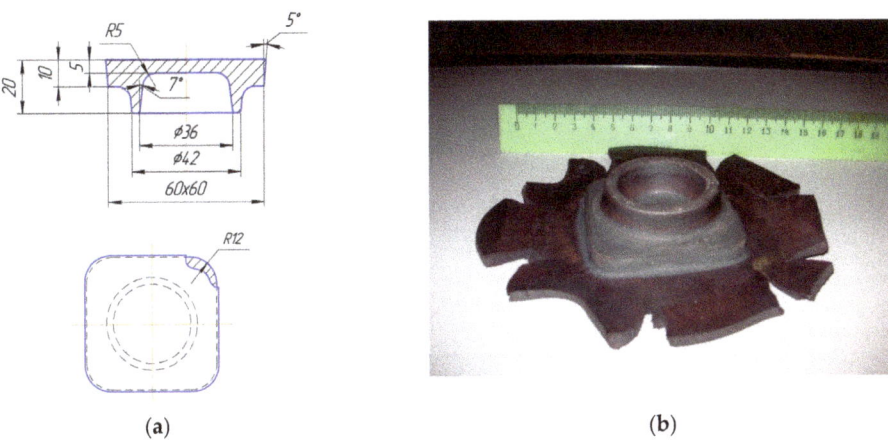

Figure 3. Sketch (a) and photo (b) of the flange forging.

Microstructural analysis was carried out on samples, the cutting scheme of which, from the fourth part of the flange's forging, is shown in Figure 4a. The measurement of the size and shape of graphite grains was carried out on layer-by-layer and end-face sections of the samples. Zones of the microstructural studies were selected on the planes of layered samples, in which the values of deformation and stress in the intensity indicator were determined using mathematical modeling. For samples №1, №3, and №7, they are given in Tables 2 and 3.

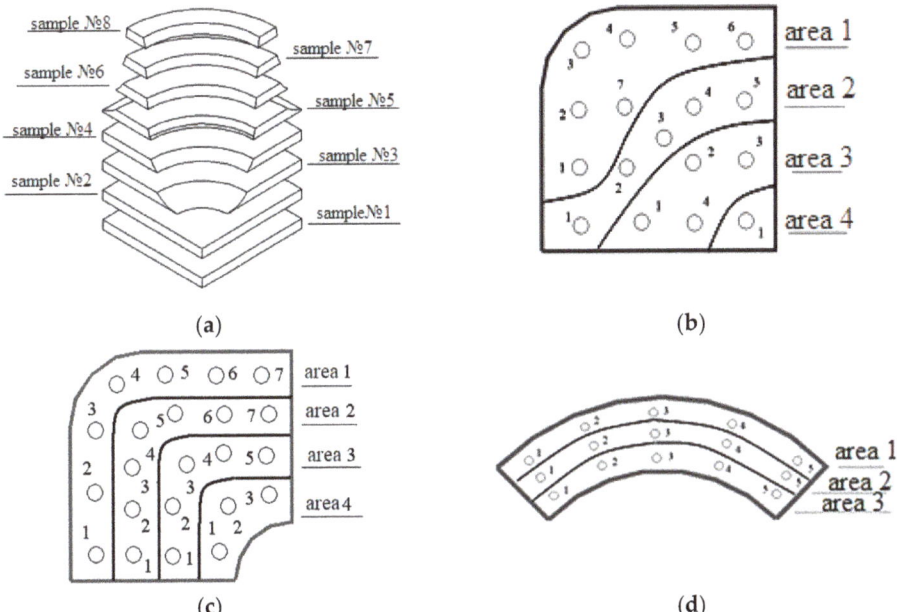

Figure 4. Diagram of cutting the fourth part of the flange forging (a) into layered samples; designation of research areas on sample №1 (b); sample №3 (c) and sample №7 (d).

Table 2. Values of the intensity of deformation of cast iron in the studied flange zones.

point	Sample №1				Sample №3				Sample №7		
	area 1	area 2	area 3	area 4	area 1	area 2	area 3	area 4	area 1	area 2	area 3
1	2.03	2.16	1.41	1.91	0.70	1.46	1.61	2.12	2.28	1.60	1.71
2	2.02	1.38	1.38		0.95	1.37	1.63	1.79	2.38	1.87	2.51
3	2.43	1.28	1.60		1.00	1.50	1.64	2.21	2.38	1.54	1.57
4	2.53	1.42	1.62		1.20	1.54	1.77		2.29	1.58	1.36
5	2.26	1.45			1.22	1.58	2.28		2.02	1.59	1.63
6	2.51				1.25	1.62					
7	1.53				1.28	1.65					

Table 3. Stress intensity values (MPa) of cast iron in the studied flange zones.

point	Sample №1				Sample №3				Sample №7		
	area 1	area 2	area 3	area 4	area 1	area 2	area 3	area 4	area 1	area 2	area 3
1	239	238	192	189	187	199	195	204	144	143	148
2	235	194	180		190	203	193	196	144	144	144
3	251	186	188		188	200	189	200	140	145	156
4	255	198	183		197	195	193		145	145	151
5	239	199			194	192	199				
6	256				197	189					
7	205				201	195					

3. Results and Discussion

One feature of the shaping of the initial blank in an open die is the fact that individual sections have different directions of metal flow due to the implementation of various mechanical deformation schemes, which forms a multidirectional structure of graphite particles.

The material of the initial blank is initially deformed in the central zone under conditions of reverse extrusion (Figure 1a). The stressed state is characterized by a comprehensive compression scheme, which is provided by an active force in the direction of precipitation and the reaction of adjacent sections of the deformable workpiece in the radial direction to its axis. The stress state scheme does not change during the entire deformation process; the maximum stress at the end of the process has a value of 239–256 MPa. The flow of metal occurs in both directions from the center in the radial direction (Figure 1).

The beginning of deformation of the material of the lateral (peripheral) zone of the workpiece has a delay in relation to the deformation of the material of the central zone (Figure 1b). In its volume, the deformation patterns change during the process. In the initial period, the stress state of the material is characterized by uneven all-round compression. On the one hand, this is due to back pressure from the side walls of the working surface of the stamp at the place of formation of the flange part of the forging, on the other, it is from the side of the upset material of the central zone of the workpiece.

In the subsequent period of stamping, when the material flows out in the form of a technological burr, in the zone of the connector plane, the state of back pressure, as well as all-round compression, disappears. The deformed state of the material, as in the central zone, shortens in the direction of compression and elongates towards the center of the workpiece and the perimeter of the stamp. The maximum voltage of 239–256 MPa corresponds to the output of the material from the engraving along the connector plane.

The formation of the cylindrical section of the product occurs last of all under conditions of reverse extrusion, which is carried out due to counter radial displacements of the material of the central and peripheral zones of the workpiece (Figure 1c). Friction forces act on the side of the walls of the cylindrical section on the extruded material, which generally provides a scheme of uneven all-round compression. The intensity of stresses reaches values of 239–256 MPa.

The implementation of various mechanical deformation schemes during forging was expressed in the unevenness of the deformed state of the material in its volume and the change in shape and size of graphite inclusions. The location of the studied characteristic zones in the forging layers is shown in Figure 5.

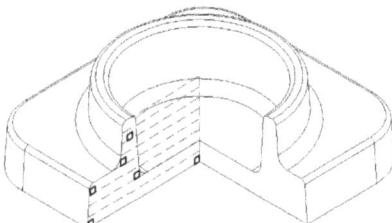

Figure 5. Diagram of the studied zones in the flange forging layers (numbering of eight layers from bottom to top).

Along the perimeter of the flange part of the forging at the place where the material enters the burr groove, in the first layer, the deformation intensity has the value ε_i = 2.02–2.52 and, in the second layer, it is ε_i = 1.64–1.9. The deformed state of the material displays shortening in the direction of compression and elongation in two other directions, primarily to the perimeter of the engraving stamp. Graphite inclusions have an elongated shape in the direction of the flow of the material into the burr groove (Figure 6).

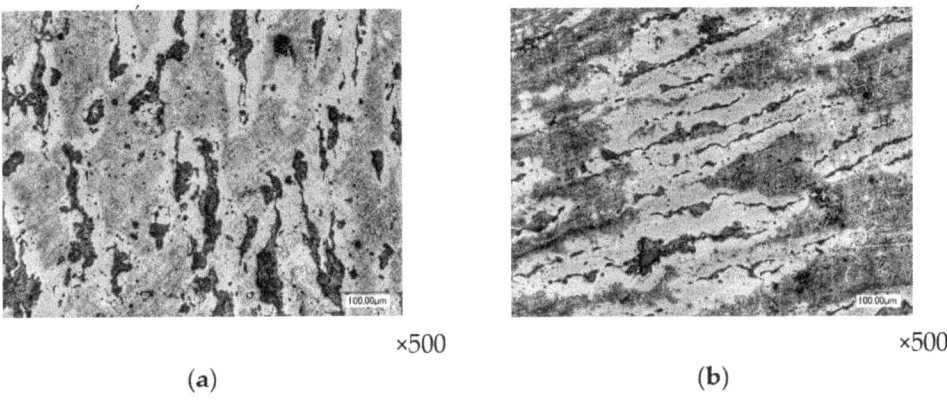

Figure 6. The shape of graphite inclusions in the side flange part of the forging (layer №1): layered section (**a**) and end section (**b**).

In the direction of the forging center, the deformed state of the first two layers decreases to the value of ε_i = 1.41–1.52 and then increases again to ε_i = 1.68–1.91. In the center, the deformed state of the material is characterized by shortening in the direction of compression and uniform elongation in the other two directions. Graphite inclusions have largely changed their spherical shape to a flattened lenticular shape, the position of which is perpendicular to the direction of precipitation (Figure 7).

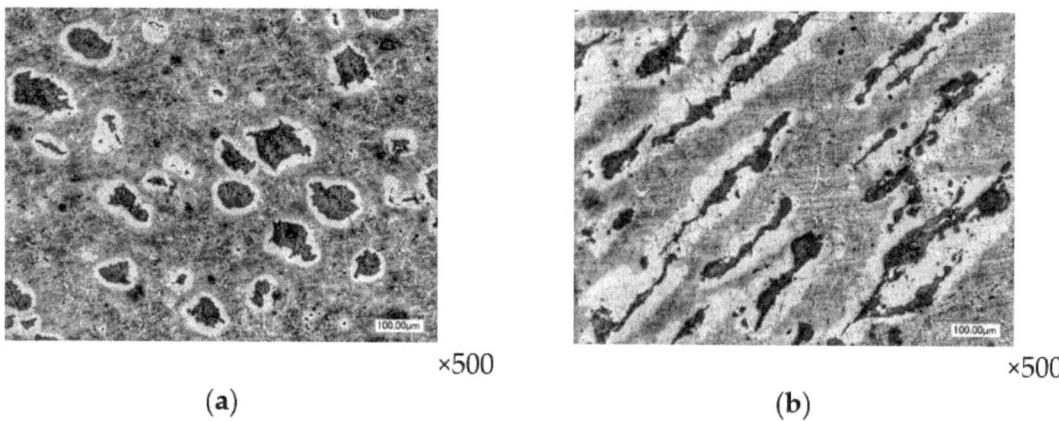

×500 ×500
(a) (b)

Figure 7. The shape of graphite inclusions in the central zone of the forging base (layer №1): layered section (a) and end section (b).

In the third and fourth layers of the perimeter of the flange part of the forging, the deformation has the value $\varepsilon_i = 0.76$–1.26, which is less than in the first two layers. As the transition to the pipe part of the forging occurs, the deformed state of these layers increases, reaching the value of $\varepsilon_i = 1.79$–2.21 in the radius section. The deformed state of the material is characterized by elongation of its height and shortening in thickness. The change in the shape of graphite inclusions is also characterized by stretching in the direction of extrusion (Figure 8).

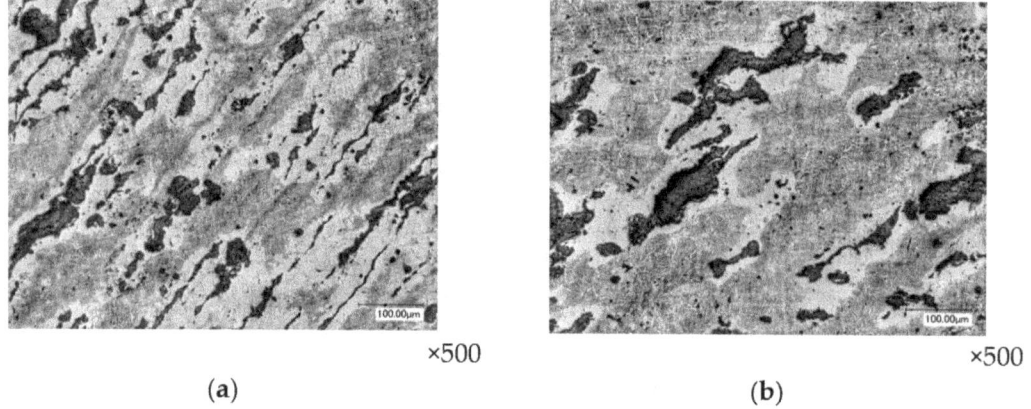

×500 ×500
(a) (b)

Figure 8. The shape of graphite inclusions on the radius section of the transition of the flange to the pipe part of the forging (layer №3): layered section (a) and end section (b).

In the fifth and sixth layers of the pipe section, the deformed state continues to increase to the value $\varepsilon_i = 2.07$–2.42. The highest value corresponds to the outer surface of the section, the smallest to the inner surface.

In the seventh and eighth layers, the deformed state of the pipe section in comparison with the previous layers decreases: in the outer layers, to the value of $\varepsilon_i = 1.5$, and in the inner layers, to the value of $\varepsilon_i = 1.35$. Graphite inclusions have slightly lost their rounded shape (Figure 9).

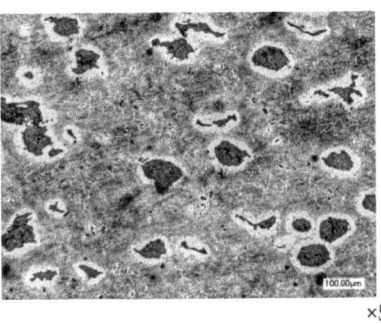 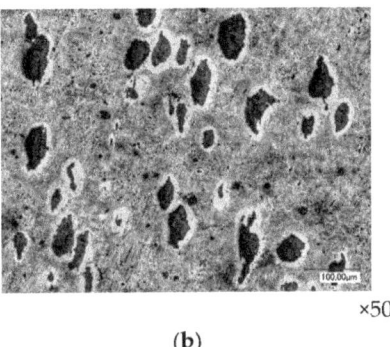

×500 ×500
(a) (b)

Figure 9. The shape of graphite inclusions in the end layers of the pipe forging section (layer №7): layered section (**a**) and end section (**b**).

Based on the result of the deformation analysis, it can be concluded that there is a significant unevenness of deformation in the forging volume, which is determined by the complex geometry of the deformation focus and, as a consequence, by different directions of the material flow [17]. The main flow occurs in the radial direction from the center of the workpiece to the burr groove. At the same time, it is impossible to fully agree with the position expressed by M.V. Storozhev, according to which, during the pre-stamping period, when the stamp figure is almost completely filled with metal, only the excess flows into the burr [19]. The results of mathematical modeling showed that during the pre-stamping period, due to the braking of the metal at the place of exit from the working area of the stamp, the direction of the main flow changes (bifurcates) and, additionally, by reverse extrusion, ensures the design of the cylindrical part of the product.

There were positive results of the stamping of the experimental batch of round forgings, in terms of the technological possibility of hot plastic deformation of ductile cast iron in open dies. In addition, the position that the deformability of the material is not a property, but the state of the material, was also confirmed and is determined by the conditions of deformation: speed, temperature, degree, and mechanical scheme of deformation. The most significant factor in increasing plasticity is considered to be deformation in a state of all-round compression. It can be created by applying special techniques that simulate hydrostatic pressure. In this work, it was created due to lateral pressure from the walls of the tool, carried out by optimizing the size of the initial workpiece. The diameter of the workpiece was selected in accordance with the dimensions of the inner contour of the flange part of the product. During deformation, the cast iron showed plasticity with a value of deformation intensity of $\varepsilon_i = 2.5$.

The study of structural changes in cast iron with spherical graphite during deformation in open dies made it possible to evaluate the ability of graphite inclusions to undergo plastic deformation under conditions of complex loading and large degrees of deformation. During forging, the material underwent the greatest deformation in three zones: along the perimeter and center of the flange part and in the transition zone of the flange part to the cylindrical section. With almost identical degrees of deformation, $\varepsilon_i = 2.1$–2.5, in all zones, graphite inclusions had different morphologies, which showed their dependence on loading conditions. The most difficult loading corresponds to the zone of the material's exit from the die cavity into the burr groove, in which there is no state of all-round compression. Unlike the other two zones, in which the scheme of uneven all-round compression is implemented, on the one hand, by a deforming tool, and on the other, by adjacent layers of metal, there is no compressive stress in the considered zone from the material outlet side. The morphology of graphite inclusions is characterized by the greatest change in shape and size, including their fragmentation.

4. Conclusions

Based on the results of the work, the following conclusions were made:

1. By the method of hot volumetric stamping in an open die, an experimental batch of forgings of the round shape of the "flange" part, with satisfactory quality of the deformed metal in macro and microstructure, grinding, and orientation of the graphite phase in the direction of the main deformations, was obtained.
2. The results of the described work made it possible to determine the technological possibility of stamping forgings of round and similar shape in open dies in terms of creating conditions for uneven all-round compression of the material, while cast iron showed plasticity with a value of deformation intensity $\varepsilon_i = 2.5$.
3. When stamping in open dies to create conditions for all-round compression, it is rational to use special techniques that simulate hydrostatic pressure, in particular, lateral pressure from the walls of the tool, carried out by selecting the size and shape of the initial workpiece.
4. The study of structural changes in ductile cast iron during deformation in open dies has shown the ability of graphite inclusions to undergo plastic deformation under conditions of complex loading with changes in shape and size, including their crushing.

Author Contributions: Conceptualization, V.G.; methodology, V.G., G.G. and A.K.; software, I.K.; formal analysis, V.G.; investigation, V.G., A.K., G.G., I.K. and E.B.; resources, V.G., A.K., G.G., I.K. and E.B.; data curation, V.G.; writing—original draft preparation, V.G., G.G. and A.K.; writing—review and editing, V.G. and A.K.; supervision, A.K.; project administration, V.G. and A.K.; funding acquisition, A.K. All authors have read and agreed to the published version of the manuscript.

Funding: This research was funded by grant 14.Z50.31.0036, awarded to Nizhny Novgorod Technical University n.a. R.E. Alekseev by Ministry of Science and Higher Education of the Russian Federation.

Institutional Review Board Statement: Not applicable.

Informed Consent Statement: Not applicable.

Data Availability Statement: The data presented in this study are available on request from the corresponding author.

Conflicts of Interest: The authors declare no conflict of interest.

References

1. Arzamasov, B.N.; Solovyova, T.V.; Gerasimov, S.A. *Handbook of Structural Materials*; The Bauman Moscow State Technical University, BMSTU: Moscow, Russia, 2005.
2. Kiani-Rashid, A.R. and Rounaghi, S.A. The New Methods of Graphite Nodules Detection in Ductile Cast Iron. *Mater. Manuf. Process.* **2011**, *26*, 242–248. [CrossRef]
3. Di Cocco, V.; Iacoviello, F.; Cavallini, M. Damaging micromechanisms characterization of a ferritic ductile cast iron. *Eng. Fract. Mech.* **2010**, *77*, 2016–2023. [CrossRef]
4. Shcherbedinsky, G.V. Iron: A Promising Material of the XXI Century. *Met. Sci. Heat Treat.* **2005**, *47*, 333–342. [CrossRef]
5. Lisovsky, A.V.; Romantsev, B.A. Formation of unique structures and properties for cast iron during hot metal forming. *Metallurgist* **2010**, *54*, 173–177. [CrossRef]
6. Nofal, A.A. Advances in metallurgy and applications of ADI. *J. Metall. Eng.* **2013**, *2*, 1–18.
7. Bell, J.F. *Experimental Foundations of Mechanics of Deformable Solids. Part II. Finite Deformations*; Nauka: Moscow, Russia, 1984.
8. Potapova, D.B. *Mechanics of Materials in a Complex Stress State*; Mechanical Engineering-1: Moscow, Russia, 2005.
9. Pachla, W.; Mazur, A.; Skiba, J.; Kulczyk, M.; Przybysz, S. Effect of hydrostatic extrusion with back pressure on mechanical properties of grey and nodular cast irons. *Arch. Metall. Mater.* **2011**, *56*, 945–953. [CrossRef]
10. Dubinskii, V.N.; Leushin, I.O.; Korovin, V.A.; Galkin, V.V.; Grachev, A.N.; Pryanichnikov, V.A. Hot deformation of cast iron with globular graphite. *Steel Transl.* **2007**, *37*, 11–13. [CrossRef]
11. Pokrovsky, A.D. *Hot Plastic Deformation of Cast Iron: Structure, Properties, Technological Bases*; Belarusian Science: Minsk, Belarus, 2010.
12. Shi, J.; Savas, M.A.; Smith, R.W. Plastic deformation of a model material containing soft spheroidal inclusions: Spheroidal graphite cast iron. *J. Mater. Process. Technol.* **2003**, *133*, 297–303. [CrossRef]

13. Chaus, A.S.; Soyka, J.; Pokrovsky, A.I. The effect of hot plastic deformation on changes in the microstructure of cast iron with spherical graphite. *Phys. Met. Metallogr.* **2013**, *114*, 4–104. [CrossRef]
14. Zhao, X.; Jing, T.F.; Gao, Y.W.; Qiao, G.Y.; Zhou, J.F.; Wang, W. Morphology of graphite in hot compressed nodular iron. *J. Mater. Sci.* **2004**, *39*, 6093–6096. [CrossRef]
15. Dmitriev, E.A.; Sobolev, B.M.; Rybalkin, A.A. Pressure treatment of cast iron with spheroidal graphite. *Sci. Notes GTU* **2012**, *1*, 83–86.
16. Ponomarev, A.S.; Sosenushkin, E.N.; Artes, A.E.; Klimov, V.N. The influence of pressure shaping on microstructure and quality of products from ductile iron. *Vestn. Mstu Stank.* **2011**, *3*, 115–120.
17. Khrol, I.N. The influence of hot plastic deformation on the structure of high–strength cast iron. In Proceedings of the International Symposium "Perspective Materials and Technologies", Brest, Belarus, 27–31 May 2019.
18. Qi, K.; Yu, F.; Bai, F.; Yan, Z.; Wang, Z.; Li, T. Research on the hot deformation behavior and graphite morphology of spheroidal graphite cast iron at high strain rate. *Mater. Des.* **2009**, *30*, 4511–4515. [CrossRef]
19. Pokrovsky, A.I. Plastic flow of inclusions of cementite and graphite during pressure treatment of cast iron. *Cast. Metall.* **2013**, *1*, 88–95.
20. Dudetskaya, L.R.; Danilchik, I.K.; Pokrovsky, A.I.; Khrol, I.N. Investigation of the parameters of the foundry-deformation technology for obtaining high-quality cast iron products. *Cast. Metall.* **2010**, *1*, 98–109.
21. Semenov, E.I. *Forging and Stamping: Handbook, Volume 2: Hot Volume Stamping*; Mashinostroenie: Moscow, Russia, 2010.
22. Grechnikov, F.V. *Theory of Plastic Deformation of Metals*; SNIU n.a. S.P. Korolev: Samara, Russia, 2021.

Disclaimer/Publisher's Note: The statements, opinions and data contained in all publications are solely those of the individual author(s) and contributor(s) and not of MDPI and/or the editor(s). MDPI and/or the editor(s) disclaim responsibility for any injury to people or property resulting from any ideas, methods, instructions or products referred to in the content.

Article

Experimental Research on Sheared Edge Formation in the Shear-Slitting of Grain-Oriented Electrical Steel Workpieces

Łukasz Bohdal *, Agnieszka Kułakowska, Radosław Patyk, Marcin Kułakowski, Monika Szada-Borzyszkowska and Kamil Banaszek

Department of Mechanical Engineering, Koszalin University of Technology, Racławicka 15–17 Street, 75-620 Koszalin, Poland
* Correspondence: lukasz.bohdal@tu.koszalin.pl

Abstract: This study sought to experimentally develop guidelines for shaping 0.3-mm-thick cold-rolled grain-oriented ET 110-30LS steel using a shear-slitting operation. Coated and non-coated steel was used for the analysis. The coated sheet had a thin inorganic C-5 coating on both sides applied to the C-2 substrate. The first part of this paper presents an analysis of the quality of the cut surface depending on the adopted machining parameters, which were the control variables on the production lines. The second part presents an analysis of the magnetic parameters of the cut samples, which allowed for the specific impact of the quality of the cut edge on the selected magnetic features. Finally, an optimization task was developed to obtain a set of acceptable solutions on the plane of controllable process variables such as slitting speed and horizontal clearance. The obtained results can be used to control the shear-slitting process on production lines and obtain high-quality workpieces.

Keywords: electrical steel; shear-slitting; sheared edge; magnetic properties; optimization

Citation: Bohdal, Ł.; Kułakowska, A.; Patyk, R.; Kułakowski, M.; Szada-Borzyszkowska, M.; Banaszek, K. Experimental Research on Sheared Edge Formation in the Shear-Slitting of Grain-Oriented Electrical Steel Workpieces. *Materials* 2022, 15, 8824. https://doi.org/10.3390/ma15248824

Academic Editor: Konstantin Borodianskiy

Received: 7 November 2022
Accepted: 6 December 2022
Published: 10 December 2022

Publisher's Note: MDPI stays neutral with regard to jurisdictional claims in published maps and institutional affiliations.

Copyright: © 2022 by the authors. Licensee MDPI, Basel, Switzerland. This article is an open access article distributed under the terms and conditions of the Creative Commons Attribution (CC BY) license (https://creativecommons.org/licenses/by/4.0/).

1. Introduction

The dynamic development of manufacturing techniques is related to the ever-increasing requirements for product quality, manufacturing operations reduction, and the assurance of high efficiency in the machining process. Accordingly, there are many difficulties involved in the correct development and proper implementation of technological processes that meet these requirements. Modern materials with promising applications are appearing on the market, but to realize their potential, they will require continuous development and investigation into the the possibilities of their processing and shaping [1,2]. The trend of continuous miniaturization in electronic and magnetic components also poses new challenges to production processes. In many cases, at the production stage, it is necessary to change the geometry and surface properties of the workpiece material by separating its fragments or applying mechanical pressure. Therefore, mechanical cutting processes are constantly used in various industries, such as the electromechanical, electrotechnical, automotive, food, and paper industries. In the electrotechnical industry, a notable challenge is to ensure not only the appropriate mechanical properties, but also the magnetic properties of the shaped products. The optimization of various methods of cutting facilitates the manufacture of a product of high technological quality in one operation. The products must have no burrs and minimum shape deviations and the deformation affected zone must be of minimum width [3–5]. In the electrotechnical industry, electrical steels and amorphous and nanocrystalline tapes are the basic magnetic materials used, among others, for the production of electrical machines such as electric motors, transformer cores, and generators [6–8]. Despite the availability of models that allow for forecasting the technological quality of products in the mechanical cutting process in terms of the state of the cut surface, the proper selection of technological parameters for the process is complicated, especially in the case of shaping magnetic materials. This is due to the lack of data in the literature regarding the settings of machining parameters, for example, for the shear-slitting of grain-oriented silicon electrical sheets in the context of final cutting surface quality and the obtained magnetic properties of the product. Thus, there are known problems that occur in the manufacture of electrical machinery such as transformers and electric motor cores.

In the case of electrical steel cutting, the difficulty in production lines is in ensuring appropriate quality in the cut edge with minimal interference with the magnetic characteristics. It is necessary to analyze the physical phenomena occurring in the contact zone of the tools with the material that is shaped during the process. Knowledge of these phenomena is crucial for proper process control and the design of new tools. Creation of built-up edges on tools, process instability, changes in physicochemical properties in the cutting zone, and low dimensional and shape accuracy of the product are typical problems on production lines related to magnetic material processing that uses cutting techniques [9,10]. The occurrence of slivers, burrs, microcracks in the material, and edge bends are the reasons for poor-quality cut surfaces and result in the generation of waste [11,12].

Previous studies have discussed the selection of an appropriate cutting technology (punching, abrasive water jet cutting, laser cutting) for shaping these steels [13,14]. There is a lack of data on the correct conditions for the implementation of mechanical cutting processes to obtain appropriate cut-surface quality and the least possible interference in the magnetic properties of electrical materials. However, there are no universal guidelines for process control to obtain high-quality products. The technological quality of electrical steel products is determined not only by the characteristic features of the sheared edge, but also by the magnetic properties, which are often degraded as a result of incorrect process performance. In two previous papers [15,16], the authors conducted research on the influence of cutting speed on the magnetic properties of amorphous tapes. The hysteresis losses were significantly influenced by the increased deformation zones of the material, which were caused by the excessive wear of the cutting tool and the formation of burrs on the intersection surface. Another paper [17] presents an analysis of the process of fine punching, which helps to ensure a higher quality in the cut surface than the standard process of punching electrical sheets due to the greater triaxial compressive stress in the cutting zone caused by the use of a wedge root and a small cutting clearance. In addition, some researchers [18] developed the FEM model to facilitate the determination of the optimal clearance for punching electrical sheets with non-oriented grain. The modeling results were experimentally verified for each clearance value obtained.

Some major challenges lie in the correct control of the cutting speed, the value of the cutting clearance, and the appropriate selection of the geometry of the cutting tools depending on the type of material being cut for obtaining a product of appropriate quality. The problem of wear in cutting tools during the forming of electrical sheets has also been analyzed [13,16–20].

It has been shown that the magnetic properties are mostly influenced, apart from concentrated stresses in the cutting zone, by grain deformation and the crystallographic structure, which depend especially on the amount of cutting clearance and speed [21]. In several papers [13,15,16], attempts were made to determine the effect of cutting speed on changes in magnetic properties. The authors of two other papers [16,22] made an attempt to investigate the influence of cutting tool wear and cutting process speed on the width of the deformation zone and the magnetic properties of electrical sheets. Local changes in the cutting edge geometry due to wear resulted in the appearance of burrs on the cut surface, an increased deformation zone, and hysteresis losses. In another two papers [13,23], hysteresis losses also increased as a result of changes in the microstructure of materials such as dislocations or stresses. The test results also showed that sharp tools caused less magnetic degradation than worn, blunt tools. In one study [24], the quality of the cut edge and negative changes in magnetic parameters were significantly influenced, apart from the processing parameters, by the methods of attaching metal sheets to punching dies. Variable boundary conditions were particularly visible in the cutting process on circular shears where the sheet metal exhibited longitudinal movement.

In shear-slitting, the material separation mechanism is often very difficult to control because the tools rotate and the material is not rigidly held [25,26]. If appropriate process conditions are not ensured, excessive tensile stresses may concentrate in the cutting zone, causing local tearing of the material. Two papers [27,28] analyzed the influence of the cutting process on selected magnetic properties of grain-oriented electrical steel cut from sheets with a width of 40 to 660 mm. A significant influence of the construction of tools on the mechanism of material deformation in the cutting zone was found, and the influence of the process conditions on the iron loss of steel was analyzed. Hubert et al. [29], using FE modeling, analyzed the stress distributions in the shearing area under the use of two rotary tools. Particular attention was paid to building an effective process model and appropriate FE discretization in the areas of contact between tools and material. It has been shown that the cutting clearance is an important factor influencing the stress values. The significance of the value of the knife rake angle on the formation of burrs on the cut surface in the case of metal alloys has also been demonstrated [30]. By controlling the rake angle, it is possible to increase the quality of the cut part of the sheet [31]. In the case of electrical sheets with a small thickness, $t = 0.1-0.3$ mm, the rake angle control alone is not sufficient, because its selection depends on the cutting speed.

The objective of this study was to research the impact of main shear-slitting technological process parameters on the cut surface formation mechanisms and the final quality of the sheared edge of grain-oriented electric steel ET 110-30LS. In the process of cutting electric steels, the selection of cutting parameters is often done by trial and error, which extends the duration of the process and the amount of waste. In the case of defects on the cut edge, smoothing and deburring operations are used, which increase production costs. Currently, the literature and industry lack information on how to minimize inter-laminar short-circuit faults between the laminations of electrical machines, which are two of the main challenges for suppliers and customers of electrical steels. In several papers [32–34], additional power losses caused by interlayer faults depend on many factors, including the location of the fault points. One of the reasons for the formation of short-circuit points is the formation of burrs on the edges of packaged sheets and the deviation of the sheet shape. In many cases, it is necessary to use deburring operations. In some critical applications, e.g., large rotating machines, a second coating layer is applied to the laminates after the punching or cutting deburring operation. However, deburring processes are usually not very thorough and can reduce the accuracy of machined parts, damage the edges or surface of the sheet, and create additional stresses in the lamination. In addition, the deburring process requires an additional machining step and additional production time, leading to additional costs. Partial interlayer failure in the stator cores of rotating machines can be more destructive than transformer cores because the layers of the stator cores are welded or held together by key bars or a casing at the base of the stator yoke.

Therefore, it is crucial to define guidelines for the proper control of the cutting process so that at the stage of its duration, the correct condition of the cut edge without defects is already obtained without interfering with the magnetic properties. The novelties of our study include the following:

- Determination of appropriate conditions for the shear-slitting process to obtain minimum burr heights and minimum shape deviations, enabling the correct packaging of sheets. Currently, there is no information on how to select process parameters and conditions to minimize these features at the same time. Knowledge on this subject will allow for the manufacture of products ready for direct packaging without the need for additional machining operations.
- Comprehensive tests for coated and uncoated sheets. Electrical engineering materials in industry are shaped by cutting techniques in different phases, often before or after coating. Currently, the literature lacks information on how to cut such materials depending on their condition. Here, we have proposed universal solutions that can be used for the cutting process of both coated and uncoated materials.

- Optimization techniques that allow for determining the areas of acceptable solutions and the optimal solution. Based on our results, these techniques can be used on production lines, enabling technologists to select appropriate conditions for the cutting process, depending on the assumptions made.
- Partial experimental research, result analysis, and multi-criteria optimization of the cutting process, which enables the development of an effective numerical model of the process for blanking parts for electric machines. On the basis of the obtained research results, a computer model was developed using the finite element method for the process of blanking parts from grain-oriented electrical sheets for the construction of an electric transformer. The developed model is aimed at verifying the results of fragmentary tests and implementing research conclusions in production.

Due to the complexity of the problem, the research was carried out in several stages. In the first stage, research was carried out on the influence of the most important technological parameters of the process on the quality of the cut edge for both sheets with an *electric insulation coating* and without the coating. This enabled observation of the geometric structure of the sheared edge and analysis of the causes of its defects in the form of: burrs, bends and rounded edges, and coating delamination. Then, selected magnetic properties of the sheet metal were tested after the cutting process for the adopted processing parameters. As a result, it was possible to determine the values of the technological parameters that could most often be controlled on production lines during the process, ensuring the highest quality of the cut edge while maintaining the appropriate magnetic properties. In the final part of the paper, we present a computer model that uses the FEA of the blanking grain-oriented electrical sheet metal process for the construction of an electric transformer. The research results were implemented in industry.

2. Materials and Methods

2.1. Material Characteristics

The tests were carried out on cold-rolled grain-oriented ET 110-30LS steel of 0.3 mm thickness. The choice of material was based on technological conditions. In the case of shaping electrical sheets, there are no guidelines for the proper control of the cutting process. A thickness of $t = 0.3$ mm was chosen due to this thickness having the greatest applicability for the construction of electrical devices. The material, despite the quality guarantees provided by the supplier, was subjected to the following laboratory tests: hardness and mechanical properties. The material we analyzed is used in the construction of power and distribution transformer cores and in the production of current transformers, voltage transformers, medium and large high-efficiency generators, reactors, magnetic screens, and coiled audio transformer cores. Coated and non-coated steel was used for the analysis. The coated sheet had a thin inorganic C-5 coating on both sides, applied to the C-2 substrate (labeled in accordance with ASTM A976-13: 2018 [35]). The C-5 coating has a very good electrical insulating quality, was 1.5–3.0 μm thick per side, and had good adhesion to the substrate. This ensured insulation resistance of $>15\ \Omega\ cm^2$, and the coating is also resistant to annealing up to 840 °C in a non-oxidizing atmosphere. Coated and uncoated sheets were selected to develop guidelines for the implementation of the process for two material variants. This choice also allowed for clarification on whether universal settings of the process parameters for these materials could be used or whether they needed to be modified to maintain appropriate quality in the workpiece.

The material hardness was tested using the Brinell and Vickers methods. Measurements were made with HPO-300 and PTM-300M hardness testers. The basic mechanical properties of the material were determined by carrying out a material tensile test on a Zwick/Roell Z400 testing machine. The samples were made in accordance with the PN-EN 10002-1 + AC standard [36]. Ten replications of the test were used for each material. The results are summarized in Tables 1 and 2.

Table 1. Mechanical properties of non-coated ET 110-30LS steel (at $T = 20\,°C$).

Density [kg/dm^3]		R$_{p0.2}$ [MPa]	R$_m$ [MPa]	F$_m$ [kN]	A$_g$ [%]	Hardness [HB]	Hardness [HV]
	7.8	318	332	1	11.4	148	157
$\bar{\sigma}$	1.1	10.88	9	0.03	2.79	2.98	2.06

Table 2. Mechanical properties of coated ET 110-30LS steel (at $T = 20\,°C$).

Density [kg/dm^3]		R$_{p0.2}$ [MPa]	R$_m$ [MPa]	F$_m$ [kN]	A$_g$ [%]	Hardness [HB]	Hardness [HV]
	7.8	314	337	0.96	10.93	157	165
$\bar{\sigma}$	1.3	12.3	9.4	0.02	0.02	0.81	2.38

2.2. Experimental Setup

To conduct experimental research, a test stand was designed that consisted of a shear-slitting machine (Prinzing Maschinenbau) and high-speed camera i-SPEED TR (iX Cameras) that was used to record the cutting process. The sheets of metal were fixed in the device with special mounts with sheet holders. Then, the machining parameters were set in accordance with the five-level rotatable experiment plan. Specially purchased additional components enabled the use of high cutting speeds, precise settings for the cutting clearance, and knife overlap (Figure 1). The machine was driven in one-stage, two-stage, or stepless mode through a motor with a gear and a brake. The drive was transmitted to the upper knife and the pressure roller made of polyurethane. The horizontal clearance h_c was adjusted by a threaded socket with a scale. The slitting velocity v was set by a knob with a scale. Five replicates were performed for each level of the study plan.

As a result of the preliminary research, the most important factors were determined. The input factors included horizontal clearance (h_c) and cutting velocity (v). The values of the process parameters are summarized in Table 3.

Table 3. The values of the studied factors.

Horizontal clearance, h_c	0.02–0.1 mm
Vertical clearance, c_v	0.1 mm
Slitting velocity, v	3–32 m/min
Rake angle, α	7°

After the cutting process, the characteristic features of the cut edge were measured, including the height of the burr, the width of the roll-over, and the width of the sheared-burnished area (Figure 2). Measurements were taken in random places along the cutting line using a measuring microscope (Kestler-Vision Engineering Dynascope) with the ND 1300 Quadra-Chek measuring system. The LEXT OLS4000 confocal laser microscope from OLYMPUS was used to measure the geometric structure of the tested steel's cut edge after cutting. This microscope made it possible to obtain images with excellent quality and accurate 3D measurements using the advanced UIS2 Optical System (infinity correction, non-destructive method). The professional TalyMap Platinum software (version 7.4) was used to analyze the measurement results. Using the data obtained from the measurements of microtopography, the program made it possible to determine the value of the parameters of the geometric structure of the surface of ET 110-30LS steel after cutting, facilitating a graphical presentation of the geometric shape of the measured areas and their profiles.

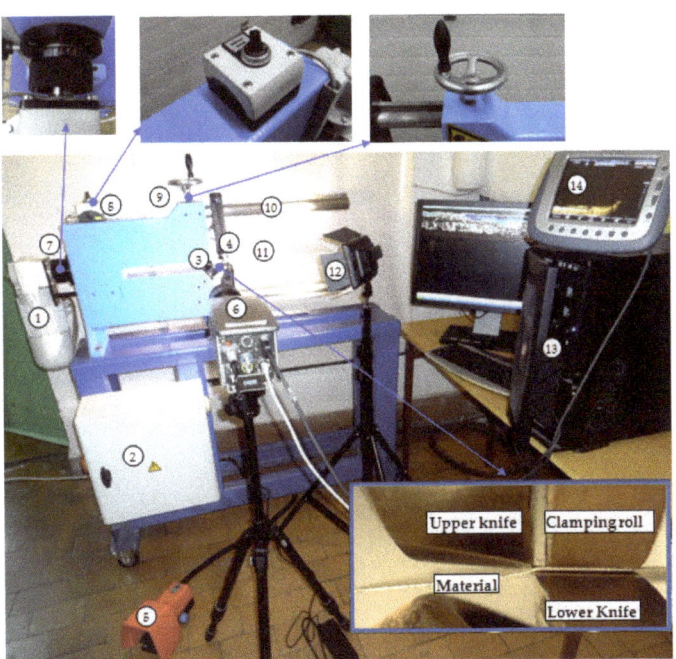

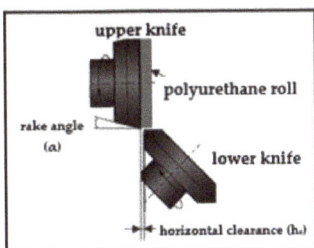
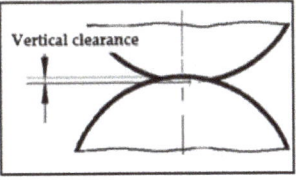

Figure 1. Experimental equipment and characteristic parameter definitions: 1—engine, 2—electrical system, 3—circular knives, 4—sheet stabilizer, 5—drive pedal, 6—high-speed camera, 7—threaded socket with a scale for adjusting the clearance, 8—cutting speed regulator, 9—knife overlap regulator, 10—scale for determining the diameter of cut discs for curvilinear contours, 11—sheet metal, 12—LED lamp, 13—PC for archiving measurement data, 14—auxiliary screen for data recording and analysis.

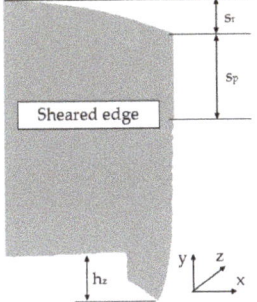

Figure 2. Typical sheared edge contour with characteristic areas, cross-sectional scheme; s_r—roll-over, s_p—sheared-burnished, h_z—burr.

3. Experimental Results and Discussion

3.1. Sheared-Edge Topography

Sample photos of the sheared edge and its surface topography maps are shown in Figures 3–10.

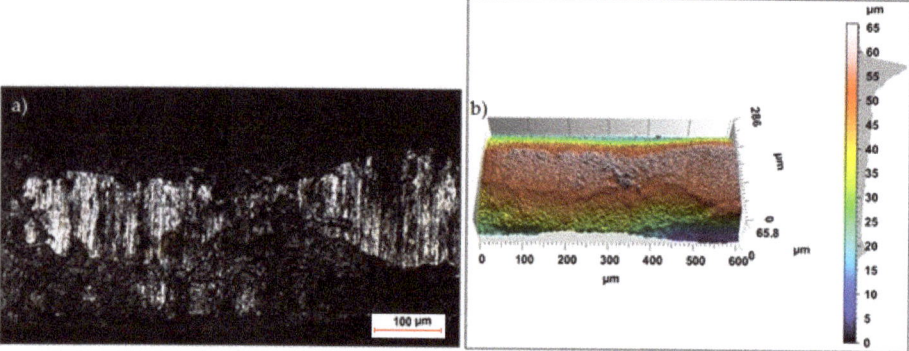

Figure 3. Views of the surface of the cut edge (h_c = 0.04 mm and v = 10.2 m/min): (**a**) surface photography, (**b**) surface topography (uncoated steel).

 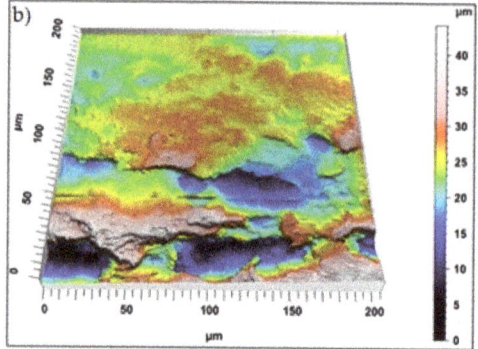

Figure 4. Views of the surface of the cut edge (h_c = 0.04 mm and v = 10.2 m/min): (**a**) surface photography, (**b**) surface topography (coated steel).

 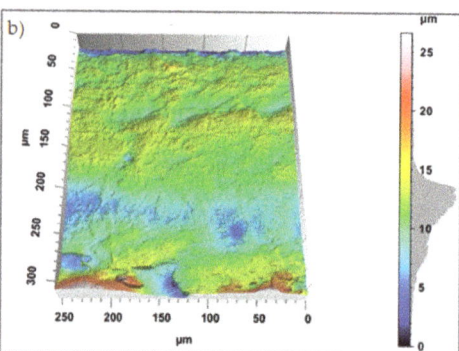

Figure 5. Views of the surface of the cut edge (h_c = 0.08 mm and v = 10.2 m/min): (**a**) surface photography, (**b**) surface topography (uncoated steel).

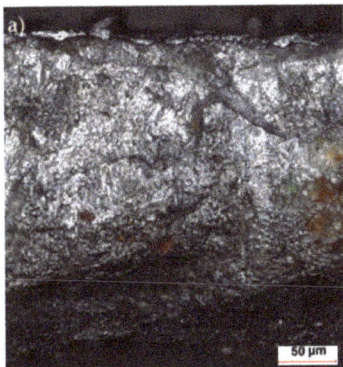
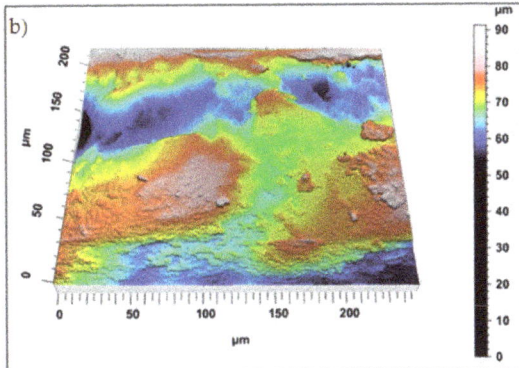

Figure 6. Views of the surface of the cut edge ($h_c = 0.08$ mm and $v = 10.2$ m/min): (**a**) surface photography, (**b**) surface topography (coated steel).

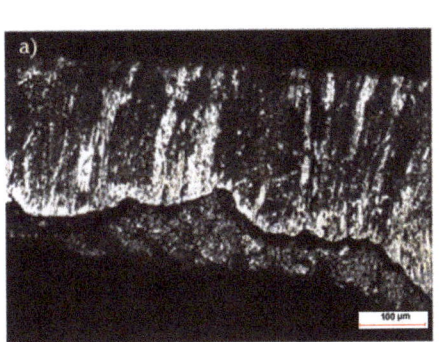
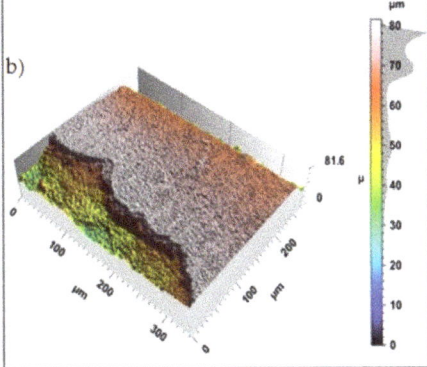

Figure 7. Views of the surface of the cut edge ($h_c = 0.02$ mm and $v = 17.5$ m/min): (**a**) surface photography, (**b**) surface topography (uncoated steel).

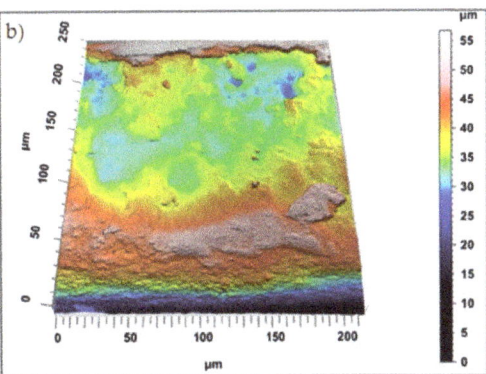

Figure 8. The view of the surface of the cut edge ($h_c = 0.02$ mm and $v = 17.5$ m/min): (**a**) surface photography, (**b**) surface topography (coated steel).

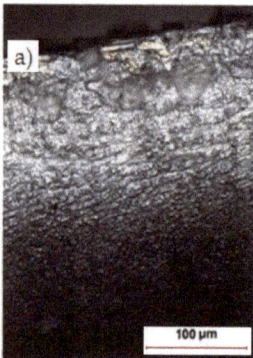 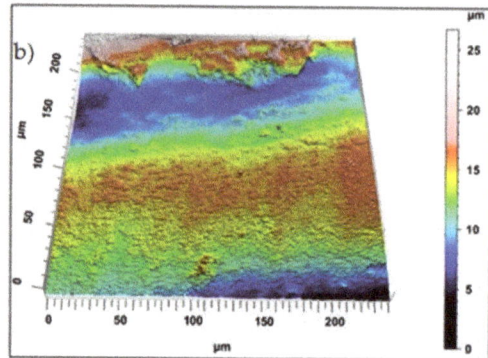

Figure 9. Views of the surface of the cut edge ($h_c = 0.06$ mm and $v = 17.5$ m/min): (**a**) surface photography, (**b**) surface topography (uncoated steel).

 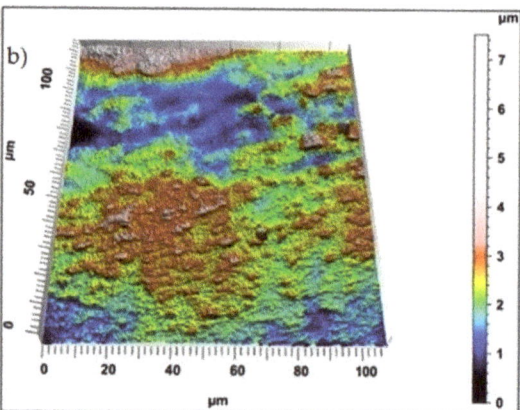

Figure 10. Views of the surface of the cut edge ($h_c = 0.06$ mm and $v = 17.5$ m/min): (**a**) surface photography, (**b**) surface topography (coated steel).

In the case of mechanical cutting processes, irregularities on the sheared surface emerge after the process in the macro and micro scales, the height and arrangement of which may affect the tribological properties of the steel after the cutting process [3,5]. The type of breakthrough may have a large impact on the operational properties of machine elements through the frictional conditions on the contact surfaces, contact stress, joint tightness, fatigue strength, or magnetic properties. Production deficiencies in surface preparation may cause mechanical damage, e.g., fatigue cracks. To improve the fatigue strength of details, rolling and burnishing processes are used. In one study [37], the authors presented a theoretical and experimental analysis of the roller-burnishing technique to achieve isotropic surface topography on cylindrical components made of austempered ductile iron (ADI) casting. Their results showed that their proposed solutions greatly improved surface roughness and eliminated the kinematic-driven roughness pattern of turning, leading to a more isotropic finishing that can reduce the negative impact of mechanical cutting on fatigue cracks.

Specimens cut with a clearance of $h_c = 0.04$ mm and cutting speed $v = 10.2$ m/min for uncoated material were characterized by a clear sheared burnished transition into fracture area in the separating material (Figure 3). The width of the sheared burnished area was greater than the width of the fractured area. The material was characterized by a slight edge-rounding and perpendicular deviation, which did not significantly reduce the quality of the cut edge. This indicates the occurrence of tensile material in selected areas, along with shear in the gap between the cutting edges, which caused local changes in the width of the sheared burnished and its unevenness. The insulating coating influenced the topography of the cut surface. Compared to the uncoated material with clearance of h_c = 0.04 mm and speed of v = 10.2 m/min, the samples were characterized by a smaller rollover in the sheared surface and smaller deviations in perpendicularity (Figure 4). The sheared burnished area was mostly dull with local shiny areas. A significant share of tensile stresses was visible in the areas of the fracture formation, indicated by the size and number of pits with a depth of more than 30 μm. On the edge of the burr, in some areas along the cutting line, fragments of the electrically insulating coating, which is a form of built-up edge on the cut surface, were visible.

Increasing the cutting clearance increased the degree of roll-over of the cut edge of the product. In the case of cutting with clearance of h_c = 0.08 mm and speed of v = 10.2 m/min the samples were characterized by a matt sheared burnished without a clear transition border in the fractured region (Figure 5).

The increase in clearance for coated material increased the area of deformation concentration and enlarged the area of the built-up edge, which also includes the coating, which was deformed and curved towards the center of the cross-section (Figure 6).

For this case, zones of irregular sliding fracture were also visible, where the arrangement and depth of the furrows indicated the occurrence of alternating phases of flow or cracking along the cut line already at the beginning of the process.

For cutting with minimum horizontal clearance of h_c = 0.02 mm and cutting speed of v = 17.5 m/min the samples had a visible transition boundary of the sheared burnished area in the fractured area (Figure 7). The width of the sheared burnished area, as in the case of cutting with clearance of h_c = 0.04 mm and speed of v = 10.2 m/min was greater than the width of the fracture area. Coated steel samples had a shiny sheared burnished area with matte areas and local depressions with a maximum depth of approximately 10 μm (Figure 8). The rounding of the cut edge and the built-up layer of the insulation coating overlapping the cut surface can be seen.

Increasing the clearance value from h_c = 0.02 mm to h_c = 0.06 mm with a cutting speed of v = 17.5 m/min resulted in a slight increase in the regularity of the sheared burnished area, which was still very smooth for uncoated material (Figure 9). The structure was slightly grainy but without the presence of peaks, which reached their maximum values in the area of the burr, creating local surface irregularities in the direction perpendicular to the cut surface. However, the burr height was minimal for both cases. The increased flow of eddy currents was caused by burrs, which contribute to increased losses due to additional electrical paths [33]. In the case of the coated material, the increase in horizontal clearance resulted in the elongation of the plastic flow phase and the emergence of a wide sheared burnished zone with a homogeneous structure and the minimum peak height. The breakthrough was grainy and shiny. A slight roll-over of the edge was visible (Figure 10).

3.2. Sheared-Edge Quality

The quality of the sheared edge after the mechanical cutting process is usually analyzed by measuring the width of characteristic areas (Figure 2) on the cut-edge profile and burr height h_z. From a technological point of view, it is very important to reduce the burr height below 20% of the thickness of the material being cut. The sheared burnished zone width should be maximized and roll-over minimized. Currently, there are no guidelines regarding the optimal values of these zones in terms of product quality nor an analysis of their influence on the magnetic properties. In a later part of the study, we analyzed the influence of the tested parameters on the selected magnetic properties of the material. The analysis of the test results showed a strong influence of cutting speed and clearance on the formation of the width of individual zones on the cut surface, both in the case of sheets with and without an insulating coating.

Figure 11 shows the effect of horizontal clearance and slitting speed on the width of the sheared-burnished zone s_p for uncoated and coated material.

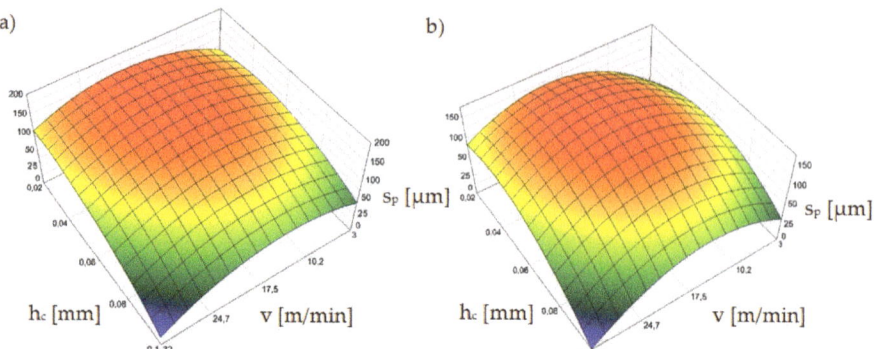

Figure 11. Graph of the dependence of the width of the sheared burnished zone (s_p) on the cutting speed (v) and the horizontal clearance (h_c): (**a**) uncoated material, (**b**) coated material.

The settings of the input parameters for which the zone of the sheared burnished area is the largest represent conditions for a proven steady state for the process, as confirmed in prior research [38–40]. In punching and blanking processes, ensuring process stability is easier than in shear slitting because the material is not displaced during the process. Therefore, the material is less exposed to local large increases in tensile stress in the cutting zone and tearing, causing an increase in the fracture areas and burr. A very important issue is the correct selection of the processing speed, which would ensure the high efficiency and appropriate technological quality of the process. In one paper [41], a mechanistic model for cutting force prediction was presented. A series of cutting trials using austenitic stainless steels and high process speeds were conducted to obtain expressions of the proper force factors. Expressions were developed for individual shear coefficients, taking into account the variable depths of cut and process speed. Deformation and strain-hardening of austenitic stainless steels at high cutting speeds were discussed. Another paper [42] presents a complete analysis of the principal beneficial aspects of mechanical surface treatment produced by the application of ball-burnishing at high speed. The use of such a treatment makes it possible to increase the fatigue strength of sheets.

Our test results for shear-slitting indicate that the lowest process stability is obtained at high cutting speeds above v = 24 m/min (Figure 11a,b). This especially applies to variants of horizontal clearance above h_c = 0.08 mm. According to other studies [30,31,38] this is probably due to the instability of the fracture process which, at high speeds, can occur more quickly even with minimal horizontal clearances in electrotechnical materials. This increases the width of the fractured zone on the cut edge. Using clearances in the range of h_c = 0.04–0.08 mm, the greatest width of the sheared burnished zone can be obtained; however, for clearances within h_c = 0.08 mm, the burr increases (Figure 12). For the v = 3 m/min cutting speed, there was a sheared burnished zone even for a clearance of h_c = 0.1 mm in the case of uncoated sheets (Figure 11a). For the coated material, when a clearance of h_c = 0.02 mm was used, it was advantageous to use a cutting speed above v = 10 m/min. (Figure 11b).

Figure 12a,b shows the influence of the analyzed parameters on the burr height on the cut edge. Significant problems occurred when the burrs electrically connected several layers, because losses increased and there was a high risk of melting the core [33]. The use of clearances above h_c = 0.06 mm caused a significant increase in the burr height for both coated and uncoated material. For clearances above h_c = 0.08 mm, the burr height was too high and may cause a problem on the production line for each analyzed cutting speed. An increase in the burr contributes to an increase in the roundness of the edge and high roll-over. The use of a clearance value below h_c = 0.06 mm allows the use of significant cutting speeds, even up to v = 24 m/min, while still obtaining an acceptable burr height value. For minimum clearance values of h_c = 0.02–0.04 mm, the width of the sheared burnished area increases, which results in a significant reduction of burrs for each speed. When using a clearance of h_c = 0.04 mm, it is more advantageous to use a cutting speed in the range of $v = 3 \div 20$ m/min for coated material and $v = 3$–28 m/min, so that the burr height does not exceed $h_z = 60$ μm.

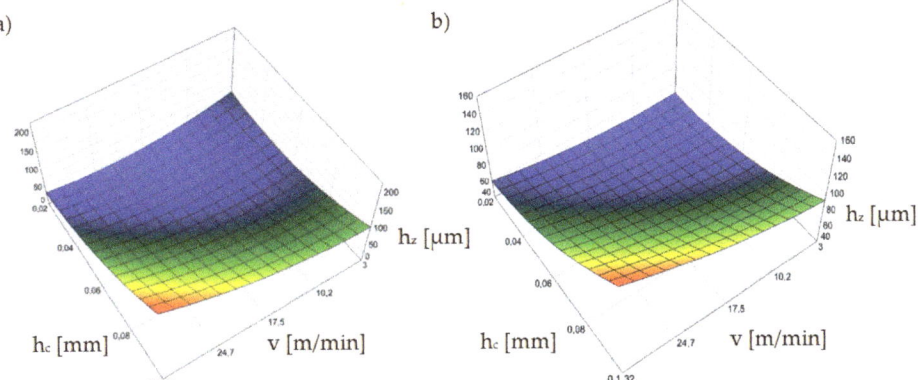

Figure 12. Graph of the dependence of the burr height (h_z) on the cutting speed (v) and the horizontal clearance (h_c): (**a**) uncoated material, (**b**) coated material.

According to a number of papers [38,43–45] in which the authors analyzed the processes of punching and blanking metal materials, the width of the roll-over zone depends mainly on the cutting clearance. One of these papers [44] presented the results of the analysis of the impact of the number of punch cycles on the height of burrs on the cut edge and roll-over formation for three values of the clearance of the blank at 5%, 10%, and 15% of the sheet thickness. It was shown that the intensity of the degradation of the working surfaces and punching edges of the punch were influenced by the friction path. However, our results indicate that it is also important to properly select the cutting speed depending on the adopted cutting clearance (Figure 13a,b). Excessive roll-over concentrates plastic strains over a larger area and contributes to the degradation of magnetic properties [45,46]. The use of cutting clearances over $h_c = 0.08$ mm was found to have an adverse affect.

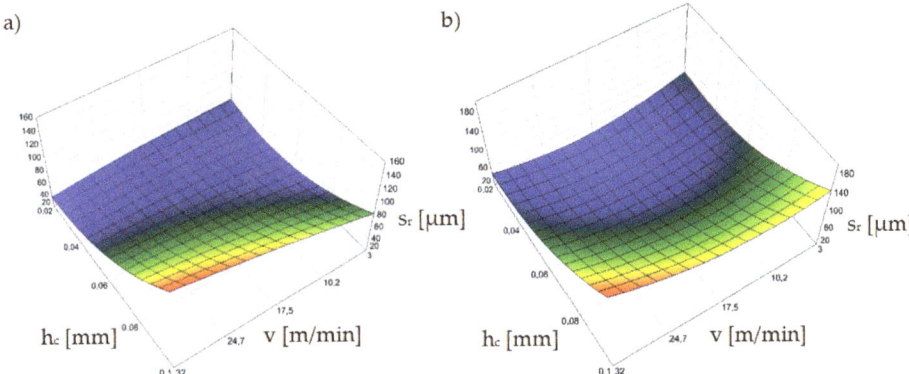

Figure 13. Graph of the dependence of the roll-over width (s_r) on the cutting speed (v) and the horizontal clearance (h_c): (**a**) uncoated material, (**b**) coated material.

In this area, the effect of the insulating coating on the width of the roll-over zone is significantly increased. For coated material, the maximum roll-over values were greater than for uncoated material for each speed value. The greatest differences occurred for the cutting speed of $v = 3$ m/min when the clearance was $h_c = 0.1$ mm. We noted that as clearances increased, ther was an excessive increase in the bending moment and local edge damage. This is related to the excessive concentration of deformations and the plastically strengthened area in the vicinity of the cut edge, as confirmed by the tests carried out in prior work [47]. The movement of defects or irregularities in the crystal, called dislocations, causes plastic deformation. This impedes the movement of the domain walls and increases hysteresis loss [47]. The use of an insulating coating results in a narrowing of the ranges of favorable parameter settings and an increase in roll-over when using minimal clearances and increased cutting speeds.

3.3. Magnetic Characteristics

There are many aspects that may have a negative impact on the magnetic properties of electrical sheets, which include excessive concentration of maximum stresses and deformations in the area of the cut edge, a wide deformation zone or the formation of burrs, built-up edges on the cutting edge, and damage to the insulating coating [31,47,48]. It was therefore important to determine to what extent the investigated process conditions affected the magnetic properties. Specially prepared samples made of laser electrotechnical steel ET 110-30LS with an insulating coating applied (closed ring samples, where outer diameter D = 120 mm, inner diameter d = 90 mm) were used for the tests. The preparation of samples was carried out according to previously described standards and procedures [46,49]. A magnetizing winding and a measuring winding were wound on each of the samples, each winding being wound uniformly to create a closed magnetic circuit and avoid magnetic flux dispersion in the material. The tests were carried out on a test stand consisting of the components shown in Figure 14.

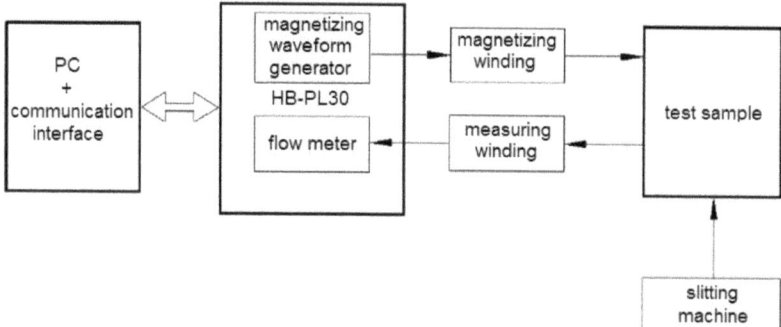

Figure 14. Block diagram of the test stand for testing magnetic characteristics.

The magnetic characteristics were measured for the determined values of the amplitude of the magnetic field strength at Hm = 250 A/m. The frequency of the demagnetizing waveform was 10 Hz. Figures 15 and 16 show the effect of the cutting speed and the clearance between the cutting tools on the hysteresis loops of ET 110-30LS steel with a thickness of t = 0.3 mm. For the tests, the rake angle of the cutting edge of the upper knife was $\alpha = 7°$, and the vertical clearance of the knives was c_v = 0.1 mm. The intensity of H_{max} and the induction of B_{max} are called saturation intensity and induction, respectively. B_s is the remanence induction. The strength of the magnetic field H_k is called the coercive force. In devices with multiple magnetization (e.g., transformer cores), hysteresis is seen as a problem because its surface area is proportional to the energy loss during one re-magnetization cycle. Given the appropriate thermal and plastic treatments and chemical composition, it is possible to minimize its surface and reduce its coercivity. In the case of soft magnetic materials, even a low range of deformation (1–10%) affects the magnetic properties [30,47,48]. In one paper [49], the authors used strength tests to show that the magnetization at saturation and remanence decrease with increases in deformation. Magnetic saturation refers to a state in which the increase of an applied external magnetic field does not increase the magnetization of the material. Remanence, or residual magnetization, is the value of the magnetic induction that remains after the removal of the external magnetic field that was magnetizing a given material.

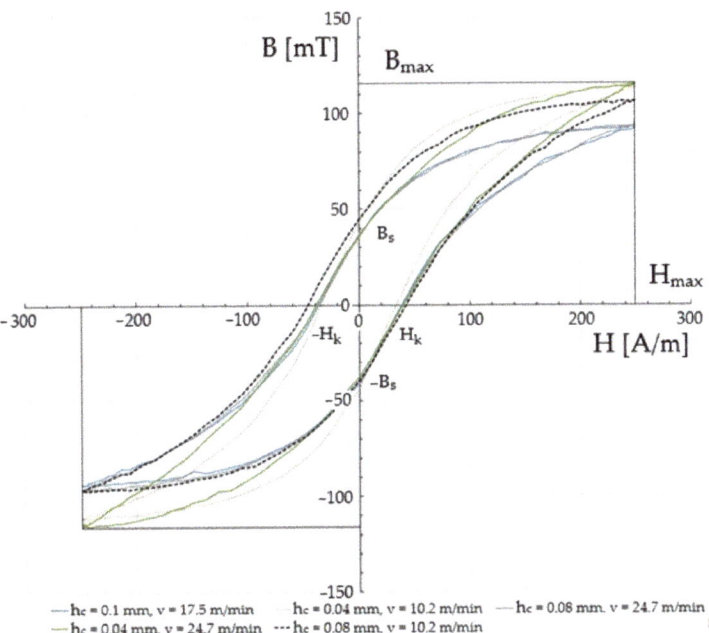

Figure 15. Influence of selected values of input factors of the cutting process on the magnetic characteristics B (H) of the ET 110-30LS sheet.

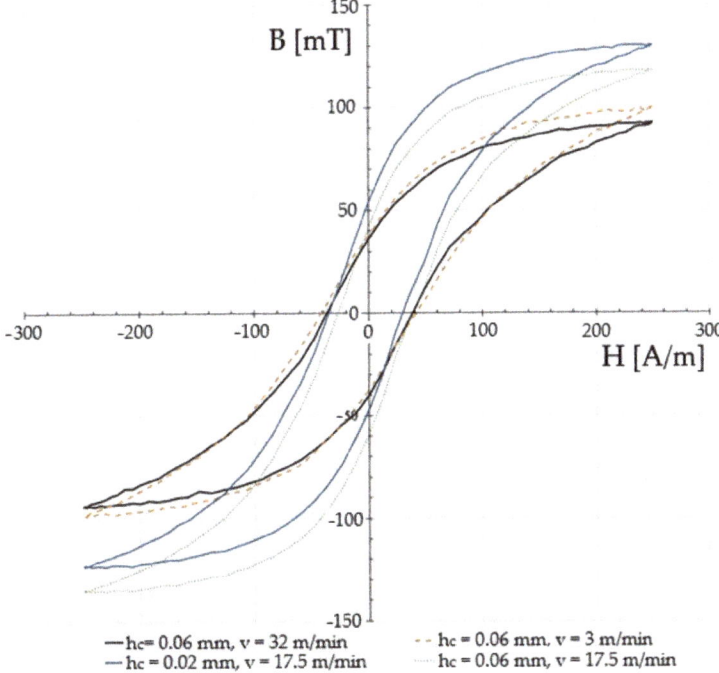

Figure 16. Influence of selected values of input factors of the cutting process on the magnetic characteristics B (H) of the ET 110-30LS sheet.

Among cutting processes, the greatest changes in the parameters of the hysteresis loop were shown in one paper to be caused by laser cutting [50]. As a result of high stresses and thermal deformations in the cutting zone, the hysteresis loops were strongly distorted and exhibited the largest internal surface area, i.e., total losses: the results for individual experiments showed that the shear-slitting process conditions had a significant influence on changes in the characteristic features of the magnetic hysteresis loop. Here, the negative impact of increased cutting clearance at high speeds on the characteristics of the hysteresis loop was particularly visible (Figures 15 and 16). Clear changes in the shapes of the hysteresis loop could be observed in the upper-bend areas of the characteristics and saturation. When using reduced cutting clearances outside of the range of h_c = 0.02–0.04 mm, the cutting speed mainly affected the characteristics of the saturation area and maximum induction (Figure 15). The parameters of the coercivity intensity and remanence induction changed slightly only for a cutting speed of v = 10.2 m/min in this range of cutting clearances. The effect of saturation induction decrease above clearance values of h_c = 0.08 mm was particularly visible.

The increase in the intensity of coercivity and remanence induction is probably due to the increased deformation zone that occurred when using increased cutting clearances and the formation of cut-edge defects. The least unfavorable changes in the parameters of the hysteresis loop occurred for the experiments performed with cutting clearances of h_c = 0.02 mm and h_c = 0.04 mm and a cutting speed of v = 17.5 m/min. The highest maximum induction and the smallest coercivity then occurred. Reducing the cutting speed to v = 3 m/min with a clearance of h_c = 0.06 mm resulted in a decrease in the value of saturation induction and an increase in coercivity. The induction of remanence also decreased. A further increase in the cutting speed value for this clearance value reduced the saturation induction, while the remanence induction and coercivity did not change significantly.

4. Optimization of the Process

On the basis of the obtained research results and the developed mathematical models of the process (regression function type II: for the magnetic hysteresis field h_f, the height of the burr on the cut edge h_z for cutting), an optimization task was developed. In industrial conditions, it is important to deliver products of planned, repeatable technological quality, and thus of functional quality. Selected features (h_f, h_z) determine both the construction quality (important, e.g., from the point of view of sheet-metal assembly) and electromagnetic quality (important, e.g., from the point of view of the efficiency of manufactured machines and electrical devices). The gradient method was used for multi-parameter optimization of the mechanical cutting process of electrical sheets. In industrial production, in addition to obtaining a high-quality product, the goal is to maximize efficiency. In the case under consideration, the efficiency of the mechanical cutting process was defined as $W = f(v)$, where v is cutting speed.

The developed type II regression function for process optimization may be an objective function or a constraint function. The optimization task was defined as follows:

$W = f(v) \rightarrow$ max	Cutting process efficiency W $\rightarrow$ maximum
h_f < 15,000 [mT·A/m]	Limiting the surface area of a magnetic hysteresis loop
h_z < 60 [μm]	Burr height limitation
3 < v < 32 [m/min]	Cutting speed limitation
0.02 < h_c < 0.1 [mm]	Clearance limitation

The developed optimization task was solved using the graphical method (Figure 17). The optimal values of the process settings were determined as follows: optimal horizontal clearance, $h_{c\ opt.}$ = 0.06 mm; optimal cutting speed, v_{opt} = 27 m/min. The designated settings guarantee high technological quality in the product with maximum efficiency in the process.

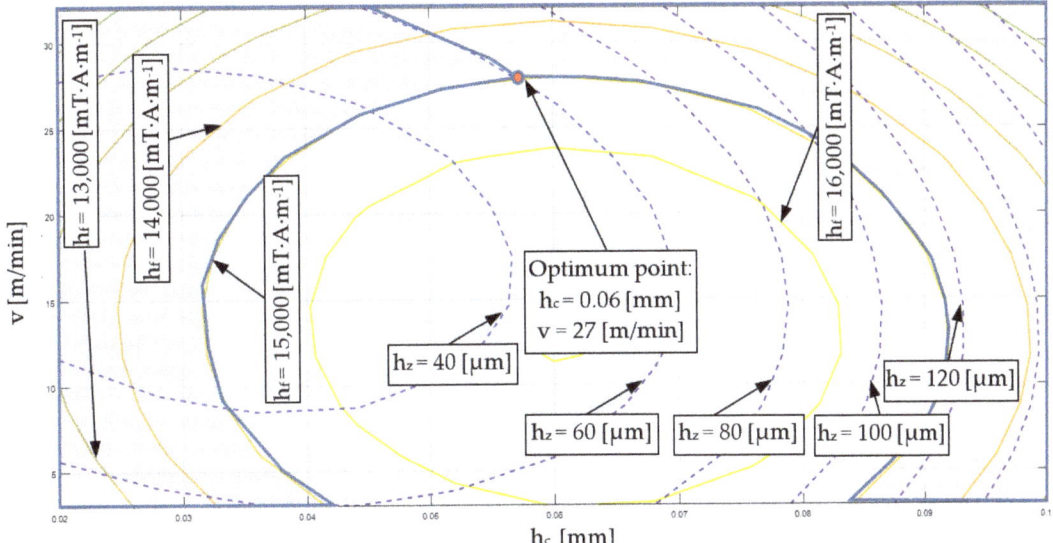

Figure 17. Graphical optimization of the cutting process.

5. Practical Application of Analysis Results

The conducted fragmentary experimental research, results analysis, and multi-criteria optimization of the cutting process enabled the development of an effective numerical model of the process for blanking parts for electric machines. Using the results, a computer model was developed with the finite element method for the process of blanking parts for the construction of an electrical transformer from grain-oriented electrical sheets. The developed model is aimed at verifying the results of fragmentary tests and implementing research conclusions into production. The industrial implementation of the research results will make it possible to obtain a polyoptimal product for mechanical (manufacturing) and electrical (efficiency of electrical devices) purposes. Figure 18 shows the FEM model of the sheet metal blanking process for the transformer core.

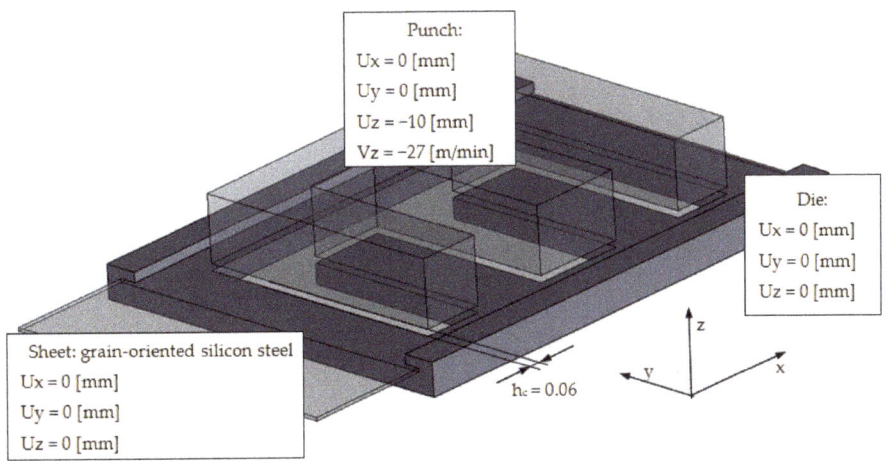

Figure 18. Geometrical model of blanking process with optimal technological parameters.

The solution of the developed equation for the motion of the object was found using the method of explicit integration (central difference) [51–53]. The results (state of reduced stresses) are shown in Figure 19.

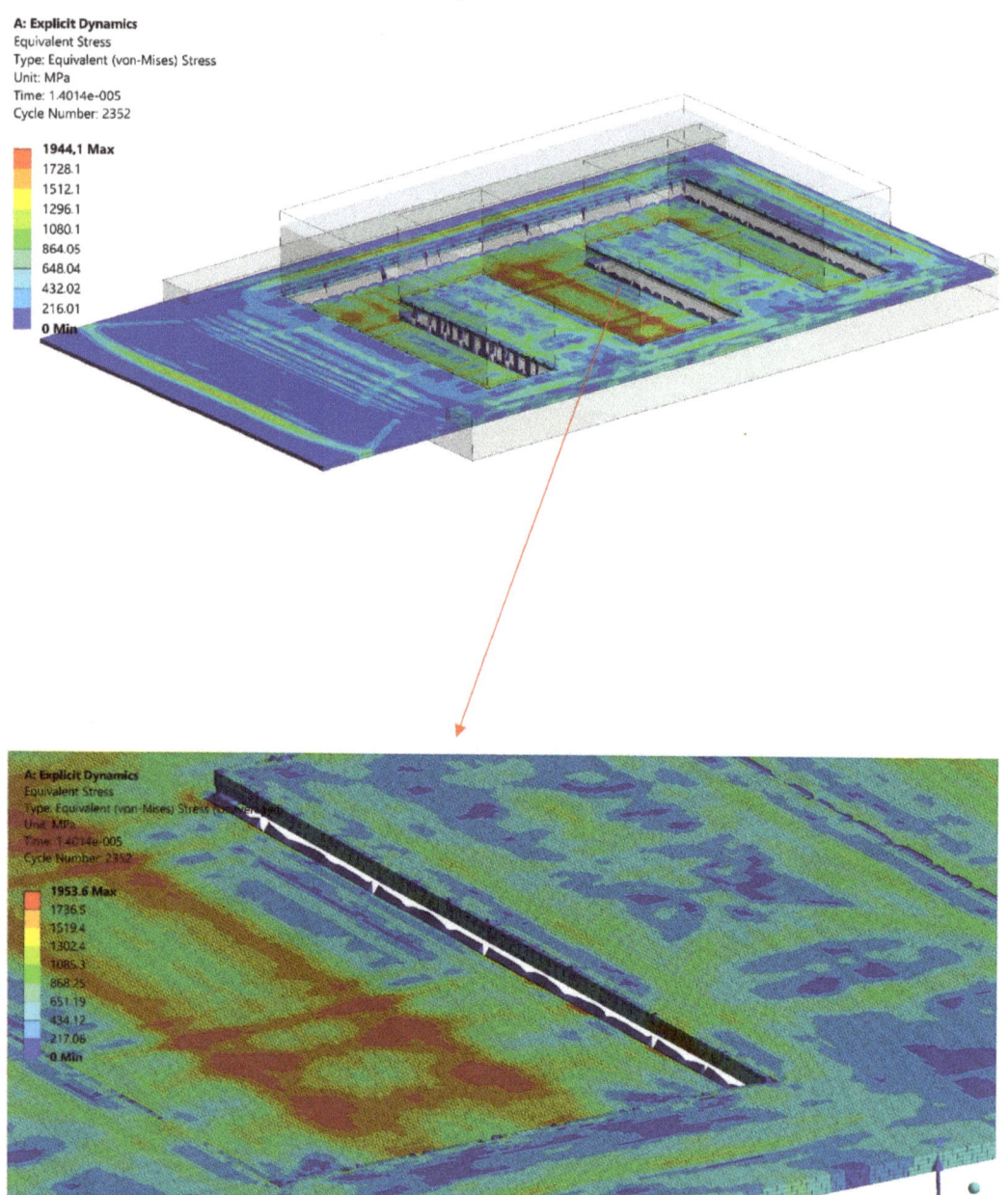

Figure 19. State of equivalent von Mises stress.

The simulation results confirm the height of the burrs obtained in experimental tests, and thus the optimal machining parameters we determined for obtaining the correct product. The last stage of our study was the implementation of technological parameters in the manufacturing process. As a result, a polyoptimal product (mechanically and electrically) was obtained, shown in Figure 20. The product confirmed previous expectations and has been implemented for production.

Figure 20. Transformer core sheet cut with optimal technological parameters from grain-oriented electrical steel.

6. Conclusions

The process of cutting electrical sheets is a very complex process. The technological quality of the product depends on many parameters related to both the condition of the workpiece and the conditions of the shearing process. The shear-slitting process is characterized by complex kinematics, which made the recommendations in the literature regarding blanking or punching processes unsuitable for use in this process. This explains the high burr formations, edge fracture, and rollover on cut surface that are observed in the shearing processes on production lines.

To enable a detailed analysis of these issues, an experimental analysis of the shear-slitting process was carried out, taking into account variable cutting conditions. As a result, the influence of the main control parameters on production lines on the quality of the cut edge and selected magnetic parameters of the material after the process was determined. Based on our results, the following conclusions can be formulated:

- Conducting tests for a material with an insulating coating, which has a composite structure, and for the same material without the coating, allowed for determining the local changes in the quality of the cut edge and identifying the appropriate cutting conditions for each of the cases. This creates new possibilities for production planning and the possibility of proper selection of machining parameters depending on whether the material will be cut with or without a coating.
- The analysis of the test results showed a strong influence of cutting speed and horizontal clearance on the formation of the width of individual zones on the cut surface, both in the case of sheets with and without an insulating coating. The test results indicate that the lowest process stability can be obtained at high cutting speeds above 24 m/min. This especially applies to variants of clearances above $h_c = 0.08$ mm.
- The use of an insulating coating reduces the effect of cutting speed on the width of the sheared burnished zone (by approx. 15%). On the other hand, the effect of horizontal clearance increases (by approx. 20%). The coating increases the stability of the plastic flow process as well as the propagation and the course of cracking for clearances in the range of $h_c = 0.02$–0.06 mm.
- In some cases, at the edge of the burr along the cut line, fragments of the insulating coating, which is a form of built-up edge on the cut surface, were observed. This can have a very negative effect on the magnetic properties and accelerate the wear of the cutting tools.

- As horizontal clearance increased, there was an excessive increase in the bending moment and local edge damage. The use of an insulating coating resulted in a narrowing of the ranges of favorable input parameters settings and an increase in roundness in the case of using minimal clearances and increased cutting speeds.
- The cutting process changes the shapes of the hysteresis loop in the areas of the upper curve of the characteristic and saturation. For reduced cutting clearances beyond the range of h_c = 0.02–0.04 mm, the cutting speed mainly affected the characteristics of the saturation area and the maximum induction. In the case of increased cutting clearances, the value of the saturation induction decreased. This was especially visible in clearance values above h_c = 0.08 mm. The increase in clearance also caused an increase in the intensity of coercivity and induction of remanence.
- Based on the research results and the use of graphical optimization, a set of acceptable solutions and an optimal solution were determined to ensure the highest quality of the cut edge and minimum disturbances in magnetic properties (v = 27 m/min, h_c = 0.06 mm). The proposed approach enables the implementation of the process for other data.

Author Contributions: Conceptualization—Ł.B., A.K., M.K.; methodology—Ł.B., M.K.; investigation—Ł.B., A.K., R.P.; software—Ł.B., M.S.-B., K.B.; data curation—R.P., K.B., M.S.-B.; validation—Ł.B., R.P.; writing—original draft preparation—Ł.B., R.P.; writing—review and editing—Ł.B., A.K., M.S.-B.; visualization—M.K., K.B.; supervision—Ł.B., A.K., M.K.; All authors have read and agreed to the published version of the manuscript.

Funding: This research received no external funding.

Institutional Review Board Statement: Not applicable.

Informed Consent Statement: Not applicable.

Data Availability Statement: Data are available from authors upon request.

Conflicts of Interest: The authors declare no conflict of interest.

References

1. Kumar, R.; Chatopadhyaya, S.; Hloch, S.; Krolczyk, G.; Legutko, S. Wear characteristics and defects analysis of friction stir welded joint of aluminum alloy 6061-T6. *Eksploat. I Niezawodn.-Maint. Reliab.* **2016**, *18*, 128–135. [CrossRef]
2. Maruda, R.W.; Feldshtein, E.; Legutko, S.; Krolczyk, G.M. Research on emulsion mist generation in the conditions of Minimum Quantity Cooling Lubrication (MQCL). *Teh. Vjesn.-Tech. Gaz.* **2015**, *22*, 1213–1218.
3. Krinninger, M.; Steinlehner, F.; Opritescu, D.; Golle, R.; Volk, W. On the influence of different parameters on the characteristic cutting surface when shear cutting aluminum. *Procedia CIRP* **2017**, *63*, 230–235. [CrossRef]
4. Gasiorek, D.; Baranowski, P.; Malachowski, J.; Mazurkiewicz, L.; Wiercigroch, M. Modelling of guillotine cutting of multi-layered aluminum sheets. *J. Manuf. Process.* **2018**, *34*, 374–388. [CrossRef]
5. Li, M. Micromechanisms of deformation and fracture in shearing aluminum alloy sheet. *Int. J. Mech. Sci.* **2000**, *42*, 889–906. [CrossRef]
6. Aggarwal, S.; Bhushan, B.; Katsube, N. Three-dimensional finite element analysis of the magnetic tape slitting process. *J. Mater. Process. Technol.* **2005**, *170*, 71–88. [CrossRef]
7. Pater, Z.; Tofil, A.; Tomczak, J.; Bulzak, T. Numerical analysis of the cross wedge rolling proces (CWR) for a stepped shaft. *Metalurgija* **2015**, *54*, 177–180.
8. Slota, J.; Kaščák, L. FEM Modeling of shear cutting of electrical steel sheets under various technological conditions. *Acta Mech. Slovaca* **2018**, *22*, 24–30. [CrossRef]
9. Ghadbeigi, H. Blanking induced damage in thin 3.2% silicon steel sheets. *Prod. Eng.* **2020**, *14*, 53–64. [CrossRef]
10. Kuo, S.K.; Lee, W.C.; Lin, S.Y.; Lu, C.Y. The influence of cutting edge deformations on magnetic performance degradation of electrical steel. In Proceedings of the 17th International Conference on Electrical Machines and Systems (ICEMS), Hangzhou, China, 22 October 2014; pp. 3041–3046.
11. Boehm, L.; Hartmann, C.; Gilch, I.; Stoecker, A.; Kawalla, R.; Wei, X.; Hirt, G.; Heller, M.; Korte-Kerzel, S.; Leuning, N.; et al. Grain size influence on the magnetic property deterioration of blanked non-oriented electrical steels. *Materials* **2021**, *14*, 7055. [CrossRef]
12. Bohdal, L. Application of a SPH coupled FEM method for simulation of trimming of aluminum autobody sheet. *Acta Mech. Autom.* **2016**, *10*, 56–61. [CrossRef]

13. Moses, A.J. Energy efficient electrical steels: Magnetic performance prediction and optimization. *Scr. Mater.* **2012**, *67*, 560–565. [CrossRef]
14. Samborski, S.; Józwik, J.; Skoczylas, J.; Kłonica, M. Adaptation of fracture mechanics methods for quality assessment of tungsten carbide cutting inserts. *Materials* **2021**, *14*, 3441. [CrossRef] [PubMed]
15. Arshad, W.; Ryckebusch, T.; Magnussen, F.; Lendenmann, H.; Eriksson, B.; Soulard, J.; Malmros, B. Incorporating lamination processing and component manufacturing in electrical machine design tools. In Proceedings of the 42nd IAS Annual Meeting, Industry Applications Conference, New Orleans, LA, USA, 23–27 September 2007; Volume 1, pp. 94–102.
16. Harstick, H.M.S.; Ritter, M.; Plath, A.; Riehemann, W. EBSD investigations on cutting edges of non-oriented electrical steel. *Metallogr. Microstruct. Anal.* **2014**, *3*, 244–251. [CrossRef]
17. Tongsrinak, P.; Panichnok, V.; Dechjarern, S. Study of blanking process with V-Ring using experiment and Finite Element Method. *Key Eng. Mater.* **2017**, *728*, 96–102. [CrossRef]
18. He, J.; Wang, Z.; Li, S.; Dong, L.; Cao, X.; Zhang, W. Optimum clearance determination in blanking coarse-grained non-oriented electrical steel sheets: Experiment and simulation. *Int. J. Mater. Form.* **2019**, *12*, 575–586. [CrossRef]
19. Zhang, H.; Xiang, Q.; Xiong, L.; Chen, C.; Jin, X.; Kang, B. Effect of processing methods on mechanical properties of non oriented high grade electrical steel. *J. Electr. Steel* **2020**, *2*, 34–38.
20. Hofmann, M.; Naumoski, H.; Herr, U.; Herzog, H.-G. Magnetic properties of electrical steel sheets in respect of cutting: Micromagnetic analysis and macromagnetic modeling. *IEEE Trans. Magn.* **2016**, *52*, 2000114. [CrossRef]
21. Baudouin, P.; De Wulf, M.; Kestens, L.; Houbaert, Y. The effect of the guillotine clearance on the magnetic properties of electrical steels. *J. Magn. Magn. Mater.* **2003**, *256*, 32–40. [CrossRef]
22. Weiss, H.A.; Leuning, N.; Steentjes, S.; Hameyer, K.; Andorfer, T.; Jenner, S.; Volk, W. Influence of shear cutting parameters on the electromagnetic properties of non-oriented electrical steel sheets. *J. Magn. Magn. Mater.* **2017**, *421*, 250–259. [CrossRef]
23. Vandenbossche, L.; Jacobs, S.; Henrotte, F.; Hameyer, K. Impact of cut edges in magnetization curves and iron losses in e-machines for automotive traction. In Proceedings of the 25th World Battery, Hybrid and Fuel Cell Electric Vehicle Symposium & Exhibition, EVS, Shenzhen, China, 7–9 November 2010.
24. Kurosaki, Y.; Mogi, H.; Fujii, H.; Kubota, T.; Shiozaki, M. Importance of punching and workability in non-oriented electrical steel sheets. *J. Magn. Magn. Mater.* **2008**, *320*, 2474–2480. [CrossRef]
25. Ma, J.; Lu, H.; Li, M.; Wang, B. Burr height in shear slitting of aluminum webs. *ASME J. Manuf. Sci. Eng.* **2006**, *128*, 46–55. [CrossRef]
26. Lu, H.; Ma, J.; Li, M. Edge trimming of aluminum sheets using shear slitting at a rake angle. *ASME J. Manuf. Sci. Eng.* **2006**, *128*, 866–873. [CrossRef]
27. Zu, G.; Zhang, X.; Zhao, J.; Cui, Y.; Wang, Y.; Jiang, Z. Analysis of the microstructure, texture and magnetic properties of strip casting 4.5 wt.% Si non-oriented electrical steel. *Mater. Des.* **2015**, *85*, 455–460. [CrossRef]
28. Godec, Z. Influence of slitting on core losses and magnetization curve of grain oriented electrical steels. *IEEE Trans. Magn.* **1977**, *13*, 1053–1057. [CrossRef]
29. Hubert, C.; Dubar, L.; Dubar, M.; Dubois, A. Finite Element simulation of the edge-trimming/cold rolling sequence: Analysis of edge cracking. *J. Mater. Process. Technol.* **2012**, *212*, 1049–1060. [CrossRef]
30. Weiss, H.A.; Trober, P.; Golle, R.; Steentjes, S.; Leuning, N.; Elfgen, S.; Hameyer, K.; Volk, W. Impact of punching parameter variations on magnetic properties of non-grain oriented electrical steel. *IEEE Trans. Ind. Appl.* **2018**, *54*, 5869–5878. [CrossRef]
31. Winter, K.; Liao, Z.; Ramanathan, R.; Axinte, D.; Vakil, G.; Gerada, C. How non-conventional machining affects the surface integrity and magnetic properties of non-oriented electrical steel. *Mater. Des.* **2021**, *210*, 110051. [CrossRef]
32. Hamzehbahmani, H.; Anderson, P.; Jenkins, K.; Lindenmo, M. Experimental study on inter-laminar short- circuit faults atrandom positions inlaminated magnetic cores. *IET Electr. Power Appl.* **2016**, *10*, 604–613. [CrossRef]
33. Mazurek, R.; Marketos, P.; Moses, A.; Vincent, J.-N. Effect of artificial burrs on the total power loss of a three-phase transformer core. *IEEE Trans. Magn.* **2010**, *46*, 638–641. [CrossRef]
34. Schulz, C.A.; Duchesne, S.; Roger, D.; Vincent, J.-N. Capacitive short circuit detection in transformer core laminations. *J. Magn. Magn. Mater.* **2008**, *320*, 911–914. [CrossRef]
35. ASTM A976; Standard Classification of Insulating Coatings for Electrical Steels by Composition. Relative Insulating Ability and Application. ASTM International (ASTM): West Conshohocken, PA, USA, 2018.
36. PN-EN 10002-1:2002; Metale—Próba rozciągania—Część 1: Metoda Badania w Temperaturze Otoczenia. Polski Komitet Normalizacyjny: Warsaw, Poland, 2002. (In Polish)
37. Rodriguez, A.; de Lacalle, L.N.; Pereira, O.; Fernandez, A.; Ayesta, I. Isotropic finishing of austempered iron casting cylindrical parts by roller burnishing. *Int. J. Adv. Manuf. Technol.* **2020**, *110*, 753–761. [CrossRef] [PubMed]
38. Subramonian, S.; Altan, T.; Campbell, C.; Ciocirlan, B. Determination of forces in high speed blanking using FEM and experiments. *J. Mater. Process. Technol.* **2013**, *213*, 2184–2190. [CrossRef]
39. Gaudilleire, C.; Ranc, N.; Larue, A.; Maillard, A.; Lorong, P. High speed blanking: An experimental method to measure induced cutting forces. *Soc. Exp. Mech.* **2013**, *53*, 1117–1126. [CrossRef]
40. Bohdal, Ł.; Kukiełka, L.; Legutko, S.; Patyk, R.; Radchenko, A.M. Modeling and experimental analysis of Shear-Slitting of AA6111-T4 Aluminum alloy sheet. *Materials* **2020**, *13*, 3175. [CrossRef] [PubMed]

41. Fernández-Abia, A.I.; Barreiro, J.; López de Lacalle, L.N.; Martínez-Pellitero, S. Behavior of austenitic stainless steels at high speed turning using specific force coefficients. *Int. J. Adv. Manuf. Technol.* **2012**, *62*, 505–515. [CrossRef]
42. Rodríguez, A.; López de Lacalle, L.N.; Celaya, A.; Lamikiz, A.; Albizuri, J. Surface improvement of shafts by the deep ball-burnishing technique. *Surf. Coat. Technol.* **2012**, *206*, 2817–2824. [CrossRef]
43. Liu, Y.; Wang, C.; Han, H.; Shan, D.; Guo, B. Investigation on effect of ultrasonic vibration on micro-blanking process of copper foil. *Int. J. Adv. Manuf. Technol.* **2017**, *93*, 2243–2249. [CrossRef]
44. Mucha, J.; Tutak, J. Analysis of the influence of blanking clearance on the wear of the punch, the change of the burr size and the geometry of the hook blanked in the hardened steel sheet. *Materials* **2019**, *12*, 1261. [CrossRef]
45. Molitor, D.A.; Kubik, C.; Hetfleisch, R.H.; Groche, P. Workpiece image-based tool wear classification in blanking processes using deep convolutional neural networks. *Prod. Eng.* **2022**, *16*, 481–492. [CrossRef]
46. Wilczyński, W. Wpływ technologii na właściwości magnetyczne rdzeni maszyn elektrycznych. *IEI Warszawa* **2003**, *215*, 6–187. (In Polish)
47. Lewis, N.; Anderson, P.; Hall, J.; Gao, Y. Power loss models in punched non-oriented electrical steel rings. *IEEE Trans. Magn.* **2016**, *52*, 7300704. [CrossRef]
48. Leuning, N.; Jaeger, M.; Schauerte, B.; Stöcker, A.; Kawalla, R.; Wei, X.; Hirt, G.; Heller, M.; Korte-Kerzel, S.; Böhm, L.; et al. Material design for low loss non-oriented electrical steel for energy efficient drives. *Materials* **2021**, *14*, 6588. [CrossRef] [PubMed]
49. Hergli, K.; Marouani, H.; Zidi, M.; Fouad, Y. Magneto-Mechanical Behavior Analysis using an Extended Jiles-Atherton Hysteresis Model for a Sheet Metal Blanking Application. *ACES J.* **2020**, *35*, 727–734.
50. Xiong, X.; Hu, S.; Hu, K.; Zeng, S. Texture and magnetic property evolution of non-oriented Fe-Si steel due to mechanical cutting. *J. Magn. Magn. Mater.* **2016**, *401*, 982–990. [CrossRef]
51. Kaldunski, P.; Kukielka, L.; Patyk, R.; Kulakowska, A.; Bohdal, L.; Chodor, J.; Kukielka, K. Study of the Influence of Selected Anisotropic Parameter in the Barlat's Model on the Drawpiece Shape. In Proceedings of the 21st International ESAFORM Conference on Material Forming: ESAFORM 2018, Palermo, Italy, 23–25 April 2018; p. 160014.
52. Kukiełka, L. New damping models of metallic materials and its application in non-linear dynamical cold processes of metal forming. In *Steel Research International, Proceedings of the 13th International Conference Metal Forming 2010, Toyohashi, Japan, 19–22 September 2010*; Special Edition 2010; Stahleisen GmbH: Düsseldorf, Germany, 2010; Volume 81, pp. 1482–1485. ISBN 978-3-514-00774-1.
53. Kukiełka, L.; Geleta, K.; Kukiełka, K. Modelling and analysis of nonlinear physical phenomena in the burnishing rolling operation with electrical current. In *Steel Research International, Metal Forming 2012*; Wiley-VCH GmbH & Co. KGaA: Weinheim, Germany, 2012; pp. 1379–1382.

Article

Investigation of Low-Pressure Sn-Passivated Cu-to-Cu Direct Bonding in 3D-Integration

Po-Yu Kung [1], Wei-Lun Huang [1], Chin-Li Kao [2], Yung-Sheng Lin [2], Yun-Ching Hung [2] and C. R. Kao [1,*]

[1] Department of Materials Science and Engineering, National Taiwan University, Taipei 10617, Taiwan
[2] Product Characterization, Corporate R&D, Advanced Semiconductor Engineering (ASE) Group, Kaohsiung City 811, Taiwan
* Correspondence: crkao@ntu.edu.tw; Tel.: +886-2-33663745

Citation: Kung, P.-Y.; Huang, W.-L.; Kao, C.-L.; Lin, Y.-S.; Hung, Y.-C.; Kao, C.R. Investigation of Low-Pressure Sn-Passivated Cu-to-Cu Direct Bonding in 3D-Integration. *Materials* **2022**, *15*, 7783. https://doi.org/10.3390/ma15217783

Academic Editor: Konstantin Borodianskiy

Received: 7 October 2022
Accepted: 2 November 2022
Published: 4 November 2022

Publisher's Note: MDPI stays neutral with regard to jurisdictional claims in published maps and institutional affiliations.

Copyright: © 2022 by the authors. Licensee MDPI, Basel, Switzerland. This article is an open access article distributed under the terms and conditions of the Creative Commons Attribution (CC BY) license (https://creativecommons.org/licenses/by/4.0/).

Abstract: Cu-to-Cu direct bonding plays an important role in three-dimensional integrated circuits (3D IC). However, the bonding process always requires high temperature, high pressure, and a high degree of consistency in height. In this study, Sn is passivated over electroplated copper. Because Sn is a soft material and has a low melting point, a successful bond can be achieved under low temperature and low pressure (1 MPa) without any planarization process. In this experiment, Sn thickness, bonding temperature, and bonding pressure are variables. Three values of thicknesses of Sn, i.e., 1 μm, 800 nm, and 600 nm were used to calculate the minimum value of Sn thickness required to compensate for the height difference. Additionally, the bonding process was conducted at two temperatures, 220 °C and 250 °C, and their optimized parameters with required pressure were found. Moreover, the optimized parameters after the Cu planarization were also investigated, and it was observed that the bonding can succeed under severe conditions as well. Finally, transmission electron microscopy (TEM) was used to observe the adhesion property between different metals and intermetallic compounds (IMCs).

Keywords: 3D integration; Cu-to-Cu bonding; low pressure; compensate height difference

1. Introduction

Recently, the shrinking of transistors has met physical limitations, because extremely small transistors cause current leakage and damage electronic devices. Therefore, 3D integration has become a promising way to implement Moore's law, as it has several advantages, such as high interconnection density, high performance, and small form factor [1,2]. There are two bonding methods in 3D IC: dielectric and metal bonding, and metal bonding plays a more important role because it determines the transmission signal and power.

In metal bonding, Cu-to-Cu direct bonding is the preferred method, owing to its low cost and good electrical conductivity [3,4]. However, the Cu-to-Cu bonding process is always conducted under high temperature and high pressure, which causes damage to electronic devices [5,6]. Moreover, the amount of Cu oxide on the surface is a crucial factor affecting the bonding quality [7]. Thus, several studies have passivated different inert materials, such as Au and Ag, over the Cu to lower the bonding temperature and pressure and to protect the Cu from oxidation [8–11]. Nonetheless, Cu-to-Cu direct bonding and metal-passivated Cu-to-Cu bonding both require high surface flatness [12], which is achieved by Chemical Mechanical Planarization (CMP) and grinding processes [13–15]. However, these are time-consuming and expensive processes. Therefore, a bonding process that could omit the CMP and grinding processes and can be conducted under low pressure and low temperature would be ideal.

Due to the motivation mentioned above, in this study, Sn was passivated over Cu to enhance the bonding. Three bonding parameters, Sn thickness, bonding temperature, and bonding pressure, were investigated in the experiment. A schematic of Cu-to-Cu direct bonding with the Sn passivation layer is shown in Figure 1. Because Sn is soft, owing to its low melting point (232 °C) [16,17], a perfect bond can be achieved under low temperature and low pressure and without CMP. This could significantly reduce the process cost, process time, and avoid high thermal budgets and compressive stress. Experiments at two bonding temperatures were conducted in this study to find the optimal parameters. These temperatures were 220 °C (<Sn melting point (Tm)) and 250 °C (>Sn Tm). At 220 °C, Sn remains solid during the bonding; this could protect the other part of electronic device which also contains Sn from melting. At 250 °C, Sn becomes a liquid, so any height difference in the metals can be better compensated for, and higher surface roughness may be tolerated. Therefore, Sn of three thickness values (1 μm, 800 nm, and 600 nm) were utilized in this study to determine the minimum required thickness and the required bonding pressure. Additionally, the present paper will compare and discuss the bonding results of Sn passivated on Cu with CMP and Sn passivated on Cu without CMP.

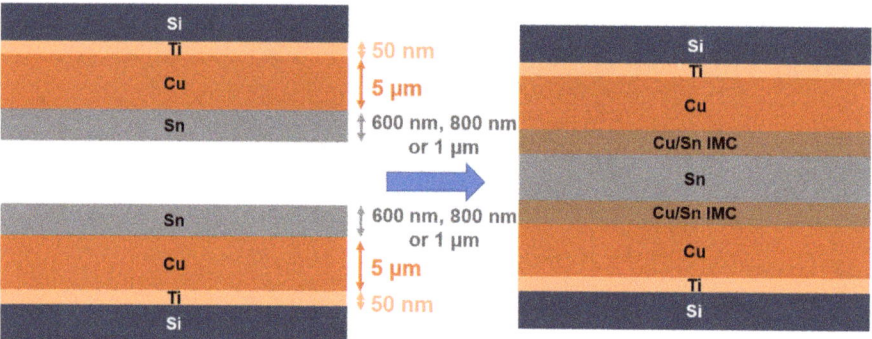

Figure 1. Schematic of the Cu-to-Cu direct bonding using Sn as the passivation layer.

2. Experimental Methods

A 50-nm Ti adhesion layer and a 300-nm Cu seed layer were sequentially sputtered onto a 4-inch silicon wafer. Subsequently, approximately 5 μm of Cu and different thicknesses of Sn were electroplated using a Cu and Sn electroplating solution from Sheng Hung Chemical Engineering Corporation in Taiwan. After electroplating, the wafer was sliced into small chips. To achieve good bonding, the chips were cleaned with acetone and 50 vol% HCl (36 wt%) followed by a DI water rinse to remove any organic materials or oxides on the sample surface.

Afterward, chips were bonded by TCB (Thermal Compressive Bonding) under a 10^{-2} torr vacuum environment. The bonding profile is shown in Figure 2. First, two chips were pressed to avoid displacement. Then, at a heating rate of 40 °C/min, the target temperature was achieved (220 °C and 250 °C) and was held for 1 min. After completion of the bonding process, the pressure was removed and a N_2 purge was used to cool down the sample to room temperature.

These samples were then mounted in epoxy and polished by SiC abrasive papers. To avoid the manual polishing of artifacts and surface impurities which could affect observations, an ion milling system (Hitachi IM4000Plus, Tokyo, Japan) with an Ar^+ ion beam was applied after the manual polishing. Then, the bonding interface was analyzed by scanning electron microscopy (SEM, Hitachi SU5000, Tokyo, Japan) to examine the cross-sectional morphology. The chemical composition of Cu–Sn intermetallic compounds (IMCs) was measured by energy-dispersive X-ray spectrometry (EDX). Some nanovoids and adhesion properties between different layers could not be observed clearly under the SEM. Therefore, transmission electron microscopy (TEM, FEI Tecnai G2 F20, Hillsboro, OR, USA) was used to analyze the interface under extremely high magnification.

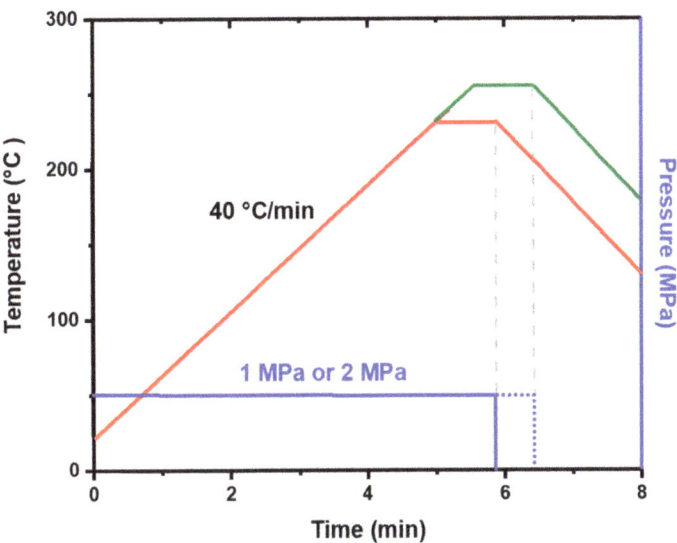

Figure 2. Pressure and temperature profile of the bonding process.

3. Results and Discussion

3.1. Bonding Parameter Optimization

Figure 3 shows the SEM image of different Sn thicknesses which were bonded at 220 °C and 250 °C under a pressure of 1 MPa for 1 min. As can be seen from Figure 3a,b, Sn bonded at both temperatures demonstrated good bonding quality, indicating that 1 µm of Sn was enough to compensate for the height difference and achieve perfect bonding at these two temperatures. It is well-known that Cu and Sn diffuse and form Cu_6Sn_5 and Cu_3Sn; Cu_3Sn is considered better than Cu_6Sn_5 due to its higher melting point and lower resistivity [18–20]. Although Cu_6Sn_5 would eventually become Cu_3Sn due to the high working temperature of the electronic device, it was necessary to analyze the adhesion properties between the different metals and IMC layers while Cu_6Sn_5 was still present in order to ensure that the bonding strength was sufficient. This phenomenon was observed in the following SEM and TEM analysis. Sn becomes a liquid at 250 °C, and it reacted quickly with Cu to form IMCs. Therefore, after bonding, only Cu_6Sn_5 and Cu_3Sn appeared at the interface, as shown in Figure 3a, but there was still some Sn remaining, as shown in Figure 3b. The bonding results when the Sn layer was reduced to 800 nm are shown in Figure 3c,d. All the IMCs became Cu_3Sn at 250 °C, and the interface showed Cu_6Sn_5 and Cu_3Sn with a low quantity of Sn at 220 °C. Furthermore, there were large gaps observed at the interface, which indicated that under these conditions, bonding did not go well, and this poor bonding might be a result of insufficient Sn. It has been reported that when two hard IMCs contact each other, the pressure is concentrated at the point of contact [21]. Thus, other points did not receive enough force, and this led to long gaps. For the 600 nm layer of Sn, because of the lower quantity of Sn, Figure 3e,f showed bigger gaps and holes at the interface. Moreover, at this thickness, no more Sn remains at 220 °C, and only Cu_6Sn_5 and Cu_3Sn can be seen at the interface. Furthermore, all the IMCs were converted to Cu_3Sn at 250 °C.

It has been reported that higher pressure could lead to better bonding [22]. Therefore, the pressure was increased from 1 MPa to 2 MPa for those parameters for which bonding was poor (refer to Figure 3). Furthermore, Figure 4 shows the result when using 2 MPa for 800 nm Sn and 600 nm Sn at 220 °C and 250 °C, respectively. From Figure 4a, the condition of 800 nm Sn bonded at 250 °C indicated a good interface when using higher pressure, and all the IMCs were converted to Cu_3Sn. Though most of the surfaces bonded well in

Figure 4b, the holes were large. This could be caused by high surface roughness, because when Sn reacted with Cu to form smooth Cu_3Sn and scallop-like Cu_6Sn_5, the depletion of Sn would make the surface rougher as shown in Figure 5 [23,24]. Thus, after two interfaces which had high surface roughness bonded with each other, the large holes finally appeared at the bonding interface. Specifically, 800 nm Sn is not thick enough to compensate for surface roughness of this magnitude. After all, the Sn remained solid at 220 °C, so it could not act like a liquid at 250 °C to flow and compensate for the height difference. For 600 nm Sn, small gaps still occurred at the interface under 2 MPa at 220 °C and 250 °C as shown in Figure 4c,d, which indicated that 2 MPa was not high enough to ensure perfect bonding. However, compared to Figure 3e,f, which are bonded under 1 MPa, the gaps were small.

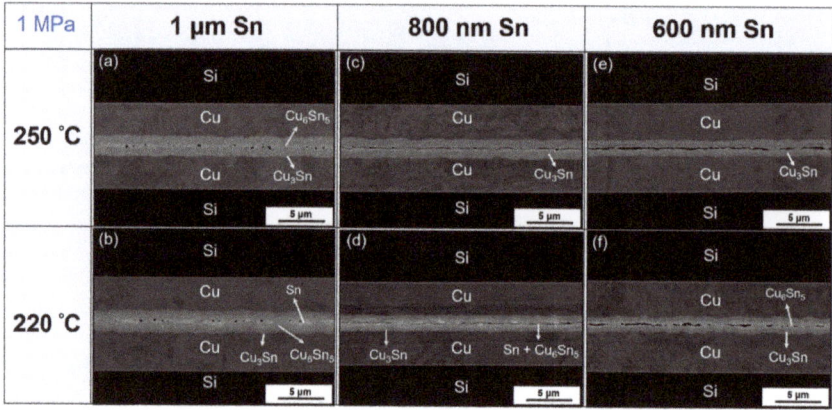

Figure 3. Cross-sectional SEM images of the bonding interface under 1 MPa for 1 min. (a) 250 °C, 1 μm Sn; (b) 220 °C, 1 μm Sn; (c) 250 °C, 800 nm Sn; (d) 220 °C, 800 nm Sn; (e) 250 °C, 600 nm Sn; (f) 220 °C, 600 nm Sn.

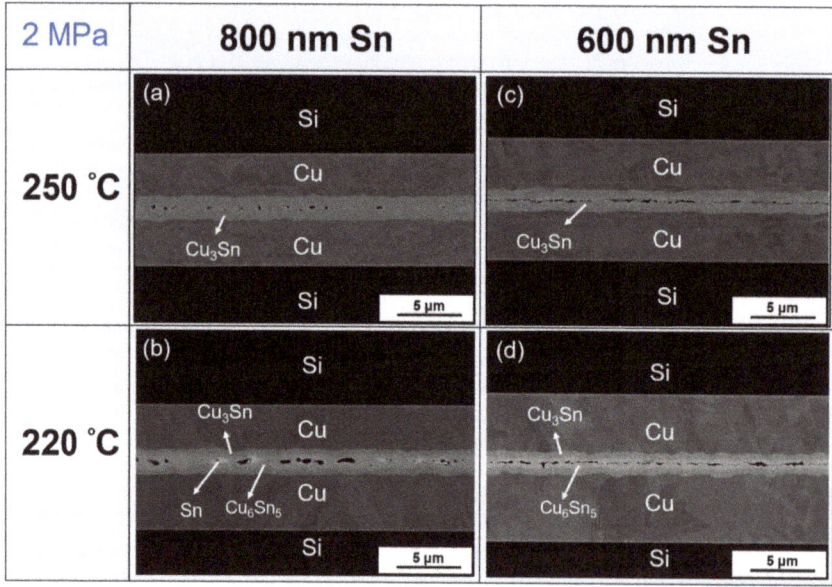

Figure 4. Cross-sectional SEM image of the bonding interface under 2 MPa for 1 min. (a) 250 °C, 800 nm Sn; (b) 220 °C, 800 nm Sn; (c) 250 °C, 600 nm Sn; (d) 220 °C, 600 nm Sn.

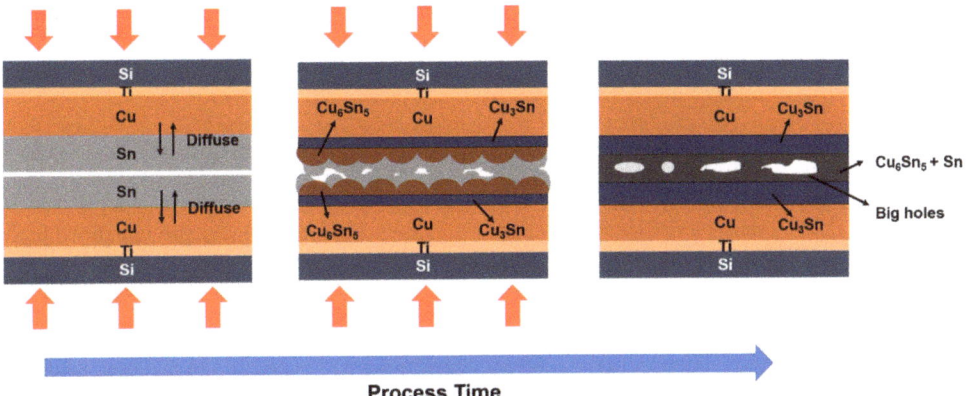

Figure 5. Schematic of IMC formation and microstructure evolution of the Cu–Sn bonding when Sn is not thick enough.

3.2. Porosity and Unbonded Interface Percentage

Although there were four parameters that indicated good bonding results, their hole percentages and the size of the holes were different. Figure 6 shows the comparison of the hole percentage and unbonded region percentage among these four parameters. These two values were calculated using the following formulae:

$$\text{Porosity }(\%) = \frac{\text{Pores Area }(\mu m^2)}{\text{Interface IMCs Area }(\mu m^2)} \tag{1}$$

$$\text{Unbonded Percentage }(\%) = \frac{\text{Unbonded Line Length }(\mu m)}{\text{Interface Length }(\mu m)} \tag{2}$$

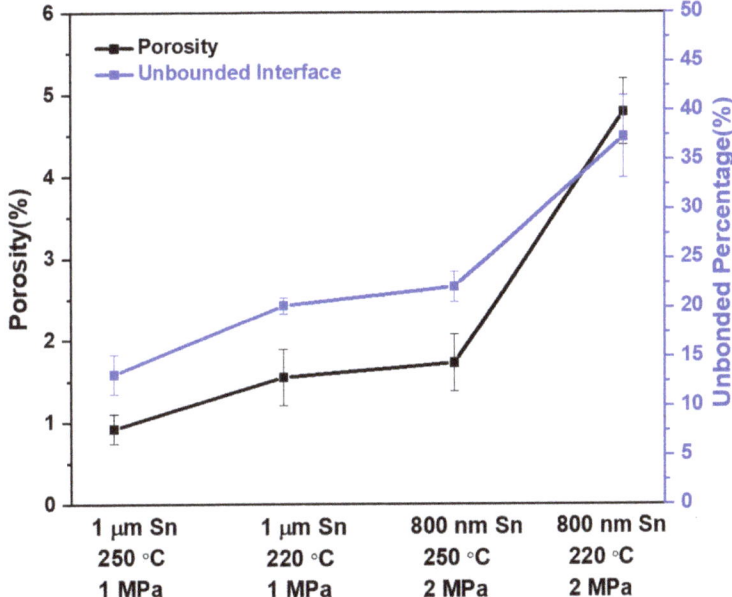

Figure 6. Comparison of porosity and unbonded percentage between four different parameters.

It was observed that the thickness of IMCs was not equal because their initial Sn thicknesses were different. Therefore, a variable called fixed interface IMCs area was set to calculate Equation (1) for each parameter. Moreover, the unbonded percentage was calculated with the unit of length in Equation (2).

In Figure 6, it can be seen that 1 μm Sn at 250 °C under 1 MPa was the best combination of parameters because it had the lowest porosity and the lowest unbonded percentage, approximately 0.9% and 13%, respectively. 1 μm Sn at a temperature of 220 °C under a pressure of 1 MPa and 800 nm Sn at a temperature of 250 °C under a pressure of 2 MPa showed similar results. Their porosities were around 1.5%, and unbonded percentages were approximately 20–25%. Finally, 800 nm Sn at 220 °C under 2 MPa had the largest porosity and unbonded percentage at 4.7% and 37.5%, respectively. Additionally, the porosity and unbonded percentage in the first three parameters showed a similar trend, as shown in Figure 6. Nevertheless, porosity increased more dramatically than the unbonded percentage for 800 nm Sn at 220 °C under 2 MPa because of the large holes caused by high surface roughness. The cause of this high surface roughness is discussed in the next paragraph.

3.3. Bonding after Grinding and CMP

In the previous section, the experiments were conducted without any planarization process. However, bonding parameter optimization after the copper planarization process is illustrated in this section. Figure 7 shows a 2D image of the electroplated Cu and Sn on electroplated copper before and after the planarization of the copper, obtained through Atomic Force Bioscopy (AFM, Bruker Bioscope resolve). As shown in Figure 7a, electroplated Cu without grinding and CMP had much higher surface undulation than electroplated Cu with grinding and CMP (Figure 7b) because it did not undergo flattening. Previous research has demonstrated that electroplated Sn contains large grains [25], so the surface had high roughness after electroplating Sn. Consequently, under the AFM 2D image, grain morphologies of Sn were apparent, which is shown in Figure 7c,d. Table 1 shows the surface roughness (Rq) of each condition. The Rq value of electroplated copper was 22.1 nm, and the surface roughness was lowered to 1.32 nm after grinding and CMP. In addition, Sn on electroplated copper also showed a smaller Rq value with copper planarization than without copper planarization, which meant that the copper which was flatter could make the surface of electroplated Sn flatter as well. Table 2 shows the maximum height difference of the samples measured by the Alpha step (Surfcoder ET3000). It has been reported that electroplated copper had a ±10% height error [26,27], and the electroplated Cu possessed a 0.96 μm height difference. These data corroborated previous research, as the copper thickness was around 5–6 μm. After grinding and CMP, the height difference was reduced to 0.28 μm. Moreover, the Sn on electroplated Cu decreased from 0.84 μm to 0.42 μm.

Table 1. AFM measurement of four different parameters.

Rq	Electroplated Cu	Sn on Electroplated Cu
Without Grinding + CMP	22.1 nm	61.1 nm
With Grinding + CMP	1.32 nm	56.2 nm

Table 2. Maximum height difference of four different parameters.

Maximum Height Difference	Electroplated Cu	Sn on Electroplated Cu
Without Grinding + CMP	0.96 μm	0.84 μm
With Grinding + CMP	0.28 μm	0.42 μm

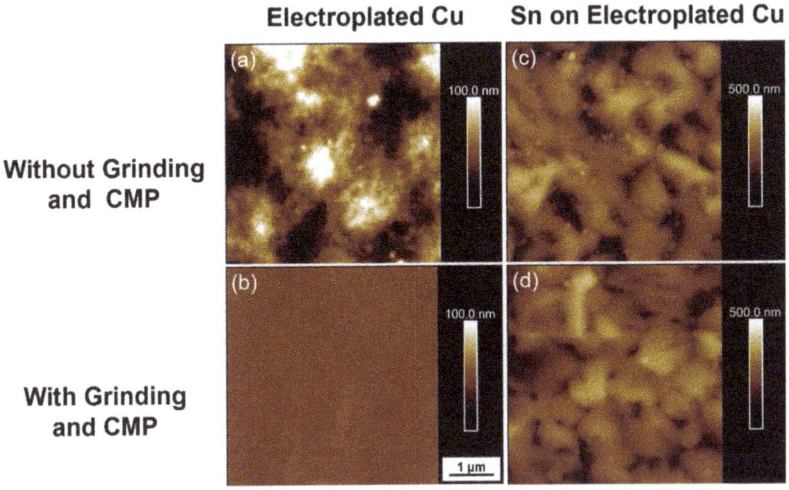

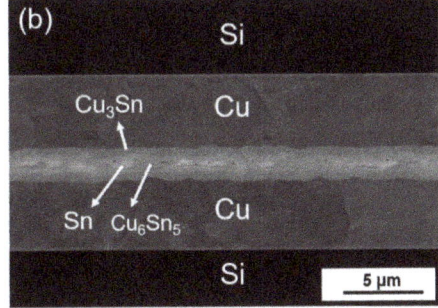

Figure 7. AFM 2D image of (**a**) Electroplated Cu without grinding and CMP; (**b**) Electroplated Cu with grinding and CMP; (**c**) Sn on electroplated Cu without grinding and CMP; (**d**) Sn on electroplated Cu with grinding and CMP.

Figure 8 shows the cross-sectional SEM bonding interface comparison of 800 nm Sn with and without the planarization process at 220 °C under 2 MPa for 1 min. Without grinding and CMP, many large holes were observed at the interface due to the high surface roughness, as shown in Figure 8a. However, after the planarization process, the bonding interface achieved perfect bonding without any apparent holes, as shown in Figure 8b. The porosity dropped from 4.8% to 0.3%, while the unbonded percentage decreased from 37.3% to 7.35%.

Figure 8. Comparison between SEM bonding interface (**a**) without and (**b**) with planarization process with 800 nm Sn at 220 °C under 2 MPa for 1 min.

Furthermore, the flatter Cu not only made the holes smaller, but it also allowed the unbonded interface to change to a bonded interface under the same parameter. Figure 9a,a1 shows the bonding interface comparisons between 600 nm Sn with and without the planarization process at 220 °C under 2 MPa for 1 min. There was a long gap that appeared in the material without grinding and CMP (Figure 9a). However, after grinding and CMP, most of the surfaces connected, with some voids appearing at the interface, as shown in Figure 9a1. The porosity was 1.25% and the unbonded percentage was 16.9%. Also, Figure 9b,b1 shows a comparison between the bonding interface before and after the planarization process, both bonded at 250 °C under 1 MPa for 1 min with 800 nm Sn. Without grinding and CMP, the interface had an apparent gap, while the interface was well-connected after flattening. The porosity and the unbonded percentage of this well-connected interface were 1.21% and 22.7%, respectively. This improvement could be attributed to the flatter surface, which prevented the two hard IMCs from contacting each other first.

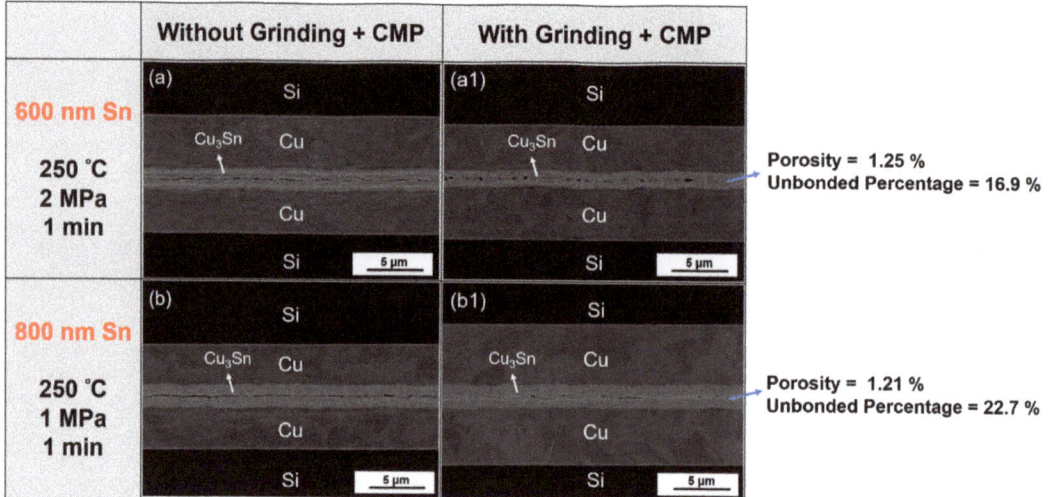

Figure 9. Comparison between (**a**,**b**) without grinding and CMP, and (**a1**,**b1**) with grinding and CMP under two bonding parameters.

3.4. Summary of Bonding Parameters and Results

This section summarizes the bonding parameters and their qualities, which were discussed in the previous paragraph. Table 3 is the summary of the porosities and unbonded percentages of optimized parameters for achieving good bonding without any planarization process. A successful bonding under 1 MPa is only feasible in the case of 1 µm Sn thickness. When using 800 nm Sn thickness, 2 MPa pressure is required. Moreover, under the same Sn thickness and pressure, the sample prepared at 250 °C bonding temperature possesses fewer defects than the one prepared at 220 °C, because the melted Sn can easily flow and fill the holes. Moreover, although the pressure exerted on 1 µm Sn is lower than on 800 nm Sn, it still shows better bonding quality without any planarization process.

Table 3. Summary of the optimized parameters without grinding or CMP.

Sn Thickness	Temperature	Pressure	Time	Porosity	Unbonded Percentage
1 µm	250 °C	1 MPa	1 min	0.9%	13.2%
	220 °C	1 MPa	1 min	1.5%	20.2%
800 nm	250 °C	2 MPa	1 min	1.7%	22.1%
	220 °C	2 MPa	1 min	4.8%	37.3%

Without Grinding or CMP.

Table 4 shows that the optimized bonding parameter changes after copper planarization when compared to no planarization. The drastic decrease in the surface roughness and maximum height difference leads to a better bonding quality and allows the bonding to succeed even under suboptimal conditions. The porosity and unbonded percentage show a drastic decrease in the 800 nm thick Sn, bonded at 220 °C under 2 MPa for 1 min. Additionally, the pressure can decrease to 1 MPa in the 800 nm thick Sn, bonded at 250 °C, and Sn thickness could be reduced to 600 nm under 2 MPa at 250 °C after grinding and CMP.

Table 4. Summary of the optimized parameters with grinding and CMP (red characters show the differences, when compared to no grinding and CMP).

Sn Thickness	Temperature	Pressure	Time	Porosity	Unbonded Percentage
800 nm	220 °C	2 MPa	1 min	4.8% → 0.3%	37.3% → 7.35%
800 nm	250 °C	2 MPa → 1 MPa	1 min	1.21%	22.7%
800 nm → 600 nm	250 °C	2 MPa	1 min	1.25%	16.9%

With Grinding + CMP.

3.5. TEM Analysis

There was only Cu_3Sn at the interface for some parameters, while the other parameters had Cu_6Sn_5 and remaining Sn after the bonding. However, these Cu_6Sn_5 and Sn were eventually converted to Cu_3Sn due to the high working temperature. It is still important to measure the adhesion properties between different layers no matter what type of IMCs were at the interface, and there might be some defects and voids which could not be observed clearly under the SEM. Therefore, TEM was required to analyze the bonding quality between different IMCs and metal layers under higher magnification. In addition, a diffraction pattern was used to perform phase identification. Figure 10a shows the bright-field TEM image of the interface bonded at 250 °C under 1 MPa for 1 min with 1 μm Sn thickness. Under this condition, the interface had a $Cu/Cu_6Sn_5/Cu_3Sn/Cu_6Sn_5/Cu$ structure. Figure 10b,c shows the selected area diffraction patterns (SADPs) of the Cu_3Sn and Cu_6Sn_5, respectively, corresponding to Figure 10a. Cu_3Sn is the columnar grain, which was in accord with the previous study [28]. Additionally, there were no voids or cracks at either the Cu–Cu_3Sn interface or the Cu_6Sn_5–Cu_3Sn interface. This phenomenon could indicate that the interface had good adhesion. The stress concentration caused by voids or cracks would not happen in this system. Moreover, the study indicated that the Sn–Cu_6Sn_5 interface is the main factor that affected the strength of the Cu–Sn system [29]. However, in this experiment, only a small amount of Sn remained at the interface. Furthermore, the remaining Sn was scattered in the Cu_6Sn_5 layer, and the Sn was depleted in most of the conditions after the bonding process. Therefore, the strength of the joints could not be determined by the Sn–Cu_6Sn_5 interface. Moreover, the previous studies revealed that the $Cu/Cu_6Sn_5/Cu_3Sn/Cu_6Sn_5/Cu$ structure had good strength [30,31], so the joint with Cu_6Sn_5 that obtained in this experiment would not be fragile.

Figure 11a shows the TEM bright-field image after all the Cu_6Sn_5 was converted to Cu_3Sn with 800 nm thick Sn bonded at a temperature of 250 °C under 2 MPa pressure for 1 min. While in this state, grain growth stopped; the Cu_3Sn grain layers on both sides came in contact with each other but did not merge together, which was in accord with the previous study [28]. Figure 11b shows the SADP of the Cu_3Sn grain shown in Figure 11a. It was reported that $Cu/Cu_3Sn/Cu$ had over 44 Mpa of shear strength [32]. No defects occurred at the Cu–Cu_3Sn interface, so the joint would also be strong when all the IMCs converted to Cu_3Sn.

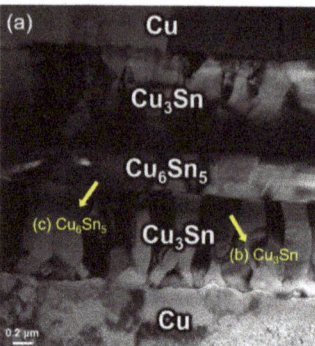 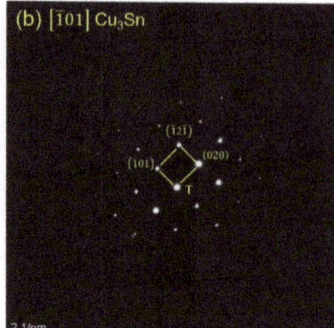 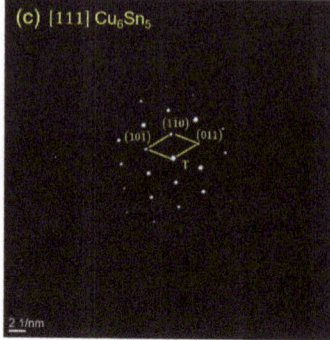

Figure 10. (**a**) Bright Field TEM of Cu/Cu$_6$Sn$_5$/Cu$_3$Sn/Cu$_6$Sn$_5$/Cu structure under the condition of 1 μm Sn, bonded at 250 °C under 1 MPa for 1 min; (**b**) SADPs of Cu$_3$Sn; (**c**) SADPs of Cu$_6$Sn$_5$.

 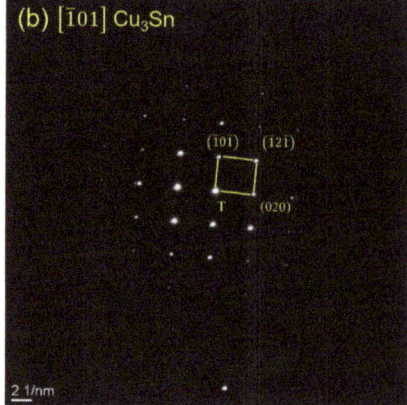

Figure 11. (**a**) Bright-field TEM of Cu–Cu$_3$Sn structure; (**b**) SADPs of Cu$_3$Sn.

4. Conclusions

To develop a Cu-to-Cu direct bonding with low temperature and low pressure and without any planarization process, passivating a Sn layer over the copper material is a promising method. Despite using the surface with a roughness of 22.1 nm Rq and 0.96 μm height difference, the bonding result is good under the conditions mentioned above. These bonding parameters were optimized, and the bonding qualities were investigated using SEM and TEM. The following conclusions can be drawn:

1. As shown in Table 3, there are three parameter changes in this study: Sn thickness, bonding temperature, and bonding pressure. From the results, higher Sn thickness (1 μm) and higher temperature (250 °C) are recommended to achieve better bonding. Moreover, if a lower Sn thickness (800 nm) is used, 2 MPa pressure can still achieve a successful bond. However, the bonding quality is not as good as the parameters of 1 μm Sn under 1 MPa pressure at both 250 °C and 220 °C. Besides, 600 nm Sn is not thick enough to bond well without the planarization process in spite of the 250 °C temperature and 2 MPa pressure.
2. Though the high surface flatness is not needed in the bonding after passivating soft Sn over Cu, the surface roughness of Cu will still affect the bonding quality. The comparison between planarization and no planarization of required bonding parameters and the bonding results are shown in Table 4.

3. As per the TEM analysis, there are no cracks or voids which occur at the Cu–Cu_3Sn, Cu_3Sn–Cu_6Sn_5, or Cu_3Sn–Cu_3Sn interfaces. This implies that there is good adhesion between different layers regardless of whether Cu_6Sn_5 is present at the interface. Therefore, the strength of this Cu–Sn joint would be sufficiently strong, because no defects occur between different layers. Thus, the defects at the bonding interface might be the dominant factor which affects the joint strength.

Finally, by passivating Sn over Cu, apart from the advantage of a low-temperature and low-pressure bonding process, the most attractive quality is that it does not require extremely low surface roughness. Because CMP is currently a time-consuming and expensive procedure and is always required before bonding in 3D IC, such as hybrid bonding, this research will provide a promising method to enhance the bonding process without grinding or CMP. Additionally, the whole bonding process can be accomplished with a cheaper and faster procedure.

Author Contributions: Conceptualization, P.-Y.K., C.-L.K. and C.R.K.; Data curation, P.-Y.K. and W.-L.H.; Formal analysis, P.-Y.K.; Funding acquisition, C.R.K.; Investigation, P.-Y.K. and W.-L.H.; Methodology, P.-Y.K., W.-L.H., C.-L.K., Y.-S.L., Y.-C.H. and C.R.K.; Project administration, C.R.K.; Supervision, C.R.K.; Validation, P.-Y.K.; Writing—original draft, P.-Y.K.; Writing—review & editing, P.-Y.K., W.-L.H., C.-L.K., Y.-S.L., Y.-C.H. and C.R.K. All authors have read and agreed to the published version of the manuscript.

Funding: This work was supported by the Ministry of Science and Technology (MOST 110-2622-E-002-016-CC1).

Institutional Review Board Statement: Not applicable.

Informed Consent Statement: Not applicable.

Data Availability Statement: Not applicable.

Acknowledgments: The authors thank Ya-Yun Yang and Ching-Yen Lin from the Precious Instrument Center of the College of Science at National Taiwan University for their assistance with ion-milling.

Conflicts of Interest: The authors declare no conflict of interest.

References

1. Das, S.; Fan, A.; Chen, K.-N.; Tan, C.S.; Checka, N.; Reif, R. Technology, performance, and computer-aided design of three-dimensional integrated circuits. In Proceedings of the 2004 International Symposium on Physical Design, Phoenix, AZ, USA, 18–21 April 2004.
2. Liang, H.-W.; Chen, H.-C.; Lin, C.-H.; Lee, C.-L.; Yang, S.-C.; Chen, K.-N. The influence of device morphology on wafer-level bonding with polymer-coated layer. In Proceedings of the IEEE International 3D Systems Integration Conference (3DIC), San Francisco, CA, USA, 8–11 November 2016.
3. Chen, K.; Tan, C. Integration schemes and enabling technologies for three-dimensional integrated circuits. *IET Comput. Digit. Tech.* **2011**, *5*, 160–168. [CrossRef]
4. Swinnen, B.; Ruythooren, W.; de Moor, P.; Bogaerts, L.; Carbonell, L.; de Munck, K.; Eyckens, B.; Stoukatch, S.; Vaes, J. 3D integration by Cu-Cu thermo-compression bonding of extremely thinned bulk-Si die containing 10 μm pitch through-Si vias. In Proceedings of the 2006 International Electron Devices Meeting, San Francisco, CA, USA, 11–13 December 2006.
5. Chen, K.N.; Fan, A.; Tan, C.S.; Reif, R. Bonding parameters of blanket copper wafer bonding. *J. Electron. Mater.* **2006**, *35*, 230–234. [CrossRef]
6. Chen, K.-N.; Fan, A.; Reif, R. Microstructure examination of copper wafer bonding. *J. Electron. Mater.* **2001**, *30*, 331–335. [CrossRef]
7. Chen, K.N.; Tan, C.S.; Fan, A.; Reif, R. Copper bonded layers analysis and effects of copper surface conditions on bonding quality for three-dimensional integration. *J. Electron. Mater.* **2005**, *34*, 1464–1467. [CrossRef]
8. Liu, D.; Chen, P.-C.; Chen, K.-N. A novel low-temperature Cu-Cu direct bonding with Cr wetting layer and Au passivation layer. In Proceedings of the IEEE 70th Electronic Components and Technology Conference (ECTC), Orlando, FL, USA, 3–30 June 2020.
9. Chen, P.-C.; Liu, D.; Chen, K.-N. Low-Temperature Wafer-Level Metal Bonding with Gold Thin Film at 100 °C. In Proceedings of the International 3D Systems Integration Conference (3DIC), Sendai, Japan, 8–10 October 2019.
10. Liu, D.; Chen, P.-C.; Tsai, Y.-C.; Chen, K.-N. Low temperature Cu to Cu direct bonding below 150° C with Au passivation layer. In Proceedings of the International 3D Systems Integration Conference (3DIC), Sendai, Japan, 8–10 October 2019.

11. Chou, T.-C.; Huang, S.-Y.; Chen, P.-J.; Hu, H.-W.; Liu, D.; Chang, C.-W.; Ni, T.-H.; Chen, C.-J.; Lin, Y.-M.; Chang, T.-C.; et al. Electrical and Reliability Investigation of Cu-to-Cu Bonding with Silver Passivation Layer in 3-D Integration. *IEEE Trans. Compon. Packag. Manuf. Technol.* **2020**, *11*, 36–42. [CrossRef]
12. Lin, P.-F.; Tran, D.-P.; Liu, H.-C.; Li, Y.-Y.; Chen, C. Interfacial Characterization of Low-Temperature Cu-to-Cu Direct Bonding with Chemical Mechanical Planarized Nanotwinned Cu Films. *Materials* **2022**, *15*, 937. [CrossRef]
13. Zantye, P.B.; Kumar, A.; Sikder, A. Chemical mechanical planarization for microelectronics applications. *Mater. Sci. Eng. R Rep.* **2004**, *45*, 89–220. [CrossRef]
14. Banerjee, G.; Rhoades, R.L. Chemical mechanical planarization historical review and future direction. *ECS Trans.* **2008**, *13*, 1. [CrossRef]
15. Krishnan, M.; Nalaskowski, J.W.; Cook, L.M. Chemical Mechanical Planarization: Slurry Chemistry, Materials, and Mechanisms. *Chem. Rev.* **2010**, *110*, 178–204. [CrossRef]
16. Zhou, J.; Sun, Y.; Xue, F. Properties of low melting point Sn–Zn–Bi solders. *J. Alloy. Compd.* **2005**, *397*, 260–264. [CrossRef]
17. Liu, Y.; Tu, K.N. Low melting point solders based on Sn, Bi, and In elements. *Mater. Today Adv.* **2020**, *8*, 100115. [CrossRef]
18. Wu, Z.; Cai, J.; Wang, Q.; Wang, J.; Wang, D. Wafer-Level Hermetic Package by Low-Temperature Cu/Sn TLP Bonding with Optimized Sn Thickness. *J. Electron. Mater.* **2017**, *46*, 6111–6118.
19. Liu, H.; Wang, K.; Aasmundtveit, K.; Hoivik, N. Intermetallic compound formation mechanisms for Cu-Sn solid–liquid interdiffusion bonding. *J. Electron. Mater.* **2012**, *41*, 2453–2462. [CrossRef]
20. Luu, T.-T.; Duan, A.; Wang, K.; Aasmundtveit, K.; Hoivik, N. Cu/Sn SLID wafer-level bonding optimization. In Proceedings of the IEEE 63rd Electronic Components and Technology Conference, Las Vegas, NV, USA, 28–31 May 2013.
21. Tang, Y.-S.; Chen, H.-C.; Kho, Y.-T.; Hsieh, Y.-S.; Chang, Y.-J.; Chen, K.-N. Investigation and Optimization of Ultrathin Buffer Layers Used in Cu/Sn Eutectic Bonding. *IEEE Trans. Compon. Packag. Manuf. Technol.* **2018**, *8*, 1225–1230. [CrossRef]
22. Das, S.; Tiwari, A.N.; Kulkarni, A.R. Thermo-compression bonding of alumina ceramics to metal. *J. Mater. Sci.* **2004**, *39*, 3345–3355. [CrossRef]
23. Yuhan, C.; Le, L. Wafer level hermetic packaging based on Cu–Sn isothermal solidification technology. *J. Semicond.* **2009**, *30*, 086001. [CrossRef]
24. Van de Wiel, H.; Vardøy, A.B.; Hayes, G.; Fischer, H.; Lapadatu, A.; Taklo, M. Characterization of hermetic wafer-level Cu-Sn SLID bonding. In Proceedings of the 4th Electronic System-Integration Technology Conference, Amsterdam, The Netherlands, 17–20 September 2012.
25. Lu, M.-H.; Hsieh, K.-C. Sn-Cu Intermetallic Grain Morphology Related to Sn Layer Thickness. *J. Electron. Mater.* **2007**, *36*, 1448–1454. [CrossRef]
26. Yang, L.; Atanasova, T.; Radisic, A.; Deconinck, J.; West, A.C.; Vereecken, P. Wafer-scale Cu plating uniformity on thin Cu seed layers. *Electrochim. Acta* **2013**, *104*, 242–248. [CrossRef]
27. Takahashi, K.M. Electroplating Copper onto Resistive Barrier Films. *J. Electrochem. Soc.* **2000**, *147*, 1414. [CrossRef]
28. Zhang, R.; Tian, Y.; Hang, C.; Liu, B.; Wang, C. Formation mechanism and orientation of Cu3Sn grains in Cu–Sn intermetallic compound joints. *Mater. Lett.* **2013**, *110*, 137–140. [CrossRef]
29. Chen, Y.; Chung, C.; Yang, C.; Kao, C.R. Single-joint shear strength of micro Cu pillar solder bumps with different amounts of intermetallics. *Microelectron. Reliab.* **2013**, *53*, 47–52. [CrossRef]
30. Kao, C.-W.; Kung, P.-Y.; Chang, C.-C.; Huang, W.-C.; Chang, F.-L.; Kao, C.R. Highly Robust Ti Adhesion Layer during Terminal Reaction in Micro-Bumps. *Materials* **2022**, *15*, 4297. [CrossRef] [PubMed]
31. Yang, C.; Song, F.; Lee, S.W.R. Effect of interfacial strength between Cu 6 Sn 5 and Cu 3 Sn intermetallics on the brittle fracture failure of lead-free solder joints with OSP pad finish. In Proceedings of the IEEE 61st Electronic Components and Technology Conference (ECTC), Lake Buena Vista, FL, USA, 31 May–3 June 2011.
32. Yao, P.; Li, X. Investigation on shear fracture of different strain rates for Cu/Cu3Sn/Cu solder joints derived from Cu–15μm Sn–Cu sandwich structure. *J. Mater. Sci. Mater. Electron.* **2020**, *31*, 2862–2876. [CrossRef]

Article

Fiber Laser Welded Cobalt Super Alloy L605: Optimization of Weldability Characteristics

B. Hari Prasad [1], G. Madhusudhan Reddy [2], Alok Kumar Das [3,*] and Konda Gokuldoss Prashanth [4,5,6,*]

1. Defence Research & Development Laboratory, Hyderabad 500058, India
2. Defence Metallurgical Research Laboratory, Hyderabad 500058, India
3. Department of Mechanical Engineering, Indian Institute of Technology (ISM), Dhanbad 826004, India
4. Department of Mechanical and Industrial Engineering, Tallinn University of Technology, 19086 Tallinn, Estonia
5. Erich Schmid Institute of Materials Science, Austrian Academy of Sciences, Jahnstrasse 12, 8700 Leoben, Austria
6. CBCMT, School of Mechanical Engineering, Vellore Institute of Technology, Vellore 630014, India
* Correspondence: alokmech@iitism.ac.in (A.K.D.); kgprashanth@gmail.com (K.G.P.)

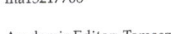

Citation: Prasad, B.H.; Madhusudhan Reddy, G.; Das, A.K.; Prashanth, K.G. Fiber Laser Welded Cobalt Super Alloy L605: Optimization of Weldability Characteristics. *Materials* 2022, 15, 7708. https://doi.org/10.3390/ma15217708

Academic Editor: Tomasz Trzepieciński

Received: 4 October 2022
Accepted: 28 October 2022
Published: 2 November 2022

Publisher's Note: MDPI stays neutral with regard to jurisdictional claims in published maps and institutional affiliations.

Copyright: © 2022 by the authors. Licensee MDPI, Basel, Switzerland. This article is an open access article distributed under the terms and conditions of the Creative Commons Attribution (CC BY) license (https://creativecommons.org/licenses/by/4.0/).

Abstract: The present study describes the laser welding of Co-based superalloy L605 (52Co-20Cr-10Ni-15W) equivalent to Haynes-25 or Stellite-25. The influence of laser welding process input parameters such as laser beam power and welding speed on mechanical and metallurgical properties of weld joints were investigated. Epitaxial grain growth and dendritic structures were visible in the weld zone. The phase analysis results indicate the formation of hard phases like CrFeNi, CoC, FeNi, and CFe in the weld zone. These hard phases are responsible for the increase in microhardness up to 321 $HV_{0.1}$ in the weld zone, which is very close to the microhardness of the parent material. From the tensile strength tests, the ductile failure of welded specimens was confirmed due to the presence of dimples, inter-granular cleavage, and micro voids in the fracture zone. The maximum tensile residual stress along the weld line is 450 MPa, whereas the maximum compressive residual stress across the weld line is 500 MPa. On successful application of Response Surface methodology (RSM), laser power of 1448.5 W and welding speed of 600 mm/min i.e., line energy or heat input equal to 144 J/mm, were found to be optimum values for getting sound weld joint properties. The EBSD analysis reveals the elongated grain growth in the weld pool and very narrow grain growth in the heat-affected zone.

Keywords: fiber laser welding; microstructure; residual stress; epitaxial growth; fractography; EBSD analysis

1. Introduction

Cobalt-based superalloy L605 is a preferred material for a wide range of high-temperature applications such as gas turbine blades, jet engine parts, aerospace systems, combustion chambers, etc. Owing to their excellent mechanical and fatigue strength; biocompatibility; corrosion and oxidation resistance, they also have wide applications in chemical, marine, and bio-medical industries [1–3]. The presence of cobalt and chromium makes it eligible for extensive biomedical applications. The presence of tungsten, chromium, and molybdenum enhances the melting point, hardness, and density of the said alloy [4]. Co-based superalloys are synthesized through a vacuum induction melting process followed by electro slag refining, which leads to very low non-metallic inclusions and hence limited crystal defects. Sometimes, it is also observed that this alloy is superior to Ni-based superalloys as it has better thermal shock resistance and anti-corrosion properties even in a hot gas environment. Different welding processes with high metal deposition rate like Gas Metal Arc Welding (GMAW), Gas Tungsten Arc Welding (GTAW) showed wide HAZ and welding defects in the joints which restricts its uses in precision aerospace components [5]. Laser welding produces minimum defects and narrow HAZ, which attracts its applications in various industries. Although there are several laser welding processes (using gas lasers and solid-state lasers), fiber laser welding is preferred over other processes due to its low maintenance cost and high reliability of the laser source.

Welding of different superalloys has been reported by many researchers. For example, Osoba et al. [6] have carried out laser welding of Haynes 282 superalloy and observed the formation of micro-segregation pattern in the fusion zone. Chen et al. [7] have investigated the grain growth phenomenon in Co-Cr-Mo alloy during laser melting and reported that the solidification starts from the epitaxial grains in the boundary zone of the melt track. The planar grain growth is not possible due to very high grain growth velocity [8]. M. Shamanian et al. [9] carried out pulsed Nd: YAG laser welding of Co-based superalloys and observed that heat input to the welding process plays a major role in the control of microstructure and grain orientation in the weldment. However, the changes in heat input had an insignificant effect on the mechanical properties. Palanivel R. et al. [10] studied the Nd: YAG laser welding of IN 800 and observed the formation of an elongated columnar but fine equiaxed dendritic structure in the fusion zone. In addition, the phase transformation occurred due to the higher cooling rate of the laser welding process. The ductile failure of welded joints was observed at higher welding speed whereas brittle failure occurred at low welding speed. Similarly, many kinds of literature describe the effects of input parameters of the laser welding process on the weld quality parameters of different materials such as the microstructure and grain orientation [11–14], bead profile [15], mechanical properties [16–18], metallurgical properties [19], corrosion resistance [20] and residual stresses [21].

In all the above-mentioned applications, the joining plays a crucial role. Some applications demand better weld geometry with the narrow heat-affected zone, which is difficult to meet through conventional welding processes [22–26]. These requirements may be met using a highly focused and concentrated heat source, as in laser and electron beam welding processes. The present investigation deals with the fiber laser welding of Co-based superalloy L605 of 2 mm thick sheets and describes the detailed analysis of the welded joints for weld bead geometry, microstructure, microhardness, tensile yield strength (yield stress), phase analysis, residual stress and EBSD study. The statistical analysis of the welding process has been carried out to develop a regression model and to correlate the dependent and independent process parameters.

2. Materials and Methods

2.1. Setup Configuration

The fiber laser welding setup consists of a 4 kW fiber laser source (make: Arnold Ravensburg, Germany), X-Y-Z CNC stages (movement range: 2500 mm × 3000 mm × 750 mm, maximum traverse speed in X-Y plane: 6000 mm/min) and an integrated CNC controller. The emission wavelength of the fiber laser is 1070 nm; the spot diameter on the focal plane is 440 µm; the beam profile is Gaussian and can be operated in continuous wave mode. The laser beam is delivered onto the worktable through an optical fiber cable. The core diameter of the optical transmission fiber is 200 µm. The welding head is equipped with an air cross jet facility to protect the lens from the deposition of metal vapors. For shielding the weld zone from atmospheric contamination, inert gas (argon) is purged from the top and bottom (above and below the workpiece) of the weld zone. The welding head is provided with X-Y-Z motion; whereas the workpiece is mounted on the machine table using a rigid fixture. The setup configuration for conducting fiber laser welding experiments is presented in Figure 1.

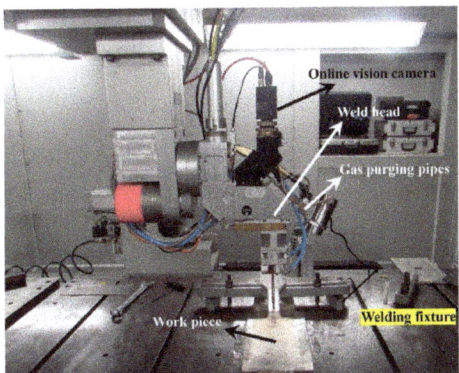

Figure 1. Setup configuration of the fiber laser welding process.

2.2. Workpiece Preparation

Cobalt-based superalloy L605 sheets having a thickness of 2 mm were used as workpieces to carry out the experiments. The samples were prepared to 60 mm × 125 mm × 2 mm size by abrasive water jet cutting, followed by milling and fine grinding of the edges. The prepared samples were cleaned with acetone to remove the oxide layer from the surfaces and the edges before conducting the welding experiments. The chemical composition of L605 material is given in Table 1.

Table 1. Chemical composition of L605 material.

Unit	Co	Cr	W	Ni	Fe	Mn	Si	C	S	P
Wt.%	Balance	20.00	15.00	10.00	3.00	1.50	0.40	0.10	0.030	0.040

2.3. Parameter Selection

Several pilot experiments were conducted by varying laser parameters such as laser beam power and welding speed at the focal plane (Figure 2) with reference to the workpiece surface for finding the ranges of different parameters to ensure full penetration welding along with the formation of acceptable bead geometry. Subsequently, the ranges of different weld parameters were selected as laser beam power: 1400–1800 W, laser beam scanning speed: 200–600 mm/min.

Figure 2. Position of the workpiece surface at the focal position of the laser beam.

2.4. Design of Experiments

In the present research, Response Surface Methodology (RSM) was followed for conducting weld experiments. Table 2 shows the levels of different input welding parameters which were decided after conducting trial experiments. Central Composite Design (CCD) was used to find the parameters setting of each experimental run. MINITAB software was used for conducting Design of Experiments (DOE). Table 3 shows CCD with parameter settings for 14 experimental runs. RSM is a statistical approach to observe the influence of simultaneous variation of any two input parameters on the responses which cannot be done in a normal parametric study, and it provides the designed model (Equation (1)) which can predict the responses with the variations of input welding parameters within the selected range.

$$Y = \alpha_0 + \alpha_1 X_1 + \alpha_2 X_2 + \alpha_3 X_3 + \alpha_{12} X_{12} + \alpha_{13} X_{13} + \alpha_{23} X_{23} + \alpha_{11} X_{11} + \alpha_{22} X_{22} + \alpha_{33} X_{33} \quad (1)$$

where, Y represents the response variable, X_1, X_2, and X_3 are the input process variables, α_0 is a constant, α_1, α_2, and α_3 are linear coefficients, α_{11}, α_{22}, and α_{33} are quadratic coefficients, and α_{12}, α_{13}, and α_{23} are interaction coefficients. The prediction ability of the developed equation is checked through ANOVA.

Table 2. The different input welding parameters used in the present study.

Parameter	Units	Level 1	Level 2	Level 3
Laser power	Watt (W)	1400	1600	1800
Welding speed	mm/min	200	400	600

Table 3. Experiment table and the measured responses.

Run Order	Pt Type	Blocks	P	WS	W1	W2	H1	H2	Y
1	0	1	1600	400	4016.83	3807.56	139.51	145.33	492
2	1	1	1800	600	3051.86	2662.39	110.45	0	509
3	0	1	1600	400	3650.61	3859.88	156.95	122.07	481
4	1	1	1400	200	4964.36	4656.27	226.71	168.58	474
5	1	1	1800	200	5958.4	4447.00	610.37	1150.99	299
6	1	1	1400	600	2336.85	1790.43	122.07	87.20	532
7	0	1	1600	400	4016.83	3731.99	244.15	198.82	486
8	−1	2	1600	200	5981.65	5830.51	412.73	494.11	540
9	0	2	1600	400	3784.31	3749.43	168.58	133.70	506
10	−1	2	1400	400	3197.19	2371.73	198.82	129.40	495
11	0	2	1600	400	3877.32	3685.49	238.34	249.96	487
12	−1	2	1600	600	3121.62	2348.48	104.64	133.70	548
13	0	2	1600	400	3830.81	3929.63	267.4	191.83	495
14	−1	2	1800	400	4150.53	3865.69	325.53	395.29	409

P: Laser power (W), WS: Scanning Speed (mm/min), W1: Top width of weld bead, W2: Bottom width of weld bead H1: Weld bead height from the parent work surface, H2: Weld undercut bottom side.

2.5. Experimentation

The prepared samples (size: 60 mm × 125 mm × 2 mm) were mounted on the welding fixture for making a butt joint without using filler material (autogenous welding). All the experiments were conducted using the parameters settings listed in Table 3. The laser source was operated in continuous mode, with a focus spot size of 0.44 mm at the focal plane. During the experimentation, the laser was irradiated perpendicular to the work surface during the welding experiments. Commercial argon gas (purity > 95%) was used as the shielding gas and purged at the rate of 10 L/min at 1 bar pressure. The gas purging was done from the top and bottom of the weld zone to prevent oxidation and other atmospheric contaminations. In this way, 14 welded samples of L605 sheets were prepared and subjected to quality evaluation. The fabricated samples were cut from the weld specimens using a wire-EDM machine and molds were prepared for mounting purposes followed by polishing to measure the microhardness.

2.6. Characterization Processes

At first, all-welded samples underwent a radiography test using a 160 kV X-ray source (model: Seram 235; make: BALTAUE NDT, Canada) to identify the internal defects (such as porosity, and internal cracks) present in the weld zone, which is required before conducting the tensile strength test. A dye penetrant test was conducted to detect surface defects. Then the tensile test specimens (prepared as per ASTM E8/E8M-15a standard) were cut from minimum defects or defect-free zones of weld samples using the wire-EDM process (Figure 3a,b). The tensile tests were carried out on Universal Tensile Testing Machine (UTM) (model: BiSS make Measure India Corporation Pvt Ltd., Secunderabad, India). The remaining part of the welded joint was kept for evaluation of other properties. The bead geometry of the weld bead was analyzed using a metallurgical microscope (make Olympus Corporation, Tokyo, Japan). The different parameters of the weld bead geometry are presented in Figure 3c. The samples were polished and etched with a chemical solution (20 mL HCl, 5 mL HNO_3, 65 gm $FeCl_3$, 150 mL distilled water, swabbing for 10–15 s) for measuring the heat-affected zone (HAZ) and microstructure analysis using Metallurgical Microscope (model: BX51M; make: Olympus Corporation, Tokyo, Japan). Microhardness tests were performed on the weld zone and on the parent material for comparison using a Digital Microhardness tester (model: MMT-X7; make MATSUZAWA, Akita, Japan). Phase and chemical composition analysis was carried out by Scanning Electron Microscope (SEM) (model EVO MA10; make ZEISS, Jena, Germany). XRD (make: Rigaku, Tokyo, Japan) analysis was done to identify the new chemical compounds formed due to welding temperature. Residual stress measurement on weld samples was carried by LXRD equipment (make: Proto, Canada). EBSD study was carried out for the welded samples in the weld zone to analyze the grain growth by FESEM with an attachment of EBSD Camera (Make: ZEISS, Jena, Germany); the results of the characterization process were analyzed and discussed in Section 3.

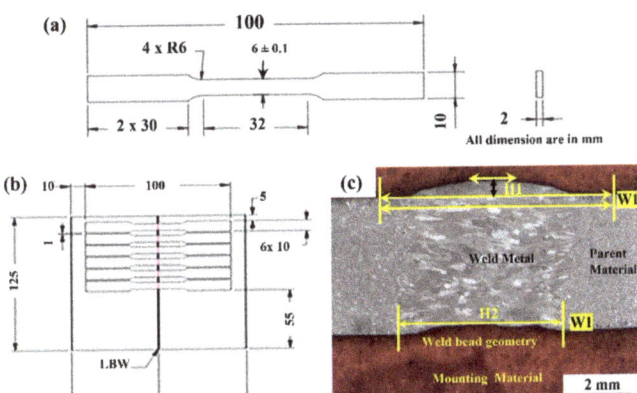

Figure 3. (a) Dimensions of the prepared tensile test specimen, (b) Tensile test specimens in the weld sample (c) Weld bead geometry on prepared weld sample (H1: weld bead height above the workpiece surface, H2: Weld undercut at the bottom surface, W1: Weld bead width at the top surface, W2: Weld bead width at the bottom surface).

3. Results and Discussion

All the prepared weld samples (14 numbers) were subjected to different tests as discussed in Section 2.6 to evaluate the characteristics of weld joints. The prominent test results and the discussions are enumerated in the following sections.

3.1. Radiography and Dye Penetrant Test

The weld zones were subjected to radiography testing as per ASTM E1742 with X-rays to find any internal defects present within them. Similarly, the dye penetrant test was carried out as per ASTM E1417 to check the surface defects. Few of the captured images of radiography and dye penetrant test are presented in Figure 4. In the radiography test, the weld samples were kept above the phosphoric image plate and then high-power X-rays (160 kV, 4 mA, exposure time: 2 min) were made to fall on the welded samples. A latent image of the sample was captured by the plate. These latent images were converted into real viewable images by a laser scanner (make GE, USA) and then images were further analyzed for internal defects. In the dye penetrant test, the liquid penetrant (Magna flux SKL-SP1) was spread over the weld zone after a thorough cleaning by the cleaner (Magna flux SKC-1) and left untouched for 30 min. Subsequently, the developer (Magna flux SKD-S2) is spread over the weld bead and left for 5 min, which pulls up the penetrant and gets accumulated in the pores and cracks. The surface turns red by which the surface defects are identified [27,28]. The results of the radiography test and die penetration tests conducted on the samples are shown in Figure 4. The presented figures show that samples are free from embedded pores and internal cracks. Only a few undercuts are visible at the edge of the welded specimen that are not the part of samples used for various tests. Tensile test specimens were cut from the weld coupons, which were having minimum or zero defects.

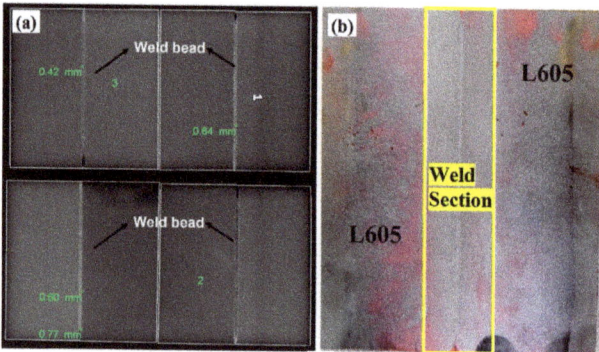

Figure 4. Captured images of (**a**) radiography test and (**b**) dye penetrant test samples.

3.2. Statistical Analysis

After the radiographic test, the other quantitative responses were measured (Table 3) and entered into the design table of RSM with CCD. All the data were analyzed to observe the effects of the combination of different process parameters on the obtained responses. Regression equations in uncoded form (Equations (2)–(6)) were developed and validated through the ANOVA (Analysis of Variance) (Table 4) which are described below. Figure 5 represents the normal probability plots for different responses. As the distribution of different points is very close to the straight line, which indicates the model has good prediction ability [29,30]. The values of R-sq and R-sq(adj) are very close to each other, which indicates the less variability of the predicted responses with respect to the input parameters. The P values in the ANOVA table for linear, square, and interaction of input parameters are found to be less than 0.05 which indicates that the developed regression model is significant [31].

$$W_1\ (\mu m) = -16422 + 28.75\ P - 15.30\ WS - 0.00807\ P^2 + 0.01387\ WS^2 - 0.00174\ P \cdot WS \qquad (2)$$

$$W_2\ (\mu m) = -34885 + 54.1\ P - 23.26\ WS - 0.01718\ P^2 + 0.00709\ WS^2 + 0.00676\ P \cdot WS \qquad (3)$$

$$H_1\ (\mu m) = 680 - 1.45\ P + 2.55\ WS + 0.000893\ P^2 + 0.000805\ WS^2 - 0.002470\ P \cdot WS \qquad (4)$$

$$H_2 \ (\mu m) = -506 - 1.54 \ P + 7.04 \ WS + 0.00162 \ P^2 + 0.00291 \ WS^2 - 0.006685 \ P \cdot WS \qquad (5)$$

$$Y \ (MPa) = -2537 + 4.44 \ P - 1.864 \ WS - 0.001580 \ P^2 + 0.000715 \ WS^2 + 0.000951 \ P \cdot WS \qquad (6)$$

Table 4. Analysis of Variance for different responses.

	W_1		W_2		H_1		H_2		Y	
Source	DF	p-Value	DF	p-Value	DF	p-Value	DF	p-Value	DF	p-Value
Linear	2	0.000	2	0.000	2	0.000	2	0.000	2	0.003
Square	2	0.001	2	0.029	2	0.137	2	0.017	2	0.026
2-Way Interaction	1	0.357	1	0.153	1	0.003	1	0.000	1	0.039
Lack-of-Fit	3	0.444	3	0.001	3	0.792	3	0.060	3	0.001
R-sq	98.11%		93.47%		92.97%		95.62%		84.97%	
R-sq(adj)	95.78%		89.39%		88.57%		92.89%		75.58%	

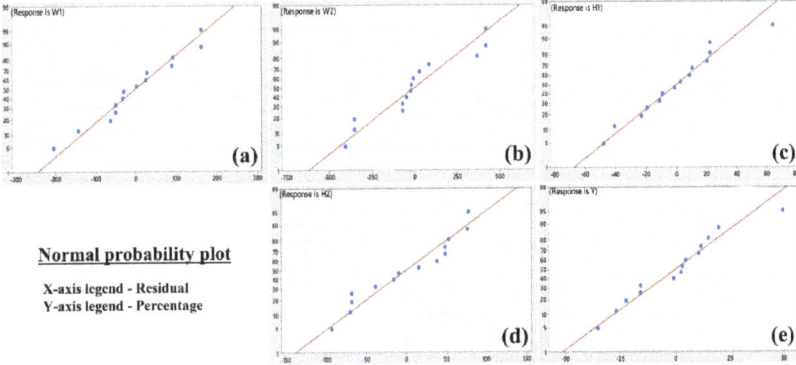

Normal probability plot

X-axis legend - Residual
Y-axis legend - Percentage

Figure 5. Normal probability plots for responses (**a**) Weld bead width at the top surface (W_1), (**b**) Weld bead width at the bottom surface (**c**) Weld bead height above the workpiece surface (H_1), (**d**) Weld undercut at the bottom surface (H_2), (**e**) Tensile yield strength (Y).

3.2.1. Weld Bead Geometry Analysis

The weld bead geometry plays a vital role in the better strength of the joint. One of the typical weld bead geometries is shown in Figure 3c for reference purpose and the measured values are presented in Table 3. Equations (2)–(6) represents the variation of geometrical parameters (responses) corresponding to the inputs which are explained with the help of the contour plots in Figure 6. The width of the weld zone is reduced with the increase in scanning speed, this might be due to laser beam passes quickly over the weld line and hence less amount of base metal is melted, however, the weld width increases reasonably with the increase in laser power. With the increase in laser power, the bead width increases due to over melting of the base material (Figure 6b). The bead width (W2) at lower side of the welding is found to be reduced with the increase in scanning speed, it may be due to the formation of keyhole in the welding process which leads to wider top and narrow bottom of the weld bead (Figure 6c). The bead height and undercut are found to be maximum at higher laser power and lower scanning speed, at the appropriate combination of laser power 1600 W and scanning speed of ~500 mm/min the bead height and undercut are found to be minimum as shown in Figure 6d,e. It shows the variation of W_1, W_2, H_1, and H_2 with respect to the laser beam power and welding speed. It has been observed that the minimum weld bead geometry (Table 3) with top width (W_1): 2336.85 µm, Bottom width (W_2): 1790.43 µm height (H_1): 104.64 µm and undercut (H_2): 0 µm were obtained at different parameter setting of laser power in the range of 1400 W to 1800 W and scanning speed 200 to 600 mm/min and the same is also confirmed from the contour plots shown in Figure 6. The optimum values of the bead geometry can be obtained at parameters settings of laser power 1448.5 W and scanning speed 600 mm/min where top width (W_1): 2594.45 µm, bottom width (W_2): 1848.62 µm, height (H_1): 126.04 µm and undercut (H_2): 130.14 µm. In view of the above, the welding parameters can be selected from the contour plots shown in Figure 6 as per the requirement of the weld bead geometry, which is in line with the reference article [32].

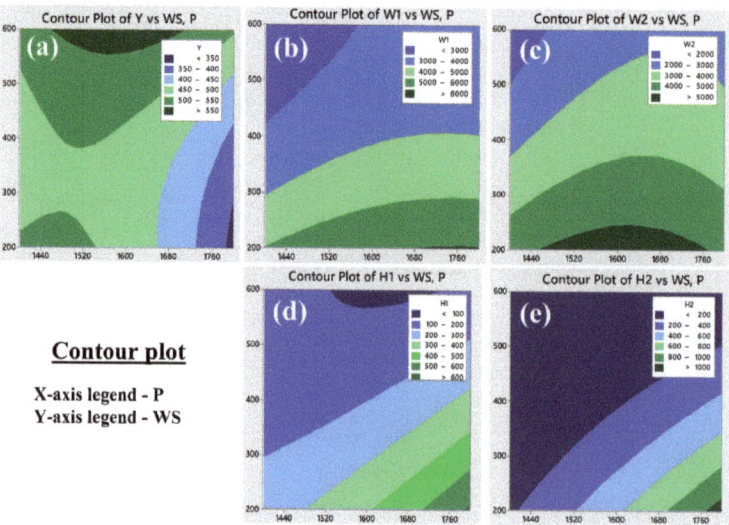

Figure 6. Contour plots for weld bead geometry: (**a**) variation of tensile yield strength (Y) (**b**) variation of top width of weld bead (W_1), (**c**) variation of bottom width of weld bead (W_1), (**d**) variation of weld bead height (H_1), (**e**) variation of undercut (H_2), w.r.t input parameters.

3.2.2. Microhardness of Weld Bead

The prepared sample for microhardness is presented in Figure 7a. The micro-hardness was measured from the centerline of the weld bead to one side towards the parent material. Figure 7b represents the microhardness profile of a typical weld sample. From the experimental data, it is observed that there is little variation in microhardness (maximum value: 321 $HV_{0.1}$) on the weld bead as compared to the parent material (average value: 305 $HV_{0.1}$) that indicates a good quality of weld joint without impacting its mechanical properties. During the laser welding, when the laser beam is irradiated on the material, considerably high temperature is developed and melts the weld sample interface. When the material solidifies, the fusion zone is created. Some parts of the heat are dissipated into the welding parts and lead to the formation of HAZ. Due to very short exposure time of the laser, the temperature of the weld zone decreases rapidly with high cooling rates. It results in the formation of austenite and austenite + carbide phases along with other different phases like CrFeNi, CoC, FeNi, and FeC that affect the microstructures of the fusion zone and microhardness is increased.

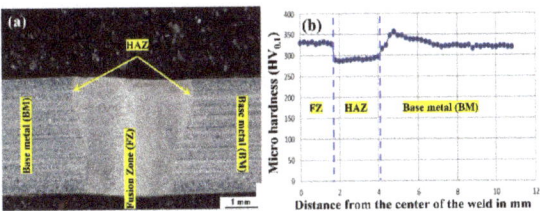

Figure 7. (a) Typical weld bead, (b) Microhardness plot of typical weld sample.

3.2.3. Tensile Yield Strength (Yield Stress)

To perform the tensile strength tests, the specimens were prepared using the wire-EDM process which is in accordance with ASTM E8 standard (Figure 3a), and the prepared tensile test specimens are shown in Figure 8a. The mounting of the specimen on UTM is shown in Figure 8c. The tensile tests for all specimen along with parent material were conducted at room temperature with a crosshead speed of 0.5 mm/min and the captured plots are shown in Figure 8d,e. The plots indicate that the amount of elongation varies between 5.21% to 25.09%, whereas the breaking load varies from 1 kN to 12 kN. During the tensile strength tests, all the samples failed in the weld zone as shown in Figure 8b and this is because of epitaxial grain formation, which was located along the axis of the weld bead [33]. The solidification of molten metal takes place at room temperature and the parent material on both sides of the weld zone acts as the chiller unit due to which the solidification starts from both sides, as a result, the end of solidification takes place at the center of the weld nugget [34]. During the solidification process, the equiaxed grain structure is formed in the weld zone due to which a parting line is observed at the center of the weld zone and that leads to the formation of the weaker part. Therefore, the optimization of the input process parameters was performed for better strength of the weld joint. It was also observed that as the heat input supply during the welding is reduced, the cooling rate increases, and it results in the formation of finer grain size and therefore results in enhanced mechanical properties [35]. In this way, laser welding parameters at high input i.e., 1800 W laser power and 200 mm/min scanning speed resulted in low yield stress of 299 MPa as compared to the laser welding parameters at low heat input i.e., 1400 W laser power and 600 mm/min, where yield stress of 546.20 MPa was obtained.

The measured values of tensile yield strength were fed to the design table and further analyzed to correlate with the input parameters presented in Equation (6). The contour plots (Figure 6a) explain the variation of yield stress w.r.t input parameters and it is observed that the tensile yield strength (yield stress) up to 541 MPa can be obtained at a parameter combination of laser power: 1448.5 W, scanning speed: 600 mm/min i.e., line energy or heat input equal to 144.44 J/mm, as shown in Figure 9.

3.2.4. Optimization of Process Parameters

In the present work, the multi-objective optimization of input process parameters was carried out using the desirability approach through the RSM technique considering responses: W_1, W_2, H_1, H_2 and Y. The desirability values of response parameters are shown in Figure 9. It is observed that the desirability value approaches 1 which indicates the good prediction ability of the developed model. In the present study, laser beam power: 1448.5W, scanning speed: 600 mm/min give response values such as W_1: 2594.45 μm, W_2: 1848.62 μm, H_1: 126.04 μm, and H_2: 130.14 μm and Y: 546.20 MPa (Figure 9). These responses were validated by conducting the confirmation experiments and the error was found to be within 5%.

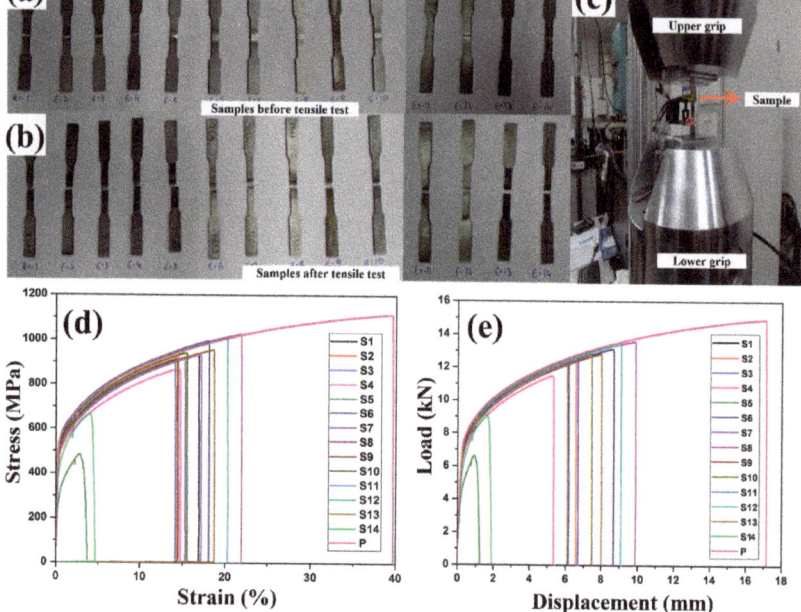

Figure 8. (a) Tensile test specimens (b) Images of tensile test specimens after failure (c) UTM configuration for the tensile test (d,e) Tensile stress-strain, load-displacement plots 3.7. Wettability studies.

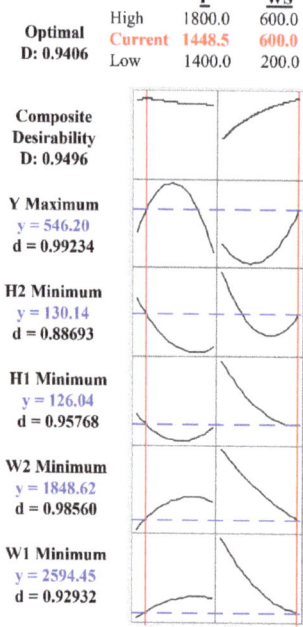

Figure 9. Optimization plot for the weld bead.

3.3. Analysis of Weld Bead

The mechanical and metallurgical properties were evaluated further to investigate the weldability characteristics of L605 alloy. The prepared weld samples which had high and low heat input were considered in this investigation, which are presented in the following sections.

3.3.1. Morphology and Microstructure

The prepared weld samples were polished to get a mirror finish surface and etched with a chemical solution to observe the microstructure by metallurgical microscope and SEM (Figure 10). Figure 11c,d represents the morphology of the cross-section of the weld bead. Small concavity is observed in the weld root as shown in Figure 10c, which may be due to high heat input during the welding process with parameters laser beam power 1800 W, scanning speed 200 mm/min. The flat root is observed in the other case as shown in Figure 10d which is due to the low heat input with parameters laser beam power 1400 W, scanning speed 600 mm/min. It is also worth stating that, the transfer of heat within the weld pool happens by convection; and therefore, the fluid dynamics predominantly determine the formation of bead shape. The major factors coming into play are surface tension, volume contraction, vapor pressure, phase transformation and gravity [36]. With the reduction in energy input, the cross-section of the weld joint transforms into almost an X-shape; whereas the V-shaped cross-section is obtained when there is partial keyhole formation [37]. This leads to a reduction in the width of the weld zone, as shown in Figure 10b. Figure 10e,f indicate the magnified images of Figure 10c,d, respectively, which affirm that granular and epitaxial grain growth takes place during the solidification process.

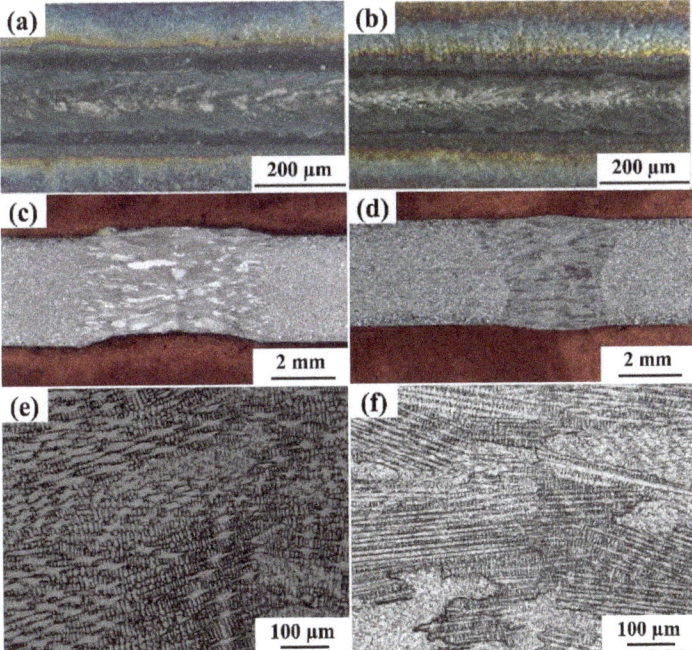

Figure 10. Optical images of surface and cross-section of welded joints (**a,c,e**) at 1800 W, 200 mm/min and (**b,d,f**) at 1400 W, 600 mm/min.

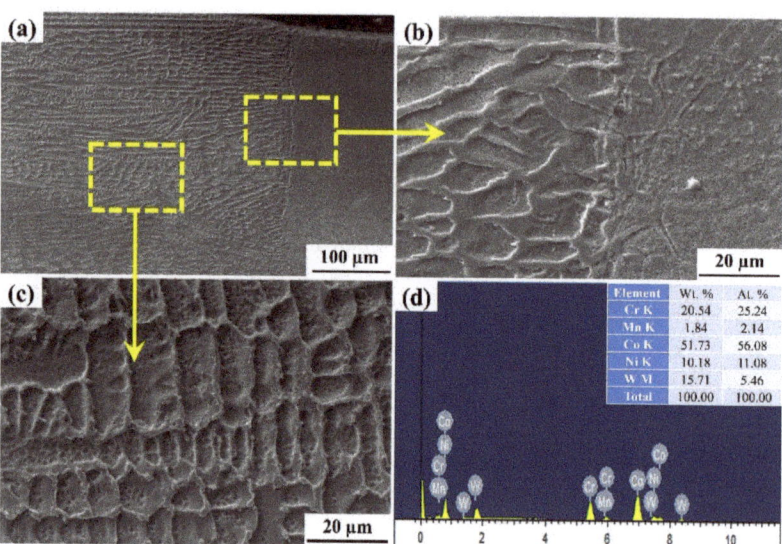

Figure 11. SEM images of the weld joint with high heat input (weld parameters: 1800 W, 200 mm/min). (**a**) Weld cross section (**b**) Interface zone (**c**) Microstructure (**d**) EDS report.

In the present work, the two typical samples fabricated at laser power of 1400 W and 200 mm/min scanning speed, and 1800 W and 200 mm/min were considered for the microstructural analysis through SEM. The detailed investigation of the weld section at the interface and weld zone has been selected. The SEM micrographs revel the long columnar dendritic structure, which is divided into cellular and columnar dendritic structures. Figures 11 and 12 represent microstructure and Energy Dispersive Spectroscopy (EDS) results of the weld bead. From Figure 11, it has been observed that the dendritic microstructures are formed having long columnar grains at the interface (Figure 11b) and cellular dendrites (Figure 11c). The fusion zone microstructure at this section revels formation of secondary dendritic arm, the magnifying image shows the formation of intermetallic phase whose presence is confirmed through the EDS report (Figure 11d), the grains are parallel and distributed uniformly with mix coarse and fine columnar dendritic (uniaxial) which might be due to the proper mixing of material at the fusion zone [38,39]. Figure 12 shows the SEM images of the weld cross section, i.e., at the interface and the fusion zone; this shows the microstructure with no directional solidification with columnar and cellular dendritic structure. The white granular structure confirms the present of Co, which is confirmed through EDS report, as shown in Figure 12d. The Variation in thermal properties like specific heat and thermal conductivity with changing temperature of the base alloys leads to the formation of microstructure obtained after welding. The presence of Co, Cr, W, Ni, and Mn was observed in the inter-dendritic structure at the interface (Figures 11b,d and 12b,d). On the other hand, the laser welding process associated with a rapid cooling rate without pores and cracks was observed at the interface zone in the prepared weld samples (Figures 11a and 12a) [40]. The columnar dendritic and cellular dendritic structures were formed in the weld zone, and it may happen due to the formation of CrFeNi, CoC, FeNi, and CFe phases in the inter-dendritic regions and this provides strength to the weld zone [41].

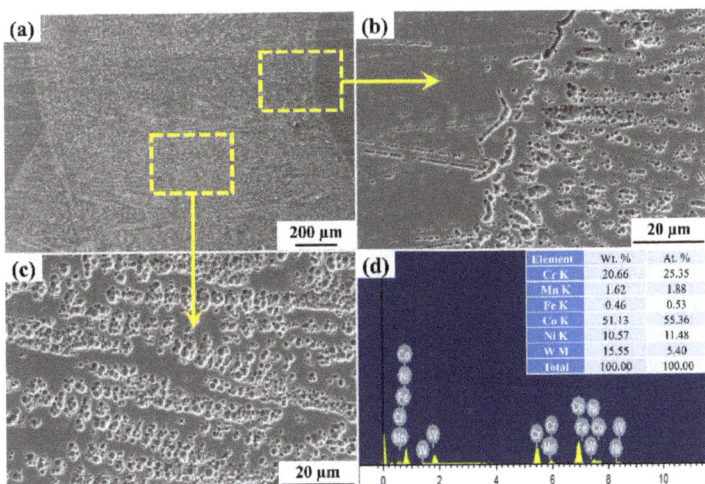

Figure 12. SEM images of welded joints with low heat input (weld parameters: 1400 W, 600 mm/min). (**a**) Weld cross section (**b**) Interface zone (**c**) Microstructure (**d**) EDS report.

3.3.2. Phase Analysis

X-ray diffraction (XRD) test was performed with laser parameter settings of 1800 W, 200 mm/min, and 1400 W and 600 mm/min on the weld bead cross-section to examine the presence of different phases (Figure 13a) along with XRD plot of parent material as shown in Figure 13b. The 2θ angle varied from 30 to 90° with a scanning speed of 4 deg/min. Figure 13 shows the XRD spectra of the weld beads and the analysis shows the presence of CrFeNi, CoC, FeNi, CFe, and Fe phases in the weld zone, unlike parent material.

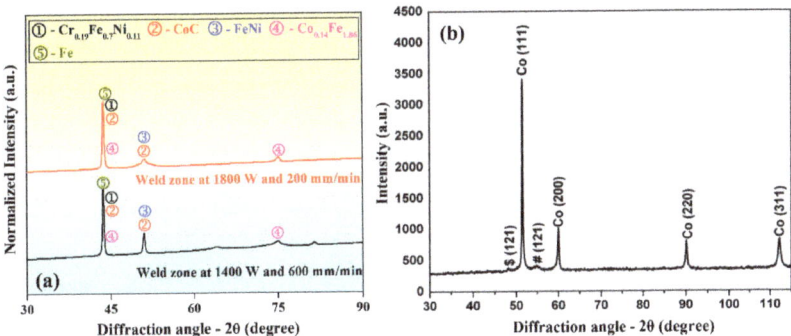

Figure 13. XRD spectra of the (**a**) weld zones processed at 1800 W, 200 mm/min, and 1400 W, 600 mm/min, respectively, and (**b**) parent material.

3.3.3. EBSD Analysis

The base metal EBSD micrograph is shown in Figure 14. The image quality map show in Figure 14 indicates a uniform fine grain size. The average grain size of the base metal is found to be ~17 μm. The fine grain size has uniformly distributed grains as shown in the Inverse pole figure map. The microstructure has around 4 % low angle grain boundaries with the remaining 96% high angle grain boundaries, as shown in Figure 14c. The pole figures and inverse pole figures show a random orientation of the grains in the base metal with no specific texture. The microstructure of the L605 weld pool region on both sides of the weld central line is shown in Figure 15. The average grain size of the HAZ has increased to 68 μm from the base metal grain size of 17 μm. This region is melted during the welding, and a dendritic/ columnar grain growth result in the formation of long elongated grains and the direction of the grains are elongated towards the base metal. The fraction of the low angle grain boundaries has increased from 4% to 15%. The resultant increase in LAB is due to the intersection for growing columnar grains. The pole figure and inverse pole figures indicate that there is a strong orientation of grains in the direction of (111) and (101), the resulting texture is like cold rolled texture.

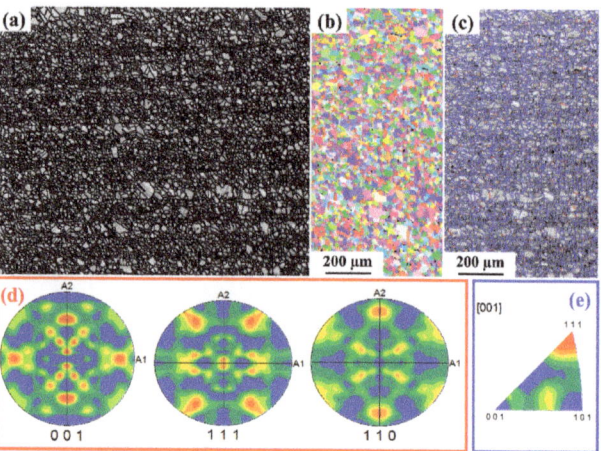

Figure 14. Base metal microstructure (a) Image quality map (b) Inverse pole figure color map (c) LAGB and HAGB distribution map (d) pole figures (e) Inverse pole figure.

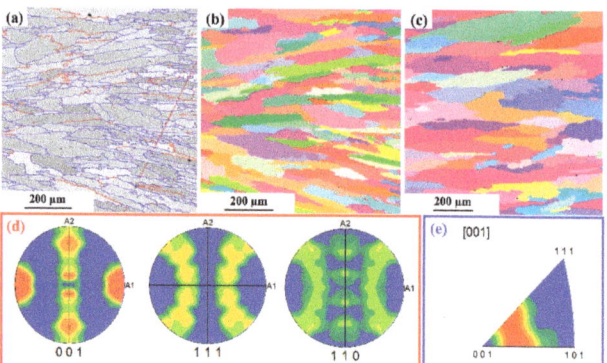

Figure 15. The Microstructure of the weld pool region (a) HAB and LAB distribution map (b) Inverse pole figure color map of weld pool region on the left side of the weld central line (c) Inverse pole figure color map of weld pool region on the right side of the weld central line. (d) Pole Figures (e) Inverse pole figures.

The EBSD maps of the complete weld bead region along with base metal are shown in Figure 16. The microstructure shows a very narrow HAZ, wherein a uniform grain size has been noticed. This region shows the grain growth compared to base metal, as this region is exposed to higher temperature below the melting point of the base material. The strain energy stored in the base material along with the temperature above recrystallization temperature has resulted in grain growth. The image quality map shows a completely dendritic and elongated grains. The pole figures indicate an overall texture orientation in <101> direction. The phase map presented indicates a fully austenitic structure with small amounts of metallic carbides. The carbide formation of $(W, Cr)_7 C_3$ was observed.

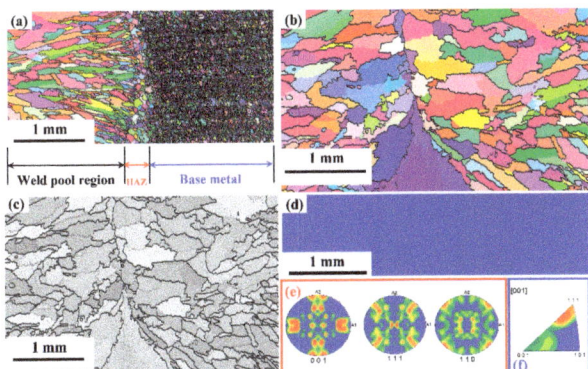

Figure 16. EBSD maps of complete L605 welds (**a**) Inverse pole figure map of the weld bead and base metal (**b**) Inver pole figure map of Weld pool region (**c**) Image quality map of weld pool region (**d**) Phase map (**e**) Pole figure of weld pool region (**f**) Inverse Pole figure of weld region.

3.3.4. Residual Stress Measurement

Residual stress was measured through the XRD technique using LXRD equipment on two prepared weld samples (input parameters 1800 W, 200 mm/min and 1400 W, 600 mm/min) at seven different points along mutually perpendicular directions (D90° and D0°) with respect to the weld line as shown in Figure 17a,b. In Figure 10c,d the grains are elongated across the weld line. It is observed the residual stress is compressive when measured along D90° (Figure 17c,d) for both parameter settings. Along D0° the compressive stress is reduced and sometimes it becomes tensile (Figure 17c) for parameters setting of 1800 W, 200 mm/min. However, at parameters setting of 1400 W, 200 mm/min, give weld bead with tensile residual stress along D0° (Figure 17d). Because of the above, the input parameters which produce the elongated grains may be preferred for the laser welding process.

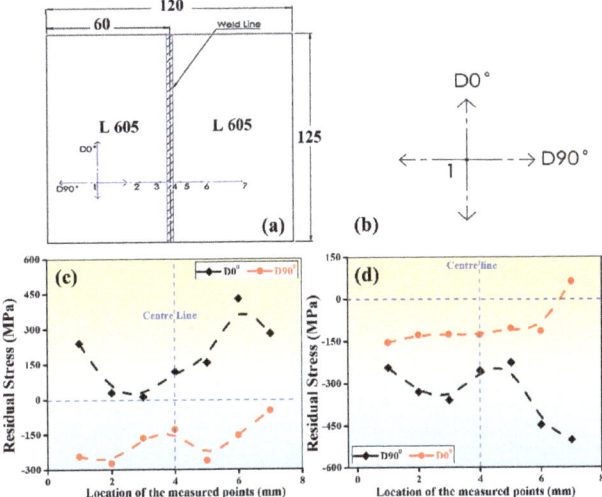

Figure 17. Residual stress measurement (**a**) Points on weld sample sketch, (point 4 represents the center of the weld bead) (**b**) Representation of D0° & D90° (**c**) Plot at D0° & D90° for weld sample with parameters 1800 W, 200 mm/min, (**d**) Plot at D0° & D90° for weld sample with parameters 1400 W, 600 mm/min.

3.3.5. Fractography Study

Figure 18 shows the SEM micrograph of the fractured surfaces of the tensile test specimens at high heat input parameters of 1800 W, 200 mm/min, and low heat input parameters of 1400 W, 600 mm/min. It is observed that tensile test specimens are subjected to ductile failure as well as cleavage failure, as illustrated in Figure 18. A significant number of dimples were observed in the fracture zone. In the inter-granular cleavage, micro-voids are seen at high and low energy input. The observed features show the ductile failure of the welded samples [42,43].

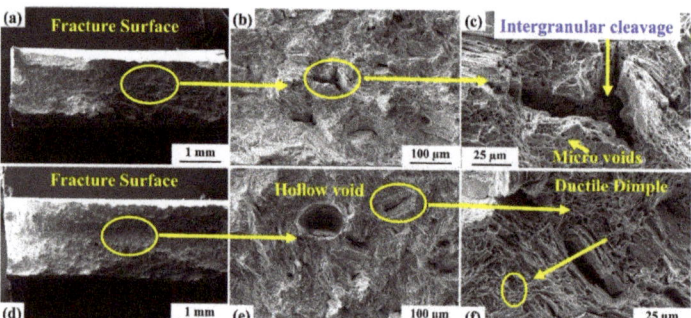

Figure 18. Fractography analysis of the specimens after conducting a tensile strength test with (**a**) high heat input parameters (exp.6), (**e**) low heat input parameters (exp. 9). (**b**) Lower magnifying image of fracture surface (exp 6) (**c**) Higher magnifying image of fracture surface (exp 6) (**d**) Overall Tensile fracture surface (exp 9) (**f**) Higher magnifying image of fracture surface (exp 9).

4. Conclusions

Fiber Laser welding of L605 cobalt superalloy sheets of a thickness of 2 mm was carried out successfully and different characterization techniques were followed for evaluation of the properties of the welded joints. From the study, the following conclusions are drawn:

- Response surface methodology was used for the development of regression equations to correlate the input and output process parameters. Within the present range of laser welding parameters, the work surface was kept at the laser beam focal plane, which gave different weld quality in the experimentation at different laser input parameters.
- Regression models were developed, and their predictability was checked through ANOVA. The best weld bead geometry of top width (W1): 2594.45 µm, bottom width (W2): 1848.62 µm, bead height (H1): 126.04µm, and undercut (H2): H2: 130.14 µm along with the highest tensile yield strength (yield stress) of 546.20 MPa were obtained at parameters setting of laser beam power: 1448.5W, scanning speed: 600 mm/min.
- The failure of tensile test samples occurred in the weld zone due to the formation of epitaxial grains, which was confirmed from the optical micrographs and to avoid this wobble mode of fiber laser welding may be adopted. The ductile failure occurred in the weld zones of the tensile test specimens, which are confirmed by the presence of dimples, intergranular cleavage, and micro voids in the fractured surfaces.
- The SEM analysis indicates the absence of micro-cracks in the weld zone. The microstructure (EBSD analysis shows a narrow HAZ, wherein a uniform grain size has been noticed. This region shows the grain growth compared to base metal, as this region is exposed to higher temperatures below the melting point of the base material. The strain energy stored in the base materials, along with the temperature above recrystallization temperature, has resulted in grain growth.
- All experiments in this work were conducted at room temperature and the material properties reported here also carried at room temperature. However, testing and characterization at high operating temperature will be considered as a part of future work.

Author Contributions: Conceptualization, A.K.D. and G.M.R.; methodology, B.H.P.; software, A.K.D. and K.G.P.; validation, B.H.P. and A.K.D.; formal analysis, B.H.P. and A.K.D.; resources, A.K.D., G.M.R. and K.G.P.; data curation, B.H.P. and K.G.P.; writing—original draft preparation, B.H.P.; writing—review and editing, B.H.P., A.K.D. and K.G.P.; visualization, A.K.D. and K.G.P.; supervision, A.K.D. and G.M.R.; project administration, K.G.P. and A.K.D.; funding acquisition, K.G.P., G.M.R. and A.K.D. All authors have read and agreed to the published version of the manuscript.

Funding: European Regional Development Grant funding ASTRA6-6 is acknowledged.

Institutional Review Board Statement: Not applicable.

Informed Consent Statement: Not applicable.

Data Availability Statement: The data may be made available on reasonable request to the authors.

Acknowledgments: Financial assistance from Defence Research Development Organisation (DRDO) is gratefully acknowledged. The authors would like to thank Shri G.A Srinivas Murthy, Director, Defence Research & Development Laboratory (DRDL) Hyderabad for his continuous encouragement and support in doing this work. The Authors also thank Shri G.V Siva Rao, P. Mastanaiah and their team, Shri D. Naveen Kumar of Defence Research & Development Laboratory, Hyderabad, P. Goshal, Samba Siva Rao, Phani Surya Kiran and their team of Defence Metallurgical Research Laboratory (DMRL), Hyderabad, Shri Satyanarayana Raju of Research Center Imarat (RCI), Shri Shakti Kumar of Indian Institute of Technology (ISM), Dhanbad, Jharkand, Shri Mukul Anand of Indian Institute of Technology (ISM), Dhanbad, Jharkand and Manowar Hussain, BIT, Sindri, for their help in conducting welding experiments, material characterization, and testing.

Conflicts of Interest: The authors declare no conflict of interest.

References

1. You, D.Y.; Gao, X.D.; Katayama, S. Review of Laser Welding Monitoring. *Sci. Technol. Weld. Join.* **2014**, *19*, 181–201. [CrossRef]
2. Sun, F.; Mantovani, D.; Prima, F. Carbides and Their Role in Advanced Mechanical Properties of L605 Alloy: Implications for Medical Devices. *Mater. Sci. Forum* **2014**, *783*, 1354–1359. [CrossRef]
3. Rathod, V.R.; Anand, R.S.; Ashok, A. Comparative Analysis of NDE Techniques with Image Processing. *Non-Destr. Test. Eval.* **2012**, *27*, 305–326. [CrossRef]
4. Wang, Z.; Tang, S.Y.; Scudino, S.; Ivanov, Y.P.; Qu, R.T.; Wang, D.; Yang, C.; Zhang, W.W.; Greer, A.L.; Eckert, J.; et al. Additive manufacturing of a martensitic Co-Cr-Mo alloy: Towards circumventing the strength-ductility trade-off. *Addit. Manuf.* **2021**, *37*, 101725. [CrossRef]
5. Shahriary, M.S.; Gorji, Y.M.; Kolagar, A.M. Gas metal arc welding in refurbishment of cobalt base superalloys. In *IOP Conference Series: Materials Science and Engineering, Proceedings of the 5th Global Conference on Materials Science and Engineering, Taichung City, Taiwan, 8–11 November 2016*; IOP Publishing: Bristol, UK, 2017; Volume 164, p. 012016. [CrossRef]
6. Osoba, L.O.; Ding, R.G.; Ojo, O.A. Microstructural analysis of laser weld fusion zone in Haynes 282 superalloy. *Mater. Charact.* **2012**, *65*, 93–99. [CrossRef]
7. Chen, Z.W.; Phan, M.A.L.; Darvish, K. Grain growth during selective laser melting of a Co–Cr–Mo alloy. *J. Mater. Sci.* **2017**, *52*, 7415–7427. [CrossRef]
8. Kurz, W.; Trivedi, R. Rapid solidification processing and microstructure formation. *Mater. Sci. Eng. A* **1994**, *179*, 46–51. [CrossRef]
9. Shamanian, M.; Valehi, M.; Kangazian, J.; Szpunar, J.A. EBSD characterization of the L-605 Co-based alloy welds processed by pulsed Nd: YAG laser welding. *Opt. Laser Technol.* **2020**, *128*, 106256. [CrossRef]
10. Nistazakis, H.E.; Stassinakis, A.N.; Muhammad, S.S.; Tombras, G.S. BER estimation for multi-hop RoFSO QAM or PSK OFDM communication systems over gamma gamma or exponentially modeled turbulence channels. *Opt. Laser Technol.* **2014**, *64*, 106–112. [CrossRef]
11. Anand, M.; Das, A.K. Issues in fabrication of 3D components through DMLS Technique: A review. *Opt. Laser Technol.* **2021**, *139*, 106914. [CrossRef]
12. Huang, Y.; Yuan, Y.; Feng, Y.; Liu, J.; Yang, L.; Cui, L. Effect of activating flux Cr_2O_3 on microstructure and properties of laser welded 5083 aluminum alloys. *Opt. Laser Technol.* **2022**, *150*, 107930. [CrossRef]
13. Wang, Z.; Ummethala, R.; Singh, N.; Tang, S.; Suryanarayana, C.; Eckert, J.; Prashanth, K.G. Selective Laser Melting of Aluminum and Its Alloys. *Materials* **2020**, *13*, 4564. [CrossRef] [PubMed]
14. Krajewski, A.; Kołodziejczak, P. Analysis of the Impact of Acoustic Vibrations on the Laser Beam Remelting Process. *Materials* **2022**, *15*, 6402.108143. [CrossRef] [PubMed]
15. Anand, M.; Das, A.K. Grain Refinement in Wire-arc Additive Manufactured Inconel 82 Alloy Through Controlled Heat Input. *J. Alloy. Compd.* **2022**, *929*, 166349. [CrossRef]
16. Selvam, J.; Raja, D.; Dinaharan, I.; Mashinini, P.M. Microstructure and mechanical characterization of Nd: YAG laser beam welded AA6061/10 wt% ZrB2 aluminum matrix composites. *Opt. Laser Technol.* **2021**, *140*, 107084. [CrossRef]

17. Zhou, S.; Wang, B.; Wu, D.; Ma, G.; Yang, G.; Wei, W. Effect of pulse energy on microstructure and properties of laser lap-welding Hastelloy C-276 and 304 stainless steel dissimilar metals. *Opt. Laser Technol.* **2021**, *142*, 107236. [CrossRef]
18. Kupiec, B.; Urbańczyk, M.; Radoń, M.; Mróz, M. Problems of HLAW Hybrid Welding of S1300QL Steel. *Materials* **2022**, *15*, 5756. [CrossRef]
19. Chatterjee, S.; Mahapatra, S.S.; Bharadwaj, V.; Upadhyay, B.N.; Bindra, K.S.; Thomas, J. Parametric appraisal of mechanical and metallurgical behavior of butt welded joints using pulsed Nd: YAG laser on thin sheets of AISI 316. *Opt. Laser Technol.* **2019**, *117*, 186–199. [CrossRef]
20. Li, G.; Lu, X.; Zhu, X.; Huang, J.; Liu, L.; Wu, Y. The interface microstructure, mechanical properties and corrosion resistance of dissimilar joints during multipass laser welding for nuclear power plants. *Opt. Laser Technol.* **2018**, *101*, 479–490. [CrossRef]
21. Fang, Z.C.; Wu, Z.L.; Huang, C.G.; Wu, C.W. Review on residual stress in selective laser melting additive manufacturing of alloy parts. *Opt. Laser Technol.* **2020**, *129*, 106283. [CrossRef]
22. Hagenlocher, C.; Weller, D.; Weber, R.; Graf, T. Reduction of the Hot Cracking Susceptibility of Laser Beam Welds in AlMgSi Alloys by Increasing the Number of Grain Boundaries. *Sci. Technol. Weld. Join.* **2019**, *24*, 313–319. [CrossRef]
23. Anand, M.; Bishwakarma, H.; Kumar, N.; Ujjwal, K.; Das, A.K. Fabrication of multilayer thin wall by WAAM technique and investigation of its microstructure and mechanical properties. *Mater. Today Proc.* **2022**, *56*, 927–930. [CrossRef]
24. Baba, N.; Watanabe, I.; Liu, J.; Atsuta, M. Mechanical Strength of Laser-Welded Cobalt-Chromium Alloy. *J. Biomed. Mater. Res. Part B Appl. Biomater.* **2004**, *69*, 121–124. [CrossRef] [PubMed]
25. Li, R.; Zhang, F.; Sun, T.; Liu, B.; Chen, S.; Tian, Y. Investigation of Strengthening Mechanism of Commercially Pure Titanium Joints Fabricated by Autogenously Laser Beam Welding and Laser-MIG Hybrid Welding Processes. *Int. J. Adv. Manuf. Technol.* **2019**, *101*, 377–389. [CrossRef]
26. Palma, D.P.D.S.; Nakazato, R.Z.; Codaro, E.N.; Acciari, H.A. Morphological and structural variations in anodic films grown on polished and electropolished titanium substrates. *Mater. Res.* **2019**, *22*. [CrossRef]
27. Shanmugarajan, B.; Padmanabham, G.; Kumar, H.; Albert, S.K.; Bhaduri, A.K. Autogenous Laser Welding Investigations on Modified 9Cr–1Mo (P91) Steel. *Sci. Technol. Weld. Joint.* **2011**, *16*, 528–534. [CrossRef]
28. Quazi, M.M.; Ishak, M.; Fazal, M.A.; Arslan, A.; Rubaiee, S.; Aiman, M.H.; Qaban, A.; Yusof, F.; Sultan, T.; Ali, M.M.; et al. A comprehensive assessment of laser welding of biomedical devices and implant materials: Recent research, development and applications. *Crit. Rev. Solid State Mater. Sci.* **2021**, *46*, 109–151. [CrossRef]
29. Dinesh Babu, P.; Buvanashekaran, G.; Balasubramanian, K.R. The Elevated Temperature Wear Analysis of Laser Surface–Hardened En25 Steel Using Response Surface Methodology. *Tribol. Trans.* **2015**, *58*, 602–615. [CrossRef]
30. Chandrasekhar, N.; Ragavendran, M.; Ravikumar, R.; Vasudevan, M.; Murugan, S. Optimization of Hybrid Laser–TIG Welding of 316LN Stainless Steel Using Genetic Algorithm. *Mater. Manuf. Process.* **2017**, *32*, 1094–1100. [CrossRef]
31. Bandhu, D.; Abhishek, K. Assessment of Weld Bead Geometry in Modified Shortcircuiting Gas Metal Arc Welding Process for Low Alloy Steel. *Mater. Manuf. Process.* **2021**, *36*, 1384–1402. [CrossRef]
32. Datta, S.; Raza, M.S.; Saha, P.; Pratihar, D.K. Effects of Process Parameters on the Quality Aspects of Weld-Bead in Laser Welding of NiTinol Sheets. *Mater. Manuf. Process.* **2019**, *34*, 648–659. [CrossRef]
33. Das Neves, M.D.M.; Lotto, A.; Berretta, J.R.; De Rossi, W.; Nilson Dias, V., Jr. Microstructure Development in Nd: YAG Laser Welding of AISI 304 and Inconel 600. *Weld. Int.* **2010**, *24*, 739–748. [CrossRef]
34. CJ, A.; Mohan, D.G. Predicting the ultimate tensile strength and wear rate of aluminium hybrid surface composites fabricated via friction stir processing using computational methods. *J. Adhes. Sci. Technol.* **2021**, *36*, 1707–1726. [CrossRef]
35. Prabakaran, M.P.; Kannan, G.R.; Pandiyarajan, R. Effects of welding speed on microstructure and mechanical properties of CO_2 laser welded dissimilar butt joints between low carbon steel and austenitic stainless steel. *Adv. Mater. Process. Technol.* **2020**, *6*, 1–12. [CrossRef]
36. Verma, A.; Natu, H.; Balasundar, I.; Chelvane, A.; Niranjani, V.L.; Mohape, M.; Mahanta, G.; Gowtam, S.; Shanmugasundaram, T. Effect of Copper on Microstructural Evolution and Mechanical Properties of Laser-Welded CoCrFeNi High Entropy Alloy. *Sci. Technol. Weld. Joint.* **2022**, *27*, 197–203. [CrossRef]
37. Khan, M.S.; Razmpoosh, M.H.; Biro, E.; Zhou, Y. A Review on the Laser Welding of Coated 22MnB5 Press-Hardened Steel and Its Impact on the Production of Tailor-Welded Blanks. *Sci. Technol. Weld. Joint.* **2020**, *25*, 447–467. [CrossRef]
38. Wu, S.C.; Hu, Y.N.; Duan, H.; Yu, C.; Jiao, H.S. On the fatigue performance of laser hybrid welded high Zn 7000 alloys for next generation railway components. *Int. J. Fatigue* **2016**, *91*, 1–10. [CrossRef]
39. Wu, S.C.; Qin, Q.B.; Hu, Y.N.; Branco, R.; Li, C.H.; Williams, C.J.; Zhang, W.H. The microstructure, mechanical, and fatigue behaviours of MAG welded G20Mn5 cast steel. *Fatigue Fract. Eng. Mater. Struct.* **2020**, *43*, 1051–1063. [CrossRef]
40. Cheepu, M.; Kumar Reddy, Y.A.; Indumathi, S.; Venkateswarlu, D. Laser Welding of Dissimilar Alloys between High Tensile Steel and Inconel Alloy for High Temperature Applications. *Adv. Mater. Process. Technol.* **2020**, *2020*, 1–12. [CrossRef]
41. Ramkumar, K.D.; Sridhar, R.; Periwal, S.; Oza, S.; Saxena, V.; Hidad, P.; Arivazhagan, N. Investigations on the structure–Property relationships of electron beam welded Inconel 625 and UNS 32205. *Mater. Des.* **2015**, *68*, 158–166. [CrossRef]

42. Deng, J.; Zhao, G.; Lei, J.; Zhong, L.; Lei, Z. Research Progress and Challenges in Laser-Controlled Cleaning of Aluminum Alloy Surfaces. *Materials* **2022**, *15*, 5469. [CrossRef] [PubMed]
43. Fernandes, F.A.; Pinto, J.P.; Vilarinho, B.; Pereira, A.B. Laser Direct Joining of Steel to Polymethylmethacrylate: The Influence of Process Parameters and Surface Mechanical Pre-Treatment on the Joint Strength and Quality. *Materials* **2022**, *15*, 5081. [CrossRef] [PubMed]

Article

Characterization of Inclusion Size Distributions in Steel Wire Rods

Pablo Huazano-Estrada [1], Martín Herrera-Trejo [1,*], Manuel de J. Castro-Román [1] and Jorge Ruiz-Mondragón [2]

1. Centro de Investigación y de Estudios Avanzados, CINVESTAV Saltillo, Av. Industria Metalúrgica No. 1062, Parque Industrial Saltillo-Ramos Arizpe, Ramos Arizpe 25900, Mexico
2. COMIMSA, Saltillo 25290, Mexico
* Correspondence: martin.herrera@cinvestav.edu.mx; Tel.: +52-844-438-9643

Citation: Huazano-Estrada, P.; Herrera-Trejo, M.; Castro-Román, M.d.J.; Ruiz-Mondragón, J. Characterization of Inclusion Size Distributions in Steel Wire Rods. *Materials* **2022**, *15*, 7681. https://doi.org/10.3390/ma15217681

Academic Editors: Konstantin Borodianskiy and Andrea Di Schino

Received: 16 September 2022
Accepted: 18 October 2022
Published: 1 November 2022

Publisher's Note: MDPI stays neutral with regard to jurisdictional claims in published maps and institutional affiliations.

Copyright: © 2022 by the authors. Licensee MDPI, Basel, Switzerland. This article is an open access article distributed under the terms and conditions of the Creative Commons Attribution (CC BY) license (https://creativecommons.org/licenses/by/4.0/).

Abstract: The control of inclusions in steel components is essential to guarantee strong performance. The reliable characterization of inclusion populations is essential not only to evaluate the quality of the components but also to allow the use of analytical procedures for the comparison and discrimination of inclusion populations. In this work, inclusion size distributions in wire rod specimens from six plant-scale heats were measured and analyzed. For the measurements, the metallographic procedure specified in the ASTM E2283 standard was used. The population density function (PDF) approach and the extreme value statistical procedure specified in the ASTM E2283 standard were used to analyze the whole size distribution and the upper tail of the size distribution, respectively. The PDF approach allowed us to identify differences among inclusion size distributions and showed that new inclusions were not formed after the liquid steel treatment process. The extreme value statistical procedure led to the prediction of the maximum inclusion length for each heat, which was used for the statistical discrimination of heats. Furthermore, the estimation of the probability of finding an inclusion larger than a given inclusion size using the extreme value theory allowed us to order the heats for different critical inclusion sizes.

Keywords: inclusion; size distribution; population density function; extreme value theory

1. Introduction

Control over inclusion cleanliness in steel products is necessary to ensure their performance under specific conditions. Inclusion size distribution is one of the parameters used to evaluate steel quality, and therefore its control during the steelmaking process is required [1,2]. Hence, the successful control of the inclusion size distribution must be based on analysis procedures that provide information on the formation and evolution of inclusions. Furthermore, predictive and discriminative analysis procedures are desirable.

The size distribution of an inclusion population is frequently estimated from metallography measurements and image analysis and represents an inclusion cleanliness parameter. Higgings [3] represented the particle size distribution by the population density function (PDF) as an alternative to the classic histogram format. The PDF approach presents the advantage of being user-independent compared to the histogram and the corresponding density function, which assumes a normal distribution [4]. Furthermore, the PDF is unique for a given inclusion population, provides information on the formation and evolution of inclusions [5] and enables the comparison of size distributions from different specimens. The logarithmic representation of PDFs is linear or quadratic and can be described by fractal (power law) and lognormal distributions, respectively [6]. Zinngrebe et al. [7] introduced the PDF approach in the analysis of inclusion size distribution and used it to study the formation of inclusions during the secondary steelmaking and casting process of Ti-alloyed Al-killed steel. The logarithmic representation of PDF showed a quadratic shape just after Al addition and became linear with time. A quadratic shape characterizes the transient deoxidation process where the transfer of the matter occurs between steel and inclusions, i.e., the formation and growth of inclusion process, whereas a linear behavior indicates that the equilibrium inclusion-steel was reached, and consequently the inclusion size distribution is the result of effective collisions and breakage phenomena and the removal of inclusions. Hence, it is expected that estimated PDFs in specimens obtained in later stages of processing can provide information on the size distribution at the end of the liquid steel treatment and its subsequent evolution. This provides evidence of the occurrence of phenomena, such as the reoxidation process, that form new inclusions after liquid treatment. It is worth mentioning recent works that use the PDF approach. Piva and Pistorius [8] used the PDF to show the evolution of the size distribution for different processing routes for steel Ca treatment, while Qifeng et al. [9] used it to analyze results obtained from the modeling of the nucleation, growth, and agglomeration of Al_2O_3 inclusions.

On the other hand, the generalized extreme value (GEV) theory based on Murakami et al.'s pioneering work [10] has been used to describe the large inclusion size tail of the size distribution, in which standard probability distributions, such as the log-normal distribution, are insufficient [4]. The method is based on measuring the maximum inclusion size in random inspection areas and fitting the Gumbel distribution to these measurements. This allows to predict the maximum inclusion length (L_{max}), i.e., the longest inclusion expected to be found in a predetermined area. The fitting of the Gumbel distribution to the size distributions of different specimens enables the statistical comparison and discrimination of L_{max}. Variations in the procedures used for the application of GEV led to the development of the ASTM E2283 standard "Standard Practice for Extreme Value Analysis of Nonmetallic Inclusions in Steel and other Microstructural Features" [11] as a guide to evaluate inclusion cleanliness. The standard was shown to be a reliable tool to assess inclusion cleanliness. Recently, Kumar and Balachandran [12] showed the standard as an effective tool for estimating the largest inclusion, which has the potential for crack nucleation that leads to fatigue failure in steel specimens. Fuchs et al. [13] confirmed the effectiveness of the standard in the evaluation of inclusion cleanliness in ultraclean gear steels. Furthermore, the standard has been used in the evaluation of the maximum inclusion size of heat-treated wire rod specimens [14]. More recently, the results using the standard showed relatively good agreement with measurements of large inclusions based on the X-ray microcomputed tomography method [15]

In this work, size distributions were measured in steel wire rod specimens from six different plant-scale heats following the metallographic procedure indicated in the ASTM E2283 standard. The PDF approach was used to analyze the whole size distribution, whereas the GEV theory was employed to describe the upper tail of the size distributions following the procedure specified in the ASTM E2283 standard and to estimate the survival probability.

2. Background
2.1. PDF Approach

The PDF concept, introduced by Higgins [3], is expressed by the following equation:

$$PDF = \frac{n_v\,(L_{XY})}{(L_Y - L_X)} \quad (1)$$

where $n_v\,(L_{XY})$ is the frequency of inclusions in a given size bin (particle number per volume), and $(L_Y - L_X)$ is the bin width with units of length. The logarithmic representation of PDFs can be linear or quadratic and can be described by fractal (power law) and lognormal distributions, respectively [6]. The probability density function of the fractal distribution is given by

$$f(x) = \frac{C}{x^D} \quad (2)$$

where C is the constant of proportionality, and D is the fractal dimension. The probability density function of the lognormal distribution is

$$f(x) = \frac{1}{x\,\sigma(2\pi)^{1/2}} \exp\left[\frac{[\ln(x) - \ln(\mu)]^2}{2\sigma^2}\right] \quad (3)$$

where m and s are the mean and standard deviation, respectively.

2.2. ASTM E2283 Standard

The ASTM E2283 standard [11] describes a procedure to statistically characterize the distribution of the largest particles in a solid matrix. It can be used in the case of inclusions in a steel matrix. Herein, essential aspects allowing us to understand the use of the standard in this work are introduced.

The ASTM E2283 procedure is based upon quantitative optical metallographic measurements and their analysis via statistical GEV theory. Six specimens per analysis are required for metallographic measurements. An area of 150 mm^2 (control area A_o) must be evaluated in four different metallographically prepared planes in each specimen, and the largest measured inclusion in each A_o is recorded. The measurements must be made using the correct magnification to ensure that the detected largest inclusion is a minimum of 20 pixels in length. The procedure provides a dataset of the 24 largest inclusions, which are listed in ascending order and can be represented as x_i. The cumulative probability P_i is calculated by the following equation:

$$P_i = \frac{x_i}{(N+1)} \quad (4)$$

where $N = 24$.

The prediction of L_{max} is based on the fitting of the Gumbel extreme value distribution to the 24 recorded inclusion lengths. The probability density function of the Gumbel distribution is expressed by

$$f_{(x)} = \frac{1}{\delta}\left[exp\left(\frac{x-\lambda}{\delta}\right)\right] \times exp\left[-exp\left(-\frac{x-\lambda}{\delta}\right)\right] \quad (5)$$

where x is the random variable (maximum inclusion length), and l and d are the location and scale parameters, respectively. The corresponding cumulative distribution is

$$F_{(x)} = exp\left(-exp\left(-\frac{x-\lambda}{\delta}\right)\right) \quad (6)$$

which can be rewritten by introducing the reduced variate y:

$$F_{(y)} = exp(-exp(-y)) \quad (7)$$

where

$$y = \frac{x-\lambda}{\delta} \quad (8)$$

Solving Equation (8) for x,

$$x = \delta\,y + \lambda \quad (9)$$

From Equation (7), y can be expressed in terms of the cumulative function as

$$y = -ln(-ln(F(y))) \quad (10)$$

Equation (10) can be related to Equation (4), and the expression for y is rewritten as

$$y = -ln(-ln(F(y))) = -\ln(-\ln(P)) \quad (11)$$

On the other hand, the return period T is used to predict how large an inclusion could be expected to be found if a reference area (A_{ref}) larger than A_0 were to be evaluated. That is,

$$T = \frac{A_{ref}}{A_0} \tag{12}$$

In the ASTM E2283 standard, A_{ref} is chosen to be 1000 times larger than A_0. Furthermore, T is statistically defined as

$$T = \frac{1}{1-P} \tag{13}$$

For a T value of 1000, the corresponding P value is 0.999 (99.9%).

Thus, for $P = 0.999$, the $y_{(P=0.99)}$ value is calculated using Equation (11), and the corresponding $x_{(P=0.99)}$ value is calculated by Equation (9). The calculated $x_{(P=0.99)}$ value corresponds to L_{max}, and Equation (9) is rewritten as

$$x_{(P=0.99)} = L_{max} = \delta\, y_{(0.99)} + \lambda \tag{14}$$

The δ and λ parameters are calculated using the maximum likelihood method ML from the log of the distribution function:

$$LL = \sum_{i=1}^{n} \ln\left(\frac{1}{\delta}\right) - \left(\frac{x_i - \lambda}{\delta}\right) - exp\left(-\frac{x_i - \lambda}{\delta}\right) \tag{15}$$

The estimated parameters are referenced as λ_{ML} and δ_{ML} and are used to construct the best-fit line through the data points using Equation (9).

The standard error SE for any inclusion of length x is

$$SE_x = \delta_{ML}\sqrt{(1.109 + 0.514y + 0.608y^2)/n} \tag{16}$$

The approximate 95% confidence interval is given by

$$95\%\ CI = \pm 2\, SE(x) \tag{17}$$

The comparison of differences in sizes of large inclusions in two steels, denoted A and B, is calculated by the approximate 95% confidence interval for $L_{max}(A) - L_{max}(B)$ by the following equation:

$$C.I = L_{max}(A) - L_{max}(B) \pm 2\sqrt{S.E_{ref}(A)^2 + S.E_{ref}(B)^2} \tag{18}$$

If the lower to upper bounds of the 95% CI include 0, then it is concluded that there is no difference in the characteristic sizes of the largest inclusions in heats A and B.

3. Materials and Methods

Measurements and analyses of inclusion size distributions in wire rod specimens from six plant-scale heats were conducted. The PDF approach was used to analyze the whole size distribution, and GEV theory was employed to analyze the upper tail of the size distributions.

Specimens from six different heats of high carbon rod wire produced at the plant scale were studied. The heats were produced by an electric arc furnace-ladle treatment-degassing-continuous casting route, and the liquid steel was deoxidized with Si. The studied steel corresponded to SAE 9254, and Table 1 shows the chemical composition expressed in weight percent (wt. %) for two specimens of each heat. The chemical analysis of C and S was performed using the infrared combustion technique, while the content of other elements was determined using the spark emission spectroscopy technique.

Table 1. Chemical composition (wt. %) of the studied heats.

Heat	Specimen	Content (wt. %)					
		C	Mn	Si	Cr	P	S
1	1	0.57	0.70	1.45	0.66	0.008	0.014
	2	0.58	0.70	1.44	0.67	0.007	0.012
2	1	0.54	0.68	1.43	0.65	0.008	0.014
	2	0.57	0.70	1.45	0.66	0.007	0.014
3	1	0.64	0.71	1.51	0.71	0.009	0.019
	2	0.65	0.72	1.52	0.71	0.009	0.019
4	1	0.59	0.70	1.5	0.70	0.008	0.019
	2	0.60	0.71	1.48	0.69	0.008	0.019
5	1	0.60	0.70	1.5	0.66	0.007	0.17
	2	0.58	0.71	1.52	0.70	0.008	0.17
6	1	0.54	0.71	1.52	0.71	0.009	0.15
	2	0.61	0.70	1.49	0.69	0.008	0.15

Inclusion size distributions were estimated using the metallographic procedure detailed in the ASTM E2283 standard, which was rigorously followed. For each heat, six specimens of wire rod in as-rolled conditions of 25 mm diameter and 200 mm length were available. Each specimen was cut into a probe of 25 mm length and sectioned longitudinally as shown in Figure 1. The plane denoted as A was progressively dry ground using 80-, 220-, 300-, 500-, 800-, 1000-, and 1200-grit SiC paper and polished with 3 and 1 μm diamond paste. The measurements were conducted under a Nikon Eclipse MA200 light optical microscope (Minato ku, Japan) at 100X magnification on 150 fields by image analysis, and the total analyzed surface was 150 mm^2. Three additional planes were analyzed, ensuring that the space between neighboring planes was at least 0.3 mm, thus avoiding inclusions being counted more than once. The total number of analyzed fields per heat was 600.

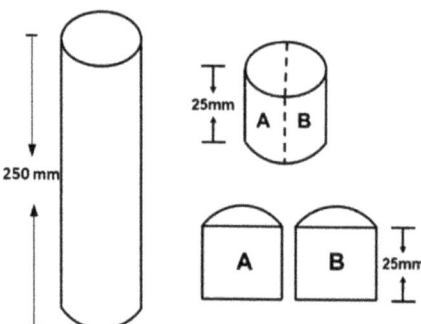

Figure 1. Obtention of probes for metallographic measurements.

The totality of the analyzed inclusions in each heat was considered to estimate the corresponding PDF using Equation (1). The frequency of inclusions per volume unit specified in that equation was estimated from the two-dimensional inclusion measurements obtained by image analysis, which were transformed into three-dimensional data using a procedure based on the Saltikov method [16,17].

By applying the ASTM E2283 standard, the measured maximum inclusion size in each of the inspected planes (24) was used for the statistical analyses described in the standard and synthetized in the previous section of this paper.

4. Results and Discussion

Figure 2 shows the measured inclusion frequency (inclusions/mm^2) and area fraction for the six specimens of each heat. In general, Heats 5 and 6 presented the highest values for both parameters, whereas Heats 1 and 2 presented the lowest values. Furthermore, Heats 5 and 6 showed more variability among specimens of the values of both parameters.

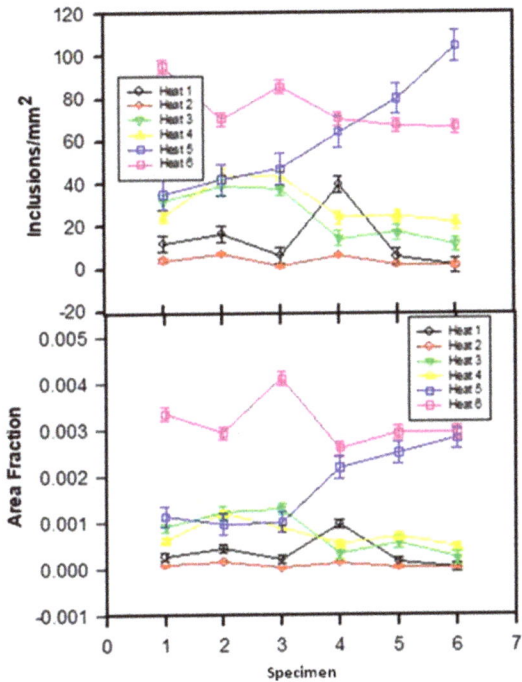

Figure 2. Evolution of inclusions/mm^2 and area fraction of inclusions.

4.1. Population Density Function PDF

Figure 3 shows the calculated PDF for each heat. The inclusion size is referred to as the inclusion equivalent diameter, which is defined as the diameter of a circle with the same area as the recorded particle [5]. Great differences were observed at small inclusion sizes, which decreased as the inclusion size increased. Below 10 µm, Heat 2 presented the lowest PDF values and was followed by Heats 1 and 3, while the rest of the heats presented similar values. In the range of 10–20 µm, Heat 2 continued to present the lowest values, followed by Heat 1, whereas Heat 6 presented the highest values; the rest of the heats presented similar values. Furthermore, the magnification of the PDF scale (inner figure) shows that at larger inclusion sizes, differences are also observed in PDF values and that Heats 2 and 6 showed the lowest and highest values of PDF, respectively. Thus, it can be stated for the studied heats that as the frequency of small inclusions increased, the population of large inclusions also increased, i.e., that the presence of large inclusions was due to the processing of liquid steel rather than to an eventual phenomenon.

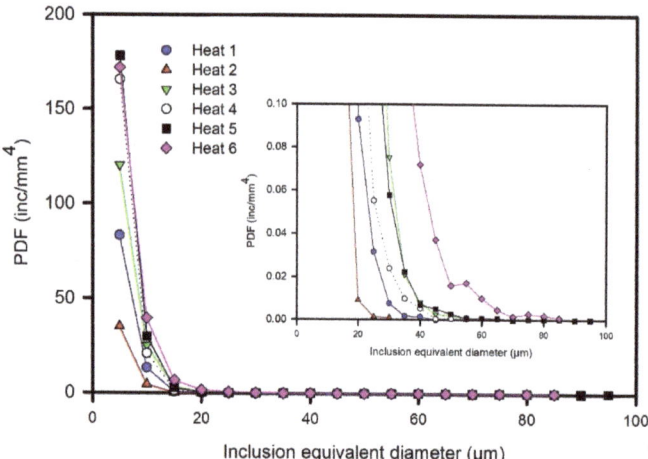

Figure 3. Evolution of the population density function PDF.

The logarithmic representation of the calculated PDFs is shown in Figure 4. A general linear power law behavior, represented by a reference straight line plotted in the figure, was observed. The reference straight line was estimated from the totality of the PDF data. The most notable deviations from the reference line corresponded to Heats 2 and 6, respectively, whereas the other heats fit better to the reference line. To clarify the observed behavior, the logarithmic representations of the PDFs of Heats 2 and 6 are shown in Figure 5, in which the data of Heat 4 were included because of its better fit to the reference line. Heat 6 presented both a higher frequency of inclusions and a larger inclusion size; in contrast, Heat 2 presented a lower frequency of inclusions and a smaller inclusion size, and Heat 4 exhibited intermediate values of both parameters. To individually analyze the heats of Figure 5, the data corresponding to the associated six samples of each heat are shown in Figure 6. As expected from a statistical point of view, more dispersion between the data were observed as the frequency decreased (Heat 2), although the linear trend continued to be observed. Thus, the heats can be ordered in decreasing order of cleanliness as follows: Heat 2, 4 and 6. Furthermore, Heats 3 and 5 had a similar behavior to Heat 4.

Then, the differences between the slopes of the straight lines of each heat can be deduced. The slope of the straight line is associated with the refining process and is specific to each process. Van Ende et al. [5] showed results for Ti-alloyed Al-killed steel produced in different plants. At the end of the refining process, a linear behavior of log PDF with a good fit was shown, and the slope values of the straight line were similar regardless of the plant of production. The slope value (−3.5) was associated with the Al-deoxidation process. Thus, the average slope value of the reference line (−0.089) in Figure 4 can be associated with the Mn–Si deoxidation process. In previous work [18], SiO_2-Al_2O_3-MnO inclusions were observed at the beginning of the ladle treatment, which evolved to SiO_2-Al_2O_3-CaO-MgO inclusions during treatment. Of note, SiO_2-Al_2O_3-CaO-MgO inclusions were observed in this work under as-rolling conditions, as shown in Figure 7 for inclusions of different sizes. This result suggests that the estimated slope of the reference line is associated with the formation of SiO_2-Al_2O_3-MnO inclusions during the Mn-Si deoxidation process and their evolution to SiO_2-Al_2O_3-CaO-MgO inclusions during ladle treatment. Furthermore, the linear behavior indicates that the new inclusions were not formed after liquid steel treatment and that inclusion populations evolved by the growth of inclusions, breakage, and removal of inclusions. In addition, the inclusions found in all samples were embedded in a matrix similar to that presented in Figure 8. The microstructure of the matrix did not vary due to the similarity of the chemical composition (Table 1) and because the inclusions were subjected to the same thermomechanical treatment.

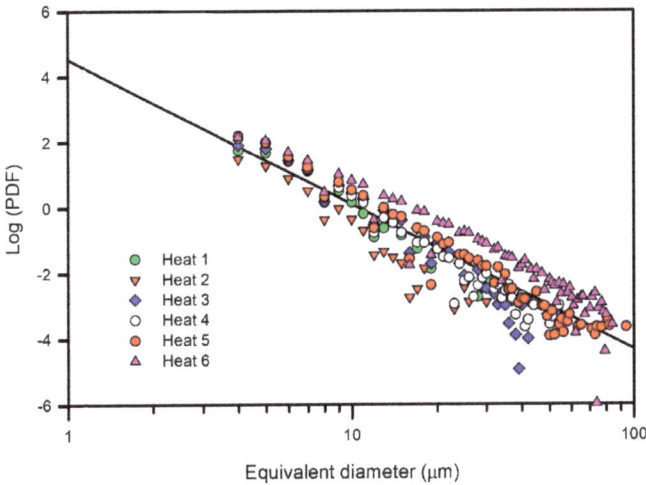

Figure 4. Log-log plots of PDFs versus inclusion equivalent diameter for all heats.

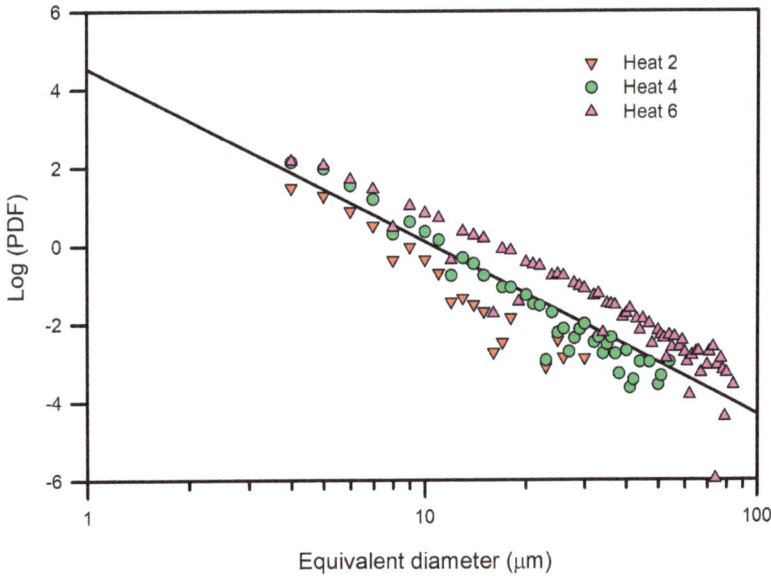

Figure 5. Log-log plot of PDFs versus inclusion equivalent diameter for Heats 2, 4 and 6.

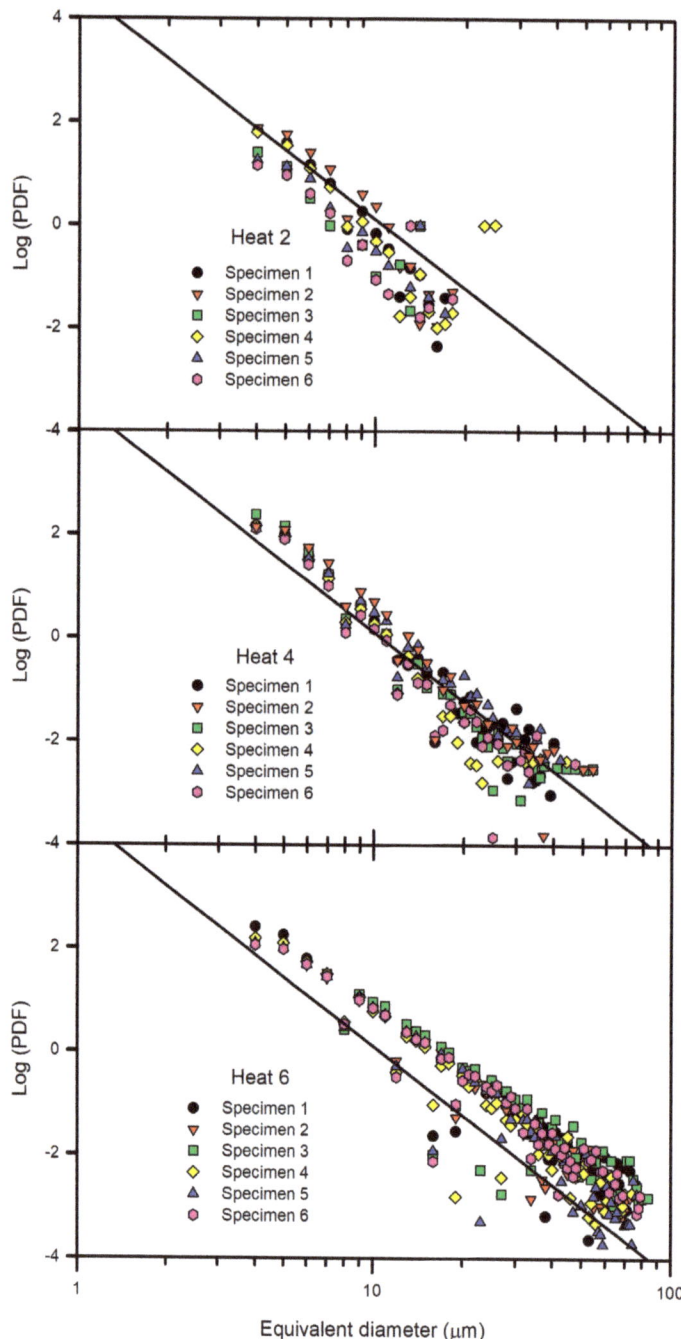

Figure 6. Log-log plot of PDFs versus inclusion equivalent diameter for specimens of Heats 2, 4 and 6.

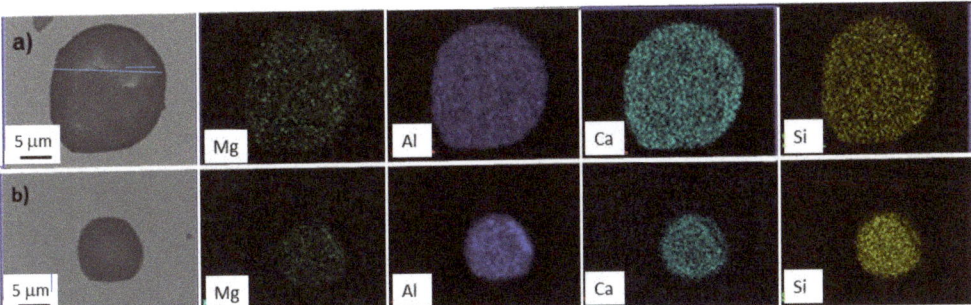

Figure 7. Elemental mapping for SiO_2-Al_2O_3-CaO-MgO inclusions of different sizes: (**a**) 20 μm and (**b**) 12 μm.

Figure 8. Typical microstructure of the steel matrix for the studied heats.

4.2. Extreme Value Distribution Analysis

Figure 9 shows a representation of the inclusion size distribution in histogram format. In this section, "length inclusion" is used to denote the inclusion size, just as the ASTM E2283 standard does. Of note, the inclusion length corresponds to the "equivalent diameter" used in the previous section. Two main observations can be highlighted. Below 20 μm, the highest frequency was shown by Heat 2; above 70 μm, however, Heat 6 had the highest frequency. For intermediate inclusion sizes, it is difficult to describe a trend.

Table 2 shows the 24 measured values of maximum inclusion length obtained for each heat from four analyzed planes in each of the six available specimens. The maximum inclusion length values are listed in increasing order as specified by the ASTM E2283 standard. The reduced variate was calculated using Equation (11) and is shown in Figure 10 for each heat as a function of the inclusion length. In general, a linear behavior for individual heats was observed. The fit of the straight lines was better at small inclusion lengths, and scattering was observed at longer inclusion lengths. It is also observed that steeper slopes correspond to lines located at shorter inclusion lengths, and consequently, the interception with the horizontal axis was different for each line.

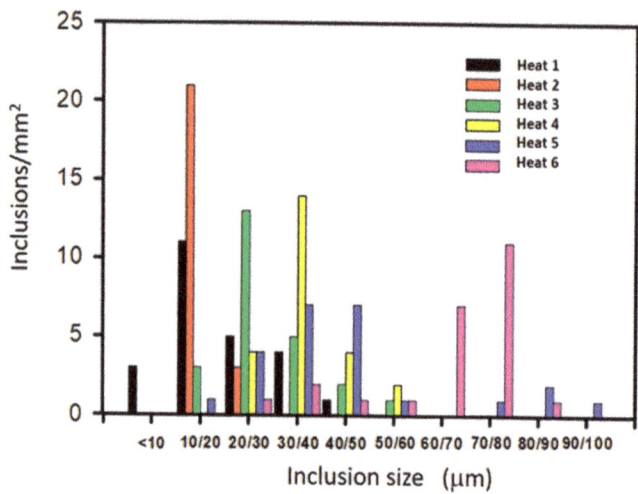

Figure 9. Inclusion size distribution.

Table 2. Inclusion maximum length L_{max} (μm) for analyzed plans in each heat.

N° Inclusion	Heat					
	1	2	3	4	5	6
1	7.5	10.4	16.7	20.2	16.7	24.8
2	8.1	11.0	19.0	20.8	21.4	36.9
3	8.7	11.5	19.6	24.8	25.4	39.2
4	10.4	11.5	20.2	26.0	26.0	43.9
5	11.6	12.1	20.2	31.2	26.5	57.1
6	12.1	12.7	20.8	31.7	31.2	60.6
7	12.1	12.7	20.8	31.7	31.7	64.0
8	12.1	13.3	21.4	32.9	34.0	64.0
9	12.7	13.3	22.5	32.9	34.6	64.6
10	14.4	13.3	23.1	32.9	35.2	65.8
11	15.0	14.4	23.7	33.45	35.2	69.8
12	15.0	14.4	24.2	34.0	39.2	69.8
13	17.3	14.4	24.8	34.6	40.4	71.0
14	19.0	15.0	25.4	35.2	45.0	72.1
15	21.9	16.2	30.0	35.8	46.2	73.3
16	22.5	16.2	30.0	35.8	46.7	75.0
17	23.7	16.2	31.7	39.8	47.9	75.0
18	26.5	17.9	32.3	39.8	48.5	76.2
19	30.0	17.9	34.0	41.5	49.0	77.9
20	31.7	17.9	35.2	43.3	50.8	77.9
21	32.9	17.9	38.1	46.2	73.0	78.5
22	38.1	22.6	42.7	49.0	80.2	79.0
23	38.7	24.2	48.5	53.1	80.8	79.6
24	40.4	29.4	52.5	53.7	93.5	83.1

Table 3 lists the values of the estimated statistical parameters according to the procedure of the ASTM E2283 standard. The values of λ, δ, and L_{max} are plotted in Figure 11, in which it is observed that the lower the values of λ and δ are, the lower are the values of L_{max}. Thus, the heats can be listed in increasing order of L_{max} values as follows: Heat 2, Heat 1, Heat 3, Heat 4, Heat 5, and Heat 6. This behavior is illustrated in Figure 12, where the probability distribution functions calculated via Equation (5) and using the values of λ and δ included in Table 3 are plotted.

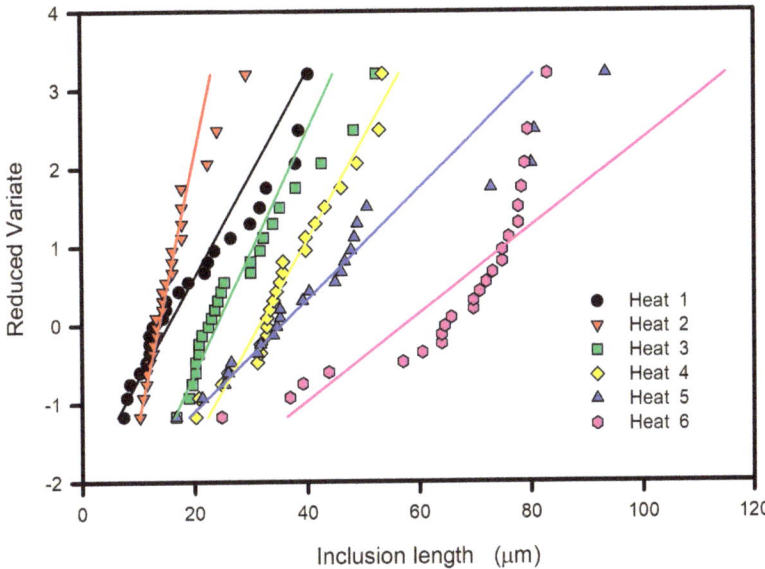

Figure 10. Reduced variate versus inclusion length.

Table 3. Values of statistical parameters (μm).

Parameters	Heat					
	1	2	3	4	5	6
λ	15.4	13.8	24.1	32.2	35.6	57.5
δ	7.6	2.9	6.5	8.8	14.1	18.0
L average	20.0	15.7	28.2	35.8	44.1	65.8
Standard deviation (Sdev)	10.4	4.6	9.6	11.8	19.8	15.3
Maximum length (L_{max})	68.2	34.2	69.0	92.7	132.9	181.9
Minimum length Y (L_{min})	6.9	6.9	6.9	6.9	6.9	6.9
Standard error (S.E.)	8.9	3.5	7.7	10.4	16.7	21.3

To determine whether the observed differences in L_{max} values were statistically significant, L_{max} values were compared according to the procedure described in the ASTM E2283 standard. The criterion of a 95% C.I. was calculated from the predicted value of L_{max} for each heat as well as the corresponding standard error. Table 4 lists the results of a round-robin comparison in matrix format, where the column and arrow titles correspond to the identification of heats. The interception of a column with an arrow denotes the range of bounds of the 95% C.I. of the corresponding heats. A C.I. range that does not include 0 is denoted in red and indicates a difference between the compared L_{max}; otherwise, it is denoted in green and indicates that there was no difference. The L_{max} of Heats 5 and 6 did not differ from each other; in contrast, they differed from the L_{max} of the rest of the heats. Furthermore, L_{max} of Heat 2 was different from other L_{max}, and L_{max} of Heats 1, 3 and 4 were not different from each other. Thus, using the L_{max} parameter as the cleanliness index, it can be stated that Heat 2 is the cleanest and significantly different from the rest of the heats; Heats 5 and 6 are the worst and significantly different from the rest of the heats; Heats 1, 3 and 4 are not different from each other, and their inclusion cleanliness is intermediate between Heat 2 and Heats 5 and 6.

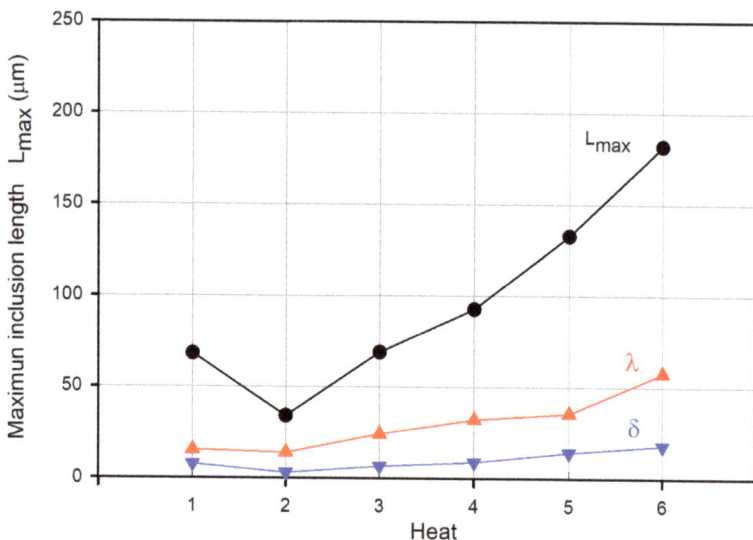

Figure 11. Maximum inclusion size L_{max}, λ, and δ for all heats.

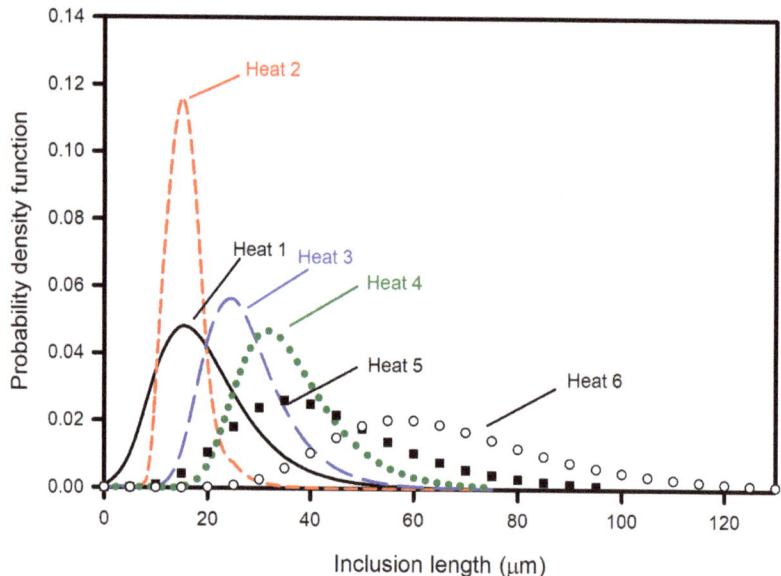

Figure 12. Probability density function for heats.

Although L_{max} represents a useful indicator that enables the comparison and discrimination of inclusion populations, there is often interest in knowing the probability of finding inclusions greater than a certain "critical" size depending on the specific application of the steel component. For example, in components subjected to fatigue, it is accepted that the size of the inclusion is important because cracks can form at the inclusion–steel interface. In this context, the survival function $S_{(x)}$ (probability of finding an inclusion larger than a given length, equal to the complement of the cumulative density function) rather than L_{max} is more convenient. Figure 13 presents $S_{(x)}$ as a function of the inclusion length for all heats. The order of heats previously stated as a function of L_{max} is confirmed for inclusion lengths greater than 35 µm, where no overlapping of curves is observed. At 35- and 10 µm inclusion lengths, the curves of Heats 3 and 4 and Heats 1 and 2 overlapped, and therefore, the order of heats was modified.

Table 4. Range of bounds of 95% confidence interval C.I.

	Heat 2	Heat 3	Heat 4	Heat 5	Heat 6
Heat 1	53.23	22.82	2.87	−26.84	−67.43
	14.79	−24.40	−51.96	−102.63	−159.96
Heat 2		−17.89	−36.64	−64.63	−104.49
		−51.70	−80.46	−132.87	−190.94
Heat 3			2.10	−27.18	−67.56
			−49.60	−100.72	−158.26
Heat 4				−0.87	−41.71
				−79.52	−136.60
Heat 5					5.21
					−103.14

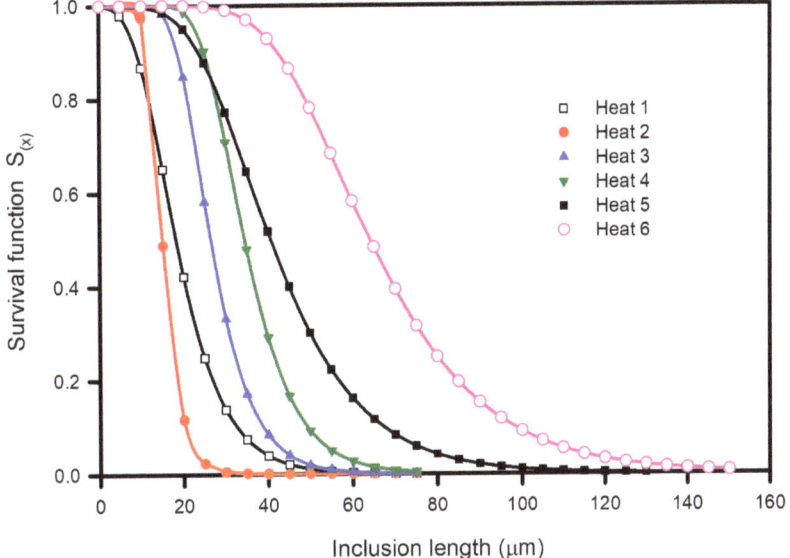

Figure 13. Survival function $S_{(x)}$ for each heat.

$S_{(x)}$ is plotted in Figure 14 from the data of Figure 13 for three critical inclusion lengths: 10, 20 and 30 µm. The heats can be ordered by increasing the $S_{(x)}$ value for each critical value; for example, for an inclusion length shorter than 10 µm, the order is Heat 1, Heat 2, Heat 3, Heat 4, Heat 5 and Heat 6. This order is altered when the critical value of 20 µm is selected, changing the order to Heat 2, Heat 1, Heat 3, Heat 5, Heat 4 and Heat 6. The observed difference is explained by the overlapping probability density curves and $S_{(x)}$ curves for Heats 1 and 2 and for Heats 4 and 5. For example, in Heat 1 and Heat 2, overlapping can be observed in the upper left corner in Figure 13, or the overlapping of probability density functions in the lower left corner in Figure 11. The same explanation can be used for the inversion of order between Heats 4 and 5. Furthermore, for critical values greater than 35 µm, where no overlapping of curves is observed, the order is Heat 2, Heat 1, Heat 3, Heat 4, Heat 5 and Heat 6. This is the same when parameter L_{max} is considered as the cleanliness parameter.

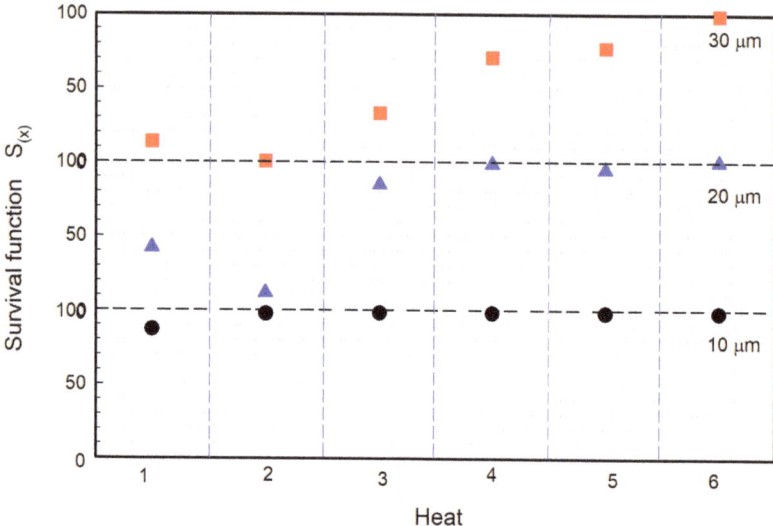

Figure 14. Survival function $S_{(x)}$ for different "critical" inclusion sizes.

5. Conclusions

Inclusion size distributions in wire rod specimens from six plant-scale heats were measured and analyzed. The measurements were taken following the metallographic procedure specified in the ASTM E2283 standard. The analysis of size distributions was developed using two approaches: (1) the estimation of PDFs according to the Higgins formalism [3] to obtain information on the whole inclusion size distribution and (2) the use of the extreme value theory to describe the upper tail of the size distributions. Thus, the following conclusions can be drawn.

- Heats were listed in decreasing order of inclusion cleanliness based on the analysis of the linear logarithmic representation of PDFs.
- No new inclusions were formed after the ladle treatment process, as inferred from the linear behavior of the logarithmic representation of PDFs, a power-law-type with time. Hence, the evolution of inclusion distribution was associated with growth, breakage, and the removal of inclusions.
- Heats were listed in decreasing order of inclusion cleanliness using the maximum inclusion length parameter L_{max}. The use of the extreme value statistics procedure specified in the ASTM E2283 standard led to a statistical comparison of L_{max}.
- Heats were ordered by considering the survival function $S_{(x)}$ values (probability of finding an inclusion larger than a "critical" inclusion length) estimated using the GEV theory. It was shown that the order can change depending on the critical value.

Author Contributions: P.H.-E. performed the experimental measurements and calculations and prepared the first draft; M.H.-T. analyzed the data and wrote the final manuscript; M.d.J.C.-R. participated in the image analysis and data analysis in the context of CONACyT Grant A1-S-44269; J.R.-M. participated in the data analysis and review of the manuscript. All authors have read and agreed to the published version of the manuscript.

Funding: P. Huazano-Estrada acknowledges the Ph.D. scholarship provided by the National Council of Science and Technology of Mexico (CONACyT) for the realization of this work.

Acknowledgments: M. Herrera-Trejo acknowledges Ternium for providing the facilities for this work.

Conflicts of Interest: The authors declare no conflict of interest.

References

1. Kaushik, P.; Lehmann, J.; Nadif, M. State of the Art in Control of Inclusions, Their Characterization, and Future Requirements. *Met. Mater. Trans. A* **2012**, *43*, 710–725. [CrossRef]
2. Pretorius, E.B.; Oltmann, H.G.; Schart, B.T. An Overview of Steel Cleanliness from an Industry Perspective. In *AISTech Proceedings 2013*; Association for Iron and Steel Technology: Warrendale, PA, USA, 2013; pp. 6–9.
3. Higgins, M. Measurement of crystal size distributions. *Am. Miner.* **2000**, *85*, 1105–1116. [CrossRef]
4. Atkinson, H.V.; Shi, G. Characterization of Inclusions in Clean Steels: A Review Including the Statistics of Extremes Methods. *Prog. Mater. Sci.* **2003**, *48*, 457–520. [CrossRef]
5. Van Ende, M.-A.; Guo, M.; Zinngrebe, E.; Blanpain, B.; Jung, I.-H. Evolution of Non-Metallic Inclusions in Secondary Steelmaking: Learning from Inclusion Size Distributions. *ISIJ Int.* **2013**, *53*, 1974–1982. [CrossRef]
6. Zinngrebe, E.; Van Hoek, C.; Visser, H.; Westendorp, A.; Jung, I.H. Inclusion Population Evolution in Ti-alloyed Al-killed Steel during Secondary Steelmaking Process. *ISIJ Int.* **2012**, *52*, 52–61. [CrossRef]
7. Bindeman, I.N. Fragmentation Phenomena in Populations of Magmatic Crystals. *Am. Miner.* **2005**, *90*, 1801–1815. [CrossRef]
8. Piva, S.; Pistorius, P. Ferrosilicon-Based Calcium Treatment of Aluminum-Killed and Silicomanganese-Killed Steels. *Metall. Mater. Trans. B* **2021**, *52*, 6–14. [CrossRef]
9. Shu, Q.; Alatarvas, T.; Visuri, V.-V.; Fabritus, T. Modelling the Nucleation, Growth and Agglomeration of Alumina Inclusion in Molten Steel by Combining Kampmann-Wagner Numerical Model with Particle Size Grouping Method. *Metall. Mater. Trans. B* **2021**, *52*, 1818–1829. [CrossRef]
10. Murakami, Y. Inclusion Rating by Statistics of Extreme Values and Its Application to Fatigue Strength Prediction and Quality Control of Materials. *Int. J. Fatigue* **1996**, *3*, 215. [CrossRef]
11. *ASTM E2283-08*; Extreme Value Analysis of Nonmetallic Inclusions in Steel and Other Microstructural Features. ASTM International: West Conshohocken, PA, USA, 2014.
12. Kumar, P.V.; Balachandran, G. Microinclusion Evaluation Using Various Standards. *Trans. Indian Inst. Met.* **2019**, *72*, 877–888. [CrossRef]
13. Fuchs, D.; Tobie TStahl, K. Challenges in Determination of Microscopic Degree of Cleanliness in Ultra-Clean Gear Steels. *J. Iron Steel Res. Int.* **2022**, *29*, 1583–1600. [CrossRef]
14. Meng, Y.; Li, J.; Wang, K.; Zhu, H. Effect of the Bloom-Heating Process on the Inclusion Size of Si-Killed Spring Steel Wire Rod. *Metall. Mater. Trans. B* **2022**, *53*, 2647–2656. [CrossRef]
15. Tian, L.; Liu, L.; Ma, B.; Zaïri, F.; Ding, N.; Guo, W.; Xu, N.; Xu, H.; Zhang, M. Evaluation of Maximum Non-Metallic Inclusion Sizes in steel by Statistics of Extreme Values Method Based on Micro-CT Imaging. *Metall. Res. Technol.* **2022**, *119*, 202. [CrossRef]
16. Gulbin, Y. On Estimation and Hypothesis Testing of the Grain Size Distribution by the Saltykov Method. *Image Anal. Stereol.* **2008**, *27*, 163–174. [CrossRef]
17. Takahashi, J.; Suito, H. Evaluation of the Accuracy of the Three-Dimensional Size Distribution Estimated from the Schwartz-Saltykov Method. *Metall. Mater. Trans. A* **2003**, *34*, 171–181. [CrossRef]
18. García-Carbajal, A.; Herrera-Trejo, M.; Castro-Cedeño, E.; Castro-Román, M.; Martínez- Enríquez, A. Characterization of Inclusion Populations in Mn-Si Deoxidized Steel. *Metall. Mater. Trans. B* **2017**, *48*, 3364–3373. [CrossRef]

Article

Heterogeneous Multiphase Microstructure Formation through Partial Recrystallization of a Warm-Deformed Medium Mn Steel during High-Temperature Partitioning

Saeed Sadeghpour *, Vahid Javaheri, Mahesh Somani, Jukka Kömi and Pentti Karjalainen

Centre for Advanced Steels Research, Materials and Mechanical Engineering, University of Oulu, 90014 Oulu, Finland
* Correspondence: saeed.sadeghpour@oulu.fi

Citation: Sadeghpour, S.; Javaheri, V.; Somani, M.; Kömi, J.; Karjalainen, P. Heterogeneous Multiphase Microstructure Formation through Partial Recrystallization of a Warm-Deformed Medium Mn Steel during High-Temperature Partitioning. *Materials* 2022, 15, 7322. https://doi.org/10.3390/ma15207322

Academic Editor: Konstantin Borodianskiy

Received: 9 September 2022
Accepted: 15 October 2022
Published: 19 October 2022

Publisher's Note: MDPI stays neutral with regard to jurisdictional claims in published maps and institutional affiliations.

Copyright: © 2022 by the authors. Licensee MDPI, Basel, Switzerland. This article is an open access article distributed under the terms and conditions of the Creative Commons Attribution (CC BY) license (https://creativecommons.org/licenses/by/4.0/).

Abstract: A novel processing route is proposed to create a heterogeneous, multiphase structure in a medium Mn steel by incorporating partial quenching above the ambient, warm deformation, and partial recrystallization at high partitioning temperatures. The processing schedule was implemented in a Gleeble thermomechanical simulator and microstructures were examined by electron microscopy and X-ray diffraction. The hardness of the structures was measured as the preliminary mechanical property. Quenching of the reaustenitized sample to 120 °C provided a microstructure consisting of 73% martensite and balance (27%) untransformed austenite. Subsequent warm deformation at 500 °C enabled partially recrystallized ferrite and retained austenite during subsequent partitioning at 650 °C. The final microstructure consisted of a heterogeneous mixture of several phases and morphologies including lath-tempered martensite, partially recrystallized ferrite, lath and equiaxed austenite, and carbides. The volume fraction of retained austenite was 29% with a grain size of 200–300 nm and an estimated average stacking fault energy of 45 mJ/m^2. The study indicates that desired novel microstructures can be imparted in these steels through suitable process design, whereby various hardening mechanisms, such as transformation-induced plasticity, bimodal grain size, phase boundary, strain partitioning, and precipitation hardening can be activated, resulting presumably in enhanced mechanical properties.

Keywords: medium Mn steel; partial recrystallization; retained austenite; partitioning; warm deformation

1. Introduction

Medium Mn steels (MMnS) containing 3–12 wt.% Mn have attracted increasing interest as potential candidates for automotive applications because of an excellent strength–ductility–toughness balance along with inexpensive alloying [1,2]. The key factor in governing a good strength–ductility combination lies in controlling properly the volume fraction and stability of retained austenite (RA). Considerable efforts have since been devoted to finetuning the balance of mechanical properties through suitable composition and process designs [3–6], thus leading to appropriate control of grain size [5,6] as well as phase fractions [7], and their morphology and heterogeneity [4].

MMnS are usually processed using intercritical annealing treatment (IAT), mostly starting from a martensitic microstructure as the dominant constituent [8]. The IAT determines the chemical composition, volume fraction, and grain size of RA [9–11]. However, in steels with a low Mn content, a long IAT time would be required to enrich austenite with an appropriate fraction of Mn to achieve a preferred fraction of stabilized RA in the final microstructure, due to the sluggish kinetics of Mn partitioning [10]. Such a prolonged IAT inevitably leads to coarsening of the recrystallized ultrafine-grained ferrite/martensite matrix and substantially deteriorates the strength of MMnS [12,13]. Therefore, a relatively high annealing temperature, usually above 600 °C (slightly below or above A$_{r1}$ temperature), is needed to accelerate Mn partitioning, while avoiding or at least delaying the loss of carbon through cementite formation. As a result, the hardness of the martensite phase decreases significantly due to extensive tempering following the reduction of both the dislocation density as well as solid solution carbon content during IAT. This makes the further strengthening of intercritically annealed MMnS difficult at a constant volume fraction of RA. Therefore, it is necessary to accelerate austenite reversion and/or partitioning using different methods.

The process of quenching and partitioning (Q&P) is normally applied in the temperature range of 250–450 °C to stabilize austenite down to room temperature (RT) through the partitioning of carbon from supersaturated martensite to untransformed austenite [14–16]. Q&P processing has successfully been applied to MMnS resulting in a microstructure comprising low-carbon martensite and carbon-enriched RA [17]. The martensite contributes to high strength, while the RA improves ductility. Recently, it has been shown [18,19] that the Q&P concept could be extended to higher temperatures to enable the partitioning of substitutional alloying elements, such as Mn in addition to C partitioning.

One effective method to accelerate the partitioning of alloying elements could be the introduction of a high density of crystal defects in the microstructure. The most common processing that is applied to MMnS includes hot-rolling followed by cold-rolling and IAT [8,20]. On the other hand, the microstructure of hot-rolled MMnS after IAT comprises lath-shaped martensite and austenite. For cold-rolled MMnS, recrystallization of deformed martensite along with austenite formation during IAT leads to the formation of an equiaxed ultrafine-grained ferritic/austenitic microstructure [21]. Kim et al. [22] reported that after cold rolling, the partially recrystallized ferrite grains and a combination of several strengthening mechanisms could result in high yield strength and enhanced work hardening of the MMnS. Bai et al. [23] suggested that the utilization of non-recrystallized austenite and recrystallized ferrite in a cold-rolled MMnS is an effective method to produce steel of ultrahigh strength (1000–1500 MPa) and excellent ductility (>40%). They showed that non-recrystallized austenite with low mechanical stability can strengthen the MMnSs through the transformation-induced plasticity (TRIP) effect, while the recrystallized ferrite can enhance ductility through sustainable plastic deformation [23]. In addition, the coupled influence of grain refinement, dislocation strengthening, and precipitation contributes to an increase in yield strength. In general, the initial microstructure strongly affects the final phase morphology, so that both the lath and globular morphologies of RA can be observed [8,20,24].

In addition to the hot and cold deformation routes, the warm rolling process can also emerge as an alternative method to achieve the required properties. There are, however, very few research works on the warm-deformed MMnS [25–29]. Chang et al. [28] and Li et al. [29] demonstrated that the warm stamped 0.1C-5Mn steel exhibited more prominent work hardening, larger ductility, and better impact toughness compared to the boron-alloyed 22MnB5 Fe-Mn steel. He et al. [26] have recently shown that warm deformation of an MMnS at 300 °C following IAT enhanced the stability of RA by creating intense dislocation networks.

The presence of prior austenite in the initial microstructure of MMnS and its fraction can also affect the partitioning process and austenite transformation from martensite during IAT and accordingly the final microstructure and mechanical properties. Liu et al. [27] reported that in a martensite–austenite dual-phase structure of 0.47C-10Mn-2Al-0.7V steel (concentrations are in wt.%), warm rolling (50% reduction at 750 °C) promoted ferrite transformation during IAT at 625 °C, in addition to austenite reversion, so that these two transformations provided a heterogeneous and multiphase microstructure susceptible to enhanced strain hardening. Tsuchiyama et al. [30] demonstrated that the presence of pre-existed austenite in 0.1C-5Mn-1.2Si steel led to the formation of fresh martensite (FM) during final cooling, which improved the tensile strength of the steel. Ding et al. [31] studied the effect of pre-existed austenite on subsequent Mn partitioning and austenite formation in a 0.2C-8Mn-2Al MMnS. They reported that 10% of pre-existed austenite accelerated the kinetics of austenite reversion and the final RA fraction could be even higher than its equilibrium value.

From the microstructure point of view, it has been shown that a heterogeneous microstructure can be very efficient in providing an enhanced strength–ductility combination in materials [27,32–34]. The hetero-structured multiphase microstructure facilitates stress and strain partitioning between the phases [35], in addition to the strain hardening induced by grain size gradient [36,37], and enables mechanisms such as deformation-induced martensite formation over a certain extended strain range depending on the stability of RA with heterogeneous sizes and morphologies [38]. Given that recrystallization is known to drastically decrease the strength of deformed materials [39], partial recrystallization has been widely used as a tool to preserve the strength to a certain extent, while improving ductility by imparting heterogeneous structures with soft and hard domains in various alloys [40]. Kim et al. [22] showed that the partially recrystallized ferrite grains contribute to the high yield strength of MMnSs due to the low mobility of screw dislocations. Accordingly, partial recrystallization provides an opportunity to achieve a heterogeneous microstructure in MMnS as well with the matrix consisting of both the recrystallized ferrite and non-recrystallized martensite that contributes to the enhancement of yield strength via heterogeneous deformation-induced hardening [22,41–43].

As concluded from the above, the design of heterogeneous microstructure in MMnS including heterogeneity in phases, morphologies, and sizes could result in a superior combination of high strength and large ductility. Though several processing strategies have since been individually investigated for accelerating the partitioning of alloying elements and promoting the heterogeneity in the final microstructure of MMnSs, incorporation of a controlled combination of these strategies, however, might seemingly be more effective in enhancing mechanical properties from the microstructure engineering perspective. In the previous work [44], a combination of factors accelerating the stabilization of austenite, including pre-existed austenite, pre-deformation at 250 °C, and high-temperature partitioning was employed to create a refined microstructure in a 0.31C-4Mn-2Ni-0.5Al-0.2Mo (in wt.%) steel. In the present work, a modified process is proposed to produce a heterogeneous multiphase microstructure consisting of martensite, recrystallized ferrite, pre-existed austenite, reverted austenite, and carbides by warm deformation at a higher temperature promoting partial recrystallization of ferrite and austenite reversion in an MMnS. The strategy is composed of three steps: (1) quenching of the austenitized specimen to a temperature below martensite start temperature (M_s) to produce a martensitic/austenitic structure, (2) warm deformation of the dual-phase alloy at a temperature above M_s to create high densities of crystal defects and potential nucleation sites for new phases in the microstructure, (3) annealing of the deformed microstructure to recover and partially recrystallize the martensite grains and to achieve the desired fraction of stabilized austenite, and finally cooling to RT.

2. Experimental Procedure

An MMnS with a chemical composition of Fe-0.31C-4Mn-2Ni-0.42Al-0.21Mo (in wt.%) was used in the present study. A 200 mm × 80 mm × 55 mm piece of the vacuum induction melted ingot was cut and homogenized at 1200 °C for 2 h and then hot-rolled to 11 mm in thickness. Cylindrical specimens with a diameter of 6 mm and a height of 9 mm were cut with the axis transverse to the rolling direction. The samples were used for processing and dilatometry tests on a Gleeble 3800 thermomechanical simulator. The specimens were first reheated at 900 °C, held for 6 min, then cooled to 120 °C to achieve a microstructure of ~75% martensite and ~25% untransformed austenite. Then, the martensitic–austenitic specimens were reheated to 500 °C where a compressive true strain of ~0.4 was applied at a constant true strain rate of 0.1. The deformed samples were immediately heated to the annealing temperatures of 600, 625, 650, and 700 °C and soaked for 20 min and subsequently cooled down to RT (hereafter referred to as samples Def-A600, Def-A625, Def-A650, and Def-A700, respectively). A constant rate of 10 °C/s was used for all the heating and cooling steps during the processing cycle. Figure 1 illustrates a schematic of the applied process.

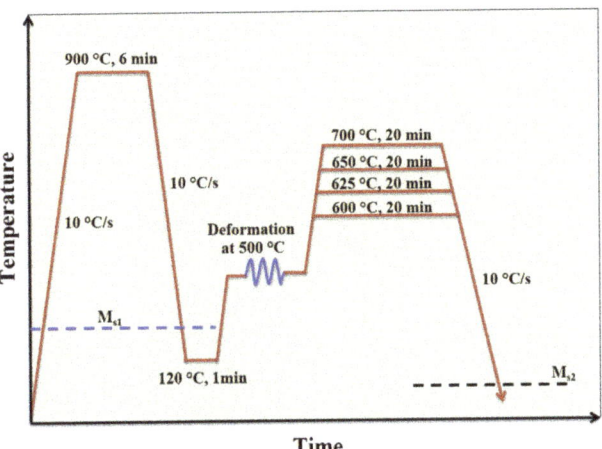

Figure 1. Sketch of the applied processing route.

To characterize the microstructures, the processed specimens were sectioned along their compression axis and prepared using standard metallographic methods. A field emission scanning electron microscope (FE-SEM) with electron backscatter diffraction (EBSD) and a scanning/transmission electron microscope (STEM/TEM) were used for microstructure observations. The specimens were finally polished with a colloidal solution of silica suspension in H_2O_2 for EBSD analysis. EBSD scans were conducted with a step size ranging from 50 to 20 nm. For STEM/TEM observations, 3 mm discs were first ground to a thickness of 100 μm and then electrochemically polished with a solution of 5% perchloric acid and 95% ethanol using a twin-jet polisher operated at −15 °C. Interesting microstructural features were analyzed in respect of chemical composition via energy-dispersive X-ray spectroscopy (EDS) mapping in the STEM. Phase fraction was determined using X-ray diffraction (XRD) with Co-Kα radiation (λ = 0.179 nm). Vickers hardness measurements were carried out on all specimens using a 5 kg load.

3. Results and Discussion

According to the dilatometry results, austenite transformation start (A_s) and finish (A_f) temperatures, and the martensite start temperature (M_s) of the studied steel were measured to be ~713 °C, ~802 °C, and 223 °C, respectively. According to XRD results shown in Figure 2a, the microstructure following reaustenitization at 900 °C and quenching to RT was mainly (93%) martensitic. The EBSD phase map of the corresponding sample confirmed that the microstructure essentially consisted of martensite and a very small amount of retained austenite, as shown in Figure 2b. The average prior austenite grain size was measured to be around 10 μm. Figure 2c shows the variation of martensite fraction as a function of quench temperature as calculated from the dilatation curve (not shown here) using the lever rule. The dashed line in Figure 2c indicates that the volume fraction of martensite is around 73% after quenching to 120 °C, so the volume fraction of pre-existed austenite is estimated to be about 27%.

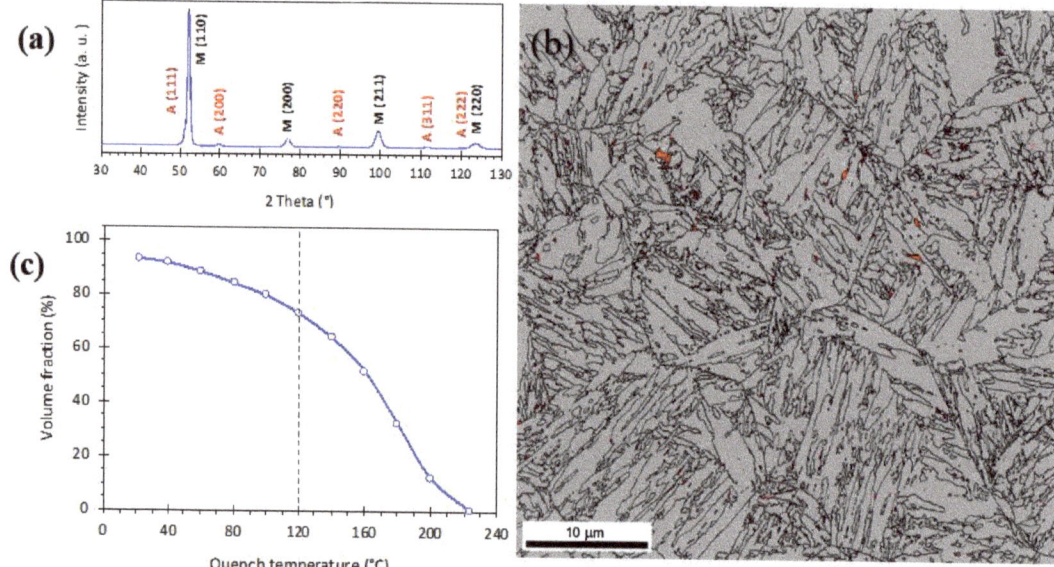

Figure 2. (a) XRD pattern of the sample quenched to RT after annealing at 900 °C, indicating martensite (M) and weak austenite (A) peaks, (b) corresponding EBSD phase map (austenite in red and martensite in grey), and (c) volume fraction of martensite at different quench temperatures.

Figure 3a presents the dilatation curves of the Def-A600, Def-A625, and Def-A650 samples. A secondary martensite formation was detected around 120 °C (M_{s2} temperature) during final cooling after annealing at 600 °C, as shown by the arrow. It is expected that after initial quenching, the M_s temperature of the untransformed austenite to be just the same as the quench-stop temperature (i.e., 120 °C). Therefore, it can be concluded that in the process of warm deformation, subsequent heating, and isothermal holding at 600 °C, the austenite was not stabilized at all. In spite of this, with increasing the annealing temperature to 625 °C and 650 °C, the M_{s2} temperatures of the untransformed austenite decreased to 90 °C and 60 °C, respectively, indicating the stabilization of austenite to some extent. Figure 3b shows the change in the diameter of the samples as a function of holding time at different annealing temperatures. The curves indicated an initial contraction up to 850 s, followed by a plateau until the end of isothermal holding. The magnitude of total contraction decreased with increasing annealing temperature from 600 °C to 650 °C. However, after a certain holding time, some microstructural changes occurred causing expansion in competition with the contraction caused by partitioning and/or austenite formation. Several processes, such as the recovery and recrystallization of deformed martensite, austenite formation, and precipitation of carbides can cause volume contraction during holding at a particular temperature. In contrast, the transformation of austenite to phases such as ferrite, pearlite, and/or bainite causes volume expansion. These phase transformations that may occur during isothermal holding can affect the final volume fraction and the chemical stability of RA. The expansions observed here are most probably associated with the austenite decomposition, which has also been investigated in the previous work [44]. The decomposition processes may reduce the austenite fraction, so that if less new austenite forms, the amount of RA will decrease despite its higher stability.

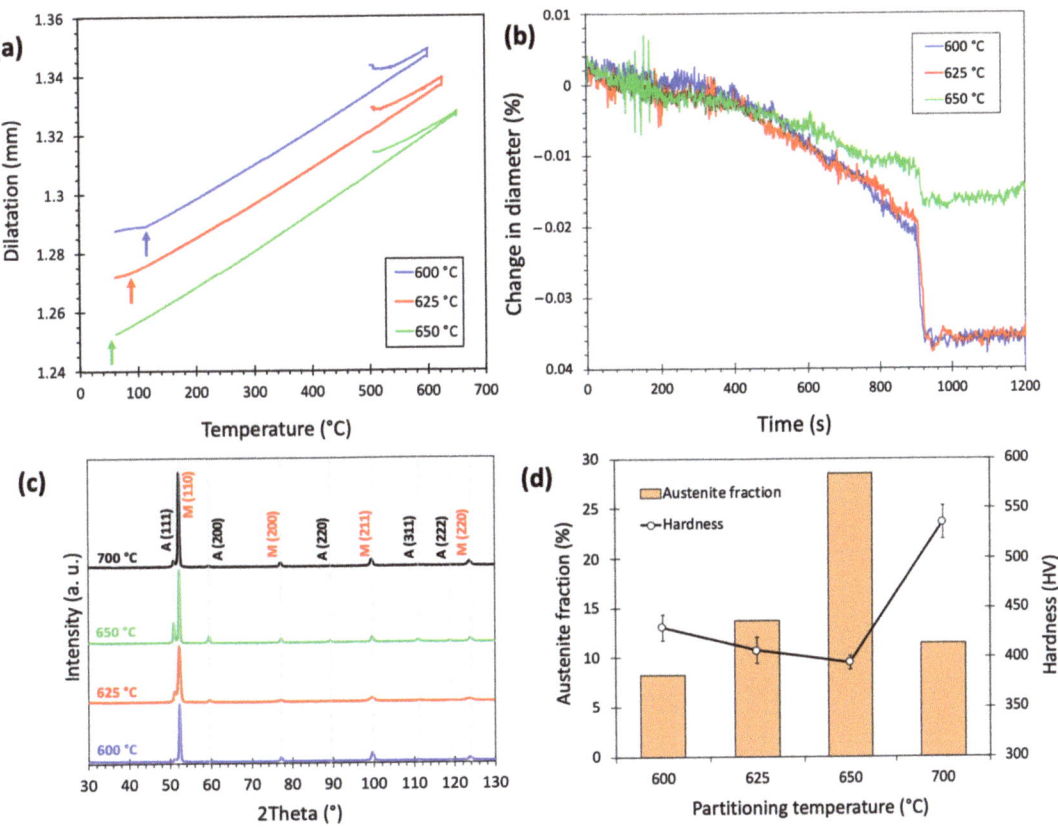

Figure 3. (a) Dilatation curves of the Def-A600, Def-A625, and Def-A650 samples, (b) corresponding curves of the change in diameter vs holding time, (c) corresponding XRD patterns showing the presence of austenite (A) and martensite (M) phases in the final microstructure and (d) corresponding austenite fractions and hardness values.

Figure 3c,d show the XRD patterns of the samples annealed at different temperatures and the corresponding volume fractions of RA calculated based on the integrated intensities of austenite and martensite peaks. According to calculations, the RA fraction increased from 8% to 29% with an increase in the annealing temperature from 600 °C to 650 °C. This increase in RA fraction, and less new fresh martensite (FM) as illustrated in the following, and a related decrease in dislocation density of tempered martensite (TM) due to higher partitioning temperature led to a decrease in hardness from 430 HV to 396 HV. Further increase in partitioning temperature to 700 °C resulted in a decreased RA fraction to 11% and an increased fraction of FM, while the hardness value increased to 536 HV. Thus, the maximum RA fraction was detected in the sample annealed at 650 °C. The change in the RA fraction and the hardness of the sample are associated with the formation of new martensite from less-stable austenite, and this is discussed based on the microstructure.

As SEM micrographs in Figure 4a–c present, the microstructure of the Def-A600 sample exhibits several features inherited from primary TM, FM, pearlite (P), RA, and carbides. The lath-shaped TM (dark areas) can be characterized thanks to the presence of various carbide precipitates. Two different morphologies of TM are detected: the equiaxed fine grains, i.e., the recrystallized ferrite with size in the range of 200–300 nm (Figure 4c), and the lath morphology. Such heterogeneity in morphology can be attributed to the partial recrystallization of martensite during annealing at 600 °C. Carbide-free FM islands are somewhat lightly etched compared to TM (Figure 4c) but can still display some substructures. Some pearlite phase constituents were also detected in the microstructure as shown by arrows in Figure 4a. In addition, two lath-type and equiaxed morphologies of RA were also observed. However, TM usually contains both large and fine carbide precipitates, unlike in the case of RA, where carbides are generally fine. Though it is difficult to distinguish RA from TM using SEM images due to their similar contrast and morphology, RA can be easily identified by the EBSD phase maps. Due to more nucleation sites created during deformation, many carbides were observed along the boundaries of partially recrystallized grains with a globular morphology, as shown in Figure 4c. As mentioned, several carbides were detected with varying sizes. Large carbides with a size in the range of 80–150 nm were observed mainly on lath boundaries, while both medium-size carbides in the range of 20–50 nm and fine ~10 nm carbides were found inside the TM.

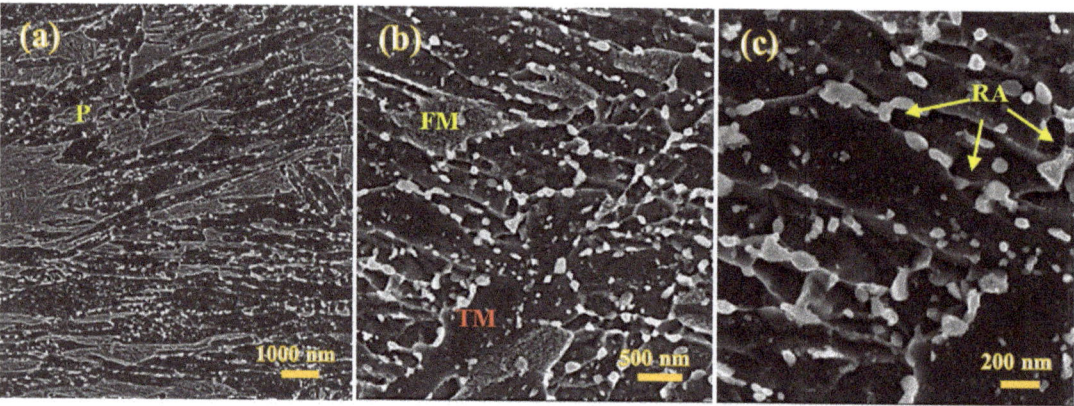

Figure 4. FE-SEM micrographs of the Def-A600 sample showing pearlite (P), tempered martensite (TM) some fresh martensite (FM), and several carbides on the lath boundaries and inside the laths (**a,b**), recrystallized ferrite and equiaxed retained austenite (RA) nucleated on carbide/ferrite boundaries and presence of very fine carbides inside the TM, ferrite, and RA grains (**c**).

Figure 5a,b presents the image quality (IQ) map overlayed by a phase map plotted from the EBSD examinations carried out on the Def-A600 specimen. The RA fraction was estimated to be ~8.5% in agreement with the XRD results presented in Figure 3d. In fact, a significant amount of austenite is consumed by transforming to P and FM during holding at 600 °C and subsequent quenching. The average grain size of RA was estimated to be around 150 nm, which is markedly smaller than that of the as-received specimen (10 μm). A considerable amount of FM was observed in the microstructure, which is in agreement with the dilatometric result (Figure 3a), giving an M_{s2} temperature of 120 °C. The IQ maps in Figure 5a,b clearly show that the TM has a higher IQ value, i.e., is brighter compared to the FM due to its lower defect density. The kernel average misorientation (KAM) map in Figure 5c shows that the TM contains a low density of geometrically necessary dislocations because of dislocation annihilation during the tempering, resulting in low KAM values. In contrast, significantly higher KAM values, i.e., dislocation densities are detected in FM areas. In addition to the differences in dislocation density, the TM contains a very low amount of C due to its partitioning into the austenite, while the FM forms from C-enriched austenite rendering it to be an untempered, high-C hard phase. Therefore, the hardness of the FM formed during final quenching is expected to be higher than that of the TM formed during initial quenching to 120 °C following austenitization (Figure 3d).

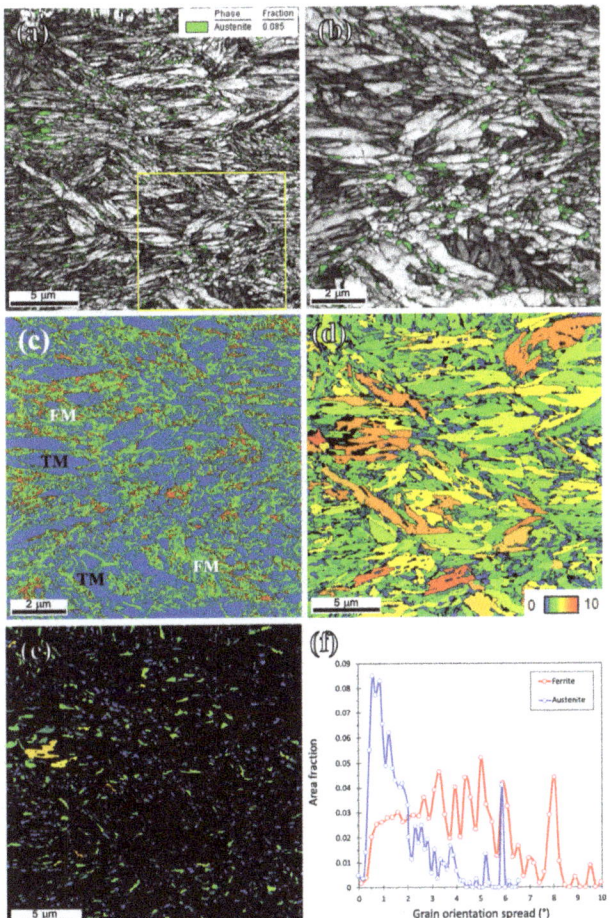

Figure 5. EBSD maps of the Def-A600 sample. (**a**) IQ map overlayed by phase map, (**b**) higher magnification of the marked area in (**a**), (**c**) KAM, (**d**) and (**e**) grain orientation spread (GOS) maps; and (**f**) GOS distribution for ferrite and austenite.

Unlike in the case of cold-rolled MMnS which usually provides a fully recrystallized microstructure after IAT [21], warm rolling only promoted partial recrystallization in the ferritic matrix as a result of a comparatively lower driving force. Further, austenite had two different types of morphologies, i.e., lamellar and equiaxed. We may assume that due to the lower driving force for recrystallization in the warm-rolled matrix, the recrystallization of ferrite and its reversion to austenite took place simultaneously. Whilst reversion of martensite laths led to the formation of lamellar RA of size <200 nm, recrystallized ferrite grains transformed to equiaxed RA with size varying in the range 150–600 nm (Figure 5a,b). In addition to the new austenite formation through reversion, partial recrystallization of lamellar pre-existed austenite may produce equiaxed or lamellar morphology in RA. Therefore, both ferrite and austenite phases showed heterogeneity in terms of size and morphology.

Figure 5d,e shows the grain orientation spread (GOS) maps of ferrite and austenite of the Def-A600 sample. As the inset in Figure 5d shows, the GOS value varies from 0 (blue) to 10 (red). This blue-to-red range was set identically so that the different conditions could be directly compared. The GOS maps comparing the local orientation of each point with respect to the orientation of the neighboring points have widely been used in evaluating the degree of recrystallization [45–47]. As seen in Figure 5d, a considerable number of ferrite grains had a high GOS value illustrating that those grains were not fully recrystallized. In contrast, most austenite grains appearing as blue or green displayed low GOS values. However, we have to notice that even with a step size of 50 nm, orientation differences are difficult to detect by GOS in grains with a size below 200 nm. The lamellar old austenite might be partly recrystallized, but 600 °C is a very low temperature for recrystallization, even though some strain partitioning might have concentrated strain in austenite. In austenitic stainless steels, the lowest recrystallization temperature is about 700 °C [48], and in duplex stainless steel, recrystallization of austenite occurs above 800 °C (though 20% cold rolling reduction, only) [49]. Therefore, despite possibly higher energy stored in austenite owing to strain partitioning between martensite and austenite during the warm deformation, there is no real evidence for the recrystallization of austenite at 600 °C, and GOS limited resolution might lead to a wrong conclusion. Fine grains may be new equiaxed austenite nucleated during annealing, being naturally dislocation-free. The formation of new dislocation-free austenite is consistent with the KAM results, though the recrystallization of pre-existed austenite cannot be ruled out and would be the subject of future studies.

Figure 5f presents a comparison of the GOS distribution for ferrite and austenite. While the ferrite grains showed a wide range of GOS owing to the presence of both recrystallized grains along with non-recrystallized grains, the austenite grains displayed a relatively narrower distribution of GOS. Two well-separated peaks for the GOS values of both ferrite and austenite grains can be identified as corresponding to the recrystallized or new grains and deformed non-recrystallized ones. Grains with a GOS less than 2° were classified as fully recrystallized ferrite which is in agreement with the reported threshold value for GOS [45–47], or dislocation-free new austenite. Based on the criterion of GOS <2°, the dislocation-free fractions were determined to be 19% and 67% for ferrite and austenite, respectively.

The microstructure of the Def-A650 sample contained several features including TM, FM, RA, and carbides, as shown in Figure 6a. It is also obvious from Figure 6b that the recrystallized grains of the specimen partitioned at 650 °C are somewhat coarser compared to those of the Def-A600 specimen. Unlike in the case of the Def-A600 sample, the absence of pearlite in the microstructure of the Def-A650 sample suggested slow kinetics of pearlite transformation at 650 °C despite holding for 20 min. The retardation of pearlitic transformation as a result of increasing annealing temperature has also been discussed in the previous study [21]. RA grains (fraction about 28%) were mainly equiaxed, in addition to the presence of a small fraction of interlath austenite in the microstructure. Additionally, the volume fraction of carbides following partitioning at 650°C was lower than that observed in the case of the sample partitioned at 600 °C, though the average size of carbides was expectedly larger at 650 °C.

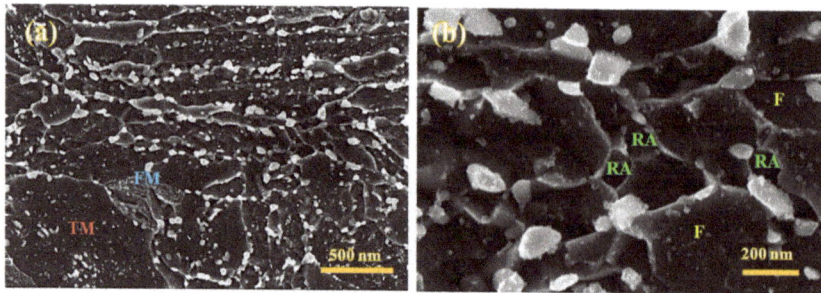

Figure 6. FE-SEM micrographs of the Def-A650 sample showing TM, FM, and carbides on the martensite lath boundaries and inside the laths (**a**), recrystallized ferrite (F), and equiaxed RA nucleated on carbide/ferrite boundaries along with very fine precipitates inside the F and RA grains (**b**).

Figure 7 illustrates the EBSD maps of the Def-A650 specimen, where the RA appears mostly as equiaxed grains, along with the TM, recrystallized ferrite, and some contents of the FM. According to Figure 7a, a significantly higher fraction of RA is observed compared to that in the Def-A600 specimen, in consistence with the XRD results shown in Figure 3c. The fraction is practically the same as that of the pre-existed austenite. Though the RA grains are somewhat coarser following partitioning at 650 °C, they are more uniform in size and distribution compared to those observed in the Def-A600 specimen, as shown in Figure 5a,b. Though the TM had both equiaxed as well as lath morphologies, the area fraction of equiaxed grains increased with increasing the annealing temperature from 600 °C to 650 °C, as a result of enhanced recrystallization rate at 650 °C. The inverse pole figure map overlayed by the high-angle boundary map, shown in Figure 7c, indicates a heterogeneous grain structure containing about 74% of high-angle boundaries. Figure 7d,e shows the GOS maps of ferrite and austenite of the Def-A650 sample. According to Figure 7d, ferrite grains were not yet fully recrystallized following 20 min holding, even though the fraction of recrystallized grains was higher than that of the Def-A600 sample (Figure 5d). The recrystallized fractions of ferrite and austenite were calculated to be around 26% and 73%, respectively. The average grain size of RA increased to 600 nm with increasing the annealing temperature to 650 °C. This corroborates that a higher amount of stored energy was available in the deformed austenite compared to the deformed martensite due to strain partitioning. As shown in Figure 7f, the ferrite grains showed a wider distribution of GOS, while the austenite grains showed a narrower distribution of GOS because of a large, recrystallized fraction.

The STEM images of the Def-A650 sample illustrate two lath-type and equiaxed morphologies for both austenite and martensite phases, as shown in Figure 8a,b. Thin interlath films of RA, surrounded by martensitic areas along with some equiaxed austenite grains can be observed as bright regions in Figure 8b. These thin RA films might be non-recrystallized pre-existed austenite or new austenite formed on non-recrystallized martensite lath boundaries. In addition to martensite and austenite, various precipitates as strings of large carbides on the lath boundaries and both spherical and rod-shaped carbides inside the TM lath were observed in the microstructure as shown in Figure 8c.

As described, the applied process resulted in a multiphase microstructure comprising TM, ferrite, RA, and carbides. Both martensite and austenite phases exhibited heterogeneous morphologies and grain structures decorated by carbides. As mentioned, these phases and morphologies could contribute to different strengthening and deformation mechanisms.

To investigate the partitioning of alloying elements in the final microstructure, STEM-EDS scans were conducted on the Def-A650 specimen. A typical STEM image along with corresponding distribution maps of C, Mn, and Ni for the sample are shown in Figure 9a–d, respectively. The contents of elements in the regions marked by identifying numbers in Figure 9c are listed in Table 1. The maps evidently show that C, Mn, and Ni have partitioned to specific regions and depletion in other regions. The Mn contents at points 1, 2, and 3 were measured to be 6.7%, 2.8%, and 15.7%, respectively. The equilibrium concentrations of Mn in austenite, ferrite, and cementite were calculated to be about 7.34%, 1.46%, and 13.85% at 650 °C, as reported elsewhere [44]. This suggests that point 1 in Figure 9c corresponds to the austenitic area, while point 2 is in the ferritic region, and point 3 lies in cementite. The corresponding SAED patterns in Figure 9e,f confirm that areas marked A_1 and M_1 in Figure 9a are distinguished as FCC and BCC structures. Since C, Mn, and to some extent Ni, partition from BCC martensite to FCC austenite, the amounts of these alloying elements in RA, especially Mn concentration, ought to be relatively higher than that in the martensitic grains. The compositions listed in Table 1 denote that there is an insignificant change in the Al and Si contents in comparison to their average value before the processing, while the C, Mn, and Ni contents are varying appreciably in different regions. This suggests that considerable partitioning of C, Mn, and Ni did take place during the process, whereas the partitioning of Al and Si to BCC ferrite was insignificant because of their lower diffusivity in FCC austenite. The partitioning of C and Mn has been reported even in a short duration of 180 s during IAT of cold-rolled MMnS in the range of 640–680 °C [50]. However, Lis et al. [51] reported no considerable partitioning of Ni, Cr, and Si in their investigation of the partitioning of various alloying elements.

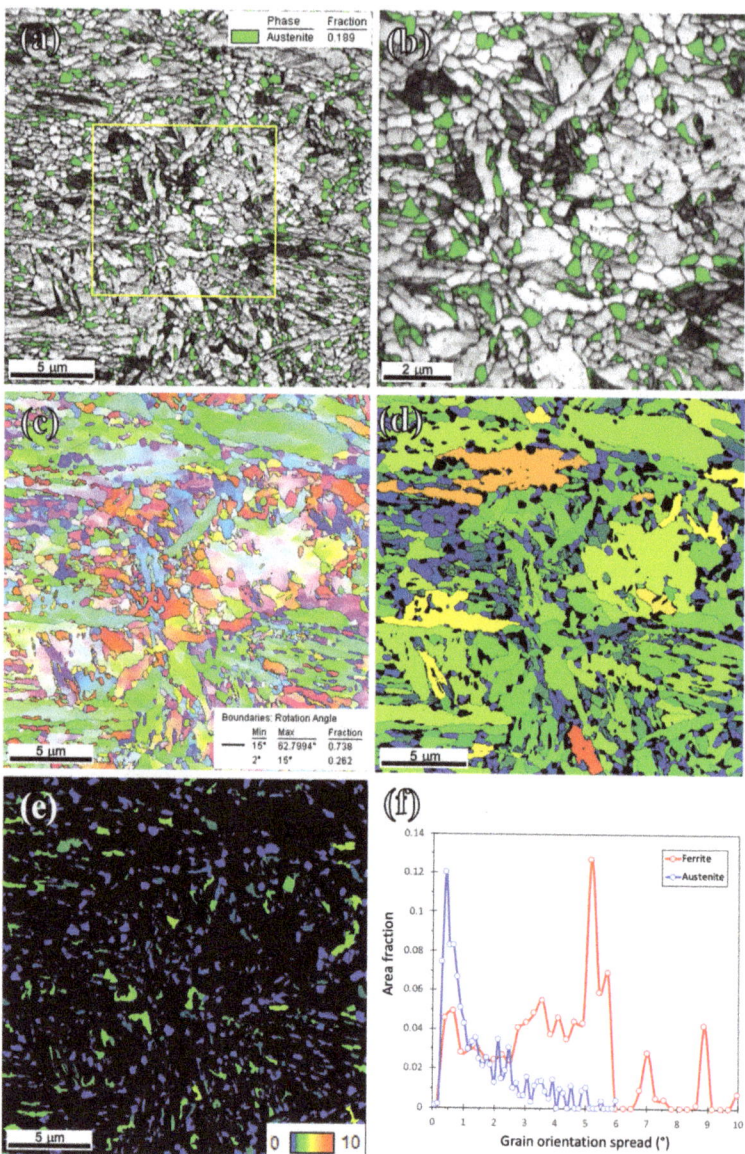

Figure 7. EBSD maps of the Def-A650 sample. (**a**) IQ map overlayed by phase map, (**b**) higher magnification of the marked area in (**a**), (**c**) IPF and (**d**,**e**) grain orientation spread (GOS) maps; and (**f**) GOS distribution for ferrite and austenite.

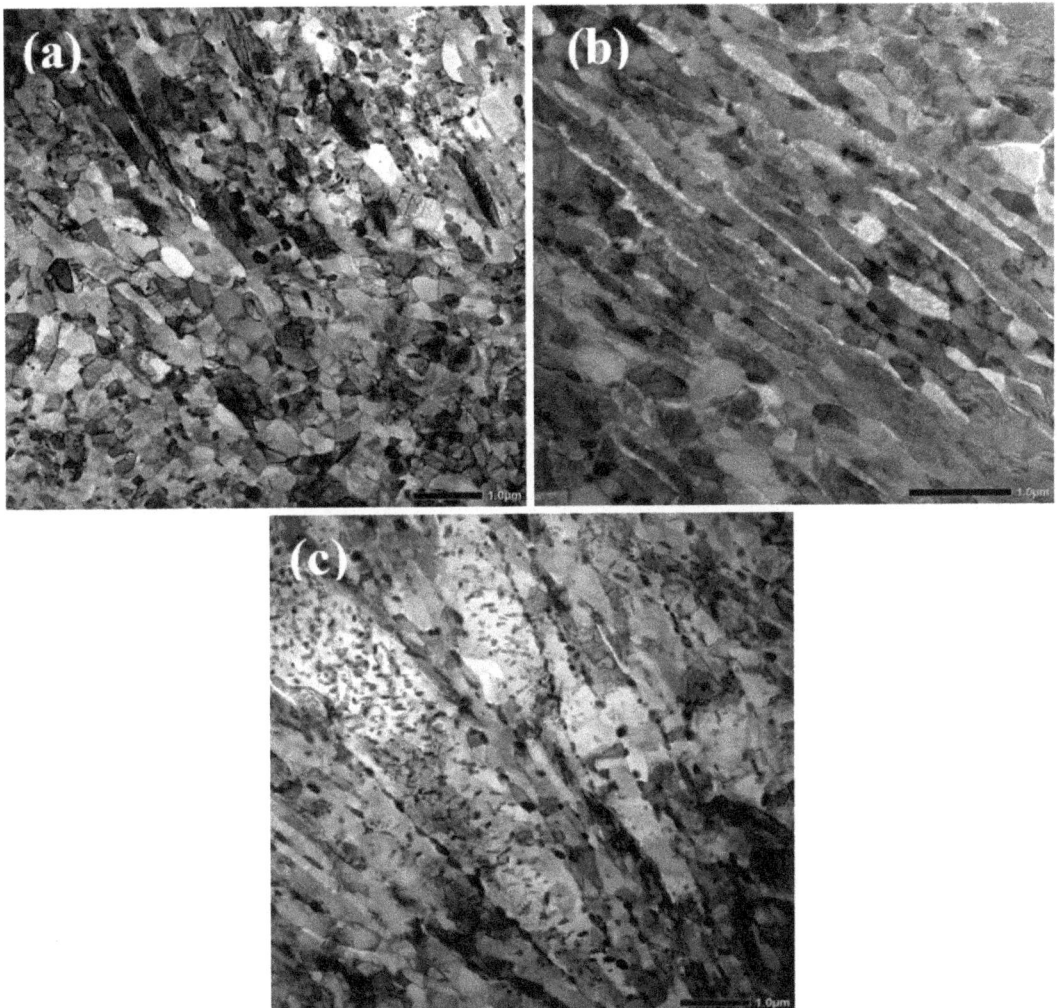

Figure 8. STEM images of the Def-A650 specimen. (**a**) Lath and equiaxed structure of martensite and RA phases, (**b**) two morphologies of retained austenite as films between the martensite laths, and some globular grains along with carbides on lath boundaries, (**c**) strings of large carbides on lath boundaries and spherical and rod-shaped carbides inside the TM laths.

Table 1. Chemical composition of the areas marked by numbers in Figure 9c measured by STEM-EDS.

Position	Mn	C	Ni	Al	Si	Mo	Phase
1	6.71	1.27	2.06	0.34	0.20	2.62	RA
2	2.84	-	0.99	0.32	0.46	0.63	F
3	15.73	9.41	0.95	0.07	0.19	0.54	Carbide
4	4.66	0.88	1.98	0.24	0.59	1.58	FM
5	7.08	1.73	2.40	0.05	0.49	-	RA

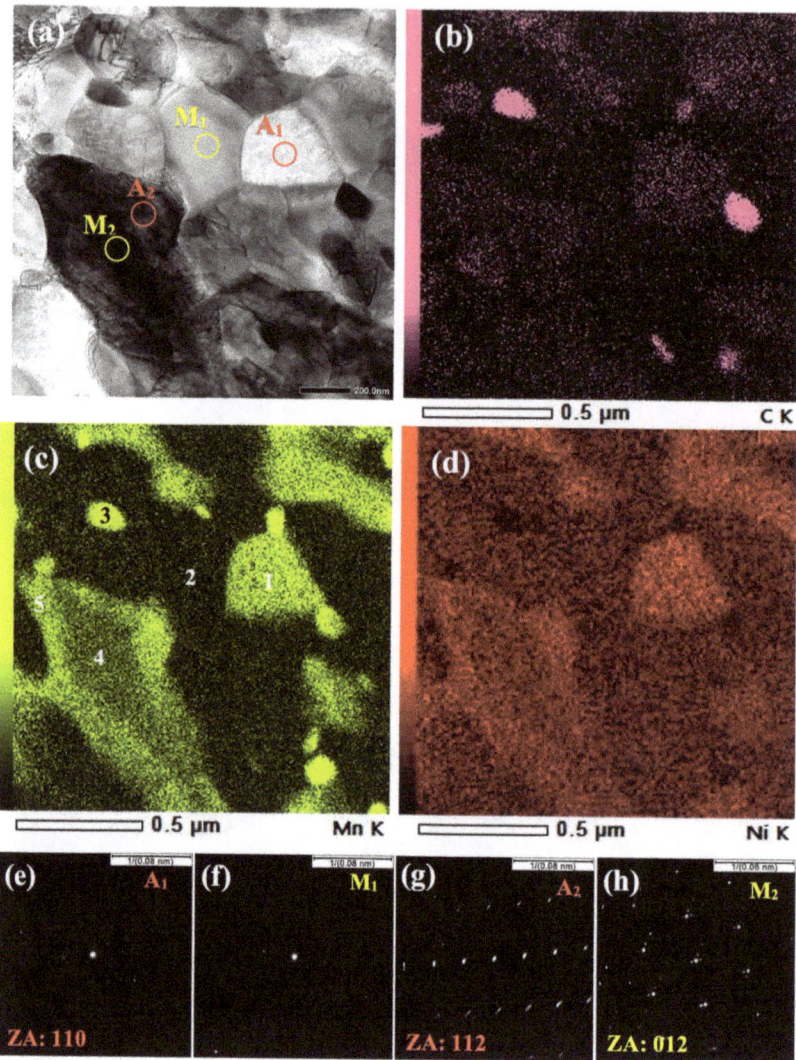

Figure 9. (a) STEM image of the Def-A650 specimen and corresponding STEM-EDS maps of (b) C, (c) Mn and (d) Ni distribution along with the electron diffraction patterns taken from the areas (e) A_1, (f) M_1, (g) A_2 and (h) M_2.

In addition to typical Mn and Ni partitioning, some martensitic regions depleted of Mn and Ni that were surrounded by thin austenitic films enriched in Mn and Ni were also observed. For example, Mn concentration at point 4 was measured to be 4.7% while Mn concentration in the outer layer, (i.e., at point 5) was measured to be 7.1%. The SAED patterns presented in Figure 9g,h confirm that the exterior area close to the boundary (i.e., A_2 in Figure 9a) has an FCC structure, while the core area (i.e., M_2 in Figure 9a) has a BCC structure. It can be concluded that the shell regions of the larger austenite grains close to the FCC/BCC interfaces, where Mn concentration was adequately high to stabilize the austenite, retained untransformed at RT. In contrast, core regions that were Mn-lean transformed to fresh martensite during the cooling to RT. It is worth mentioning that the STEM samples were prepared at −15 °C indicating that RA was stable even until this temperature. The average thickness of the Mn-enriched austenite shell was measured to be around 150–200 nm. Given that the contents of C, Mn, and Ni in FM were measured to be higher than those in TM but lower than those in RA, it was possible to distinguish the TM, FM, and RA in the microstructure, as shown in Table 1 for the selected locations.

A similar formation of martensite inside austenite has also been reported in 5Mn-0.1C steel during cooling from the intercritical region [30]. It has been shown that the content of Mn inside the core region, can be close to its average level in the material [31], which is in good agreement with the present observations. The present results indicate that despite the higher Mn concentration in the core of the prior austenite (4.7%) in comparison to its average value in the bulk material (4%), the partitioning of Mn and Ni was still inadequate to fully stabilize austenite and only a thin shell of austenite with a thickness of 150–200 nm, where the Mn content is higher (7.1%), was stabilized. This is ascribed to the fact that the annealing at 650 °C for 20 min was too short for adequate partitioning of Mn to fully stabilize austenite because of the slow diffusion of Mn in the austenite phase. In contrast, a large area of ferrite was depleted of Mn (e.g., 2.5–3% Mn) due to the higher diffusion rate in ferrite.

The authors recently showed [44] that the prior deformation can effectively improve the Mn partitioning distance by creating high densities of dislocations as easy diffusion paths. However, due to 20 min partitioning at 650 °C, some of the grains coarsened beyond 500 nm suggesting that Mn partitioning may not be enough to stabilize large austenite grains, as their core regions transformed to FM during the cooling to RT. According to previous results [44], the critical Mn concentration to stabilize austenite in the studied steel was found to be around 5.5%. However, to consider the partitioning effect of other alloying elements on austenite stabilization, the M_s temperature was calculated for different grains. On the other hand, C diffuses several hundred times faster than substitutional elements such as Mn and Ni, and its equilibration is completed at the very beginning of the heating so that the effect of deformation on carbon diffusion rate is not detectable irrespective of the partitioning duration.

During IAT, the recovery and recrystallization of martensite occur before its reversion to austenite and the recrystallization is enhanced with high stored energy provided by deformation. The effect of deformation on the accelerated recrystallization of martensite before its reversion has already been observed in dual-phase steels [52] and MMnS [53]. It has also been reported that recrystallization of ~75% cold-deformed martensite occurs in less than 5 s at a temperature around A_{c1} [54]. In the present work, a smaller deformation strain of 40% at a higher temperature of 500 °C was applied, so the stored energy is much lower compared to that generated in cold deformation. Because of the lower driving force in warm deformation, the recrystallization process is slower. The recrystallization of martensite results in fine ferrite grains with low dislocation density. However, with reduced dislocation density, accelerated diffusion processes cannot operate to transport adequate Mn atoms to desired diffusion distances in austenite [44].

It has been shown that the austenite reversion in martensite–austenite structure can start at a temperature below A_{c1} [44,55,56]. The equilibrium thermodynamic calculations also indicate that austenite might form at temperatures as low as 492 °C [44]. This can be ascribed to the pre-existed austenite so that the increase in austenite fraction results from the growth of pre-existed austenite rather than by the nucleation of new grains. According to Ding et al. [31], in the presence of prior austenite, both the growth of pre-existed austenite and the nucleation of new austenite grains may take place during intercritical annealing.

To clarify the influence of partial recrystallization on microstructural evolutions, some salient microstructural characteristics obtained in the present study (the sample deformed at 500 °C) and in the previous work [44] (a non-deformed and a sample deformed at 250 °C) are compared as listed in Table 2. The main difference concerns the recrystallization fraction of ferrite which was much smaller in the specimen deformed at 500 °C. The lower recrystallized fraction is obviously due to the reduced stored energy following deformation at the high temperature (500 °C), also resulting in a heterogeneous structure with a broad grain size distribution. Though there is uncertainty about the recrystallization of austenite due to new austenite formation, still the sample deformed at 500 °C showed a lower fraction of dislocation-free grains. The sample deformed at 500 °C exhibited a comparatively larger average grain size and a higher number fraction of grains with core–shell structure following partial recrystallization resulting in a slightly lower fraction of stabilized austenite and the formation of more FM. The average M_s temperatures and stacking fault energies (SFE) of RA in various samples were estimated using the following equations [57,58]:

$$M_s = 517 - 423C - 30.4Mn - 7.5Si + 30Al \quad (1)$$

$$SFE = 1.2 + 1.4 \, (Ni + 0.5Mn + 0.3Cu + 30C) + 0.6 \, (Cr + 2Si + 1.44B) \quad (2)$$

Table 2. Microstructural characteristics of the present steel after deformation at 500 °C (def-500) and 250 °C (def-250) and without deformation (non-def) followed by annealing at 650 °C for 20 min.

Sample	Recrystallized Ferrite Fraction (%)	Recrystallized Austenite Fraction (%)	RA Average and Range of Grain Size (μm)	Ferrite Average and Range of Grain Size (μm)	RA Fraction (%)	M_s (°C)	RA SFE (mJ/m^2)
def-500	26	73	0.23 (0.07–0.99)	0.31 (0.11–5.51)	29	−70	45
def-250	56	78	0.24 (0.07–1.32)	0.24 (0.09–2.97)	32	−89	48
non-def	0	0	0.2–0.3 *	0.8–2 *	34	−105	48

* The lath thickness was used for this sample.

The M_s and SFE values listed in Table 2 indicate that the RA in the sample deformed at 500 °C had lower stability in comparison to that in the sample deformed at 250 °C and non-deformed sample. This comparison indicates that by varying the deformation temperature, it is possible to vary the microstructure stability and heterogeneity in terms of size, distribution, phase fractions, morphology, and its stability.

Figure 10 schematically illustrates the progress of microstructural evolution during the processing of present MMnS under the application of warm deformation and subsequent partitioning annealing. The initial microstructure, which is mainly martensitic, transforms to a fully austenitic structure with a large grain size after annealing at 900 °C for 6 min (step I). Following quenching to 120 °C, a microstructure consisting of 73% martensite and 27% austenite is achieved (step II). As a result of warm deformation at 500 °C, a high dislocation density is created both in austenite as well as martensite phases (step III). Strain partitioning to the softer austenite phase makes austenite grains far more strained meaning acquisition of a higher dislocation density. During subsequent IAT at 650 °C for 20 min, both the formation of new austenite and partial recrystallisation of martensite take place along with partial carbide dissolution (step IV). During the final cooling to RT, some austenitic regions with lower chemical stability transform to FM creating a core–shell structure in addition to the bulky FM regions (step V). The applied processing route results in a heterogeneous multiphase microstructure comprising coarse-grained TM, fine recrystallized ferrite, ultrafine RA, FM, and fine carbides. The grain refinement along with C, Mn, and Ni partitioning promotes higher stability of RA grains that are envisaged to provide pronounced TRIP and/or TWIP effects and consequently enhance the strain hardening rate and ductility.

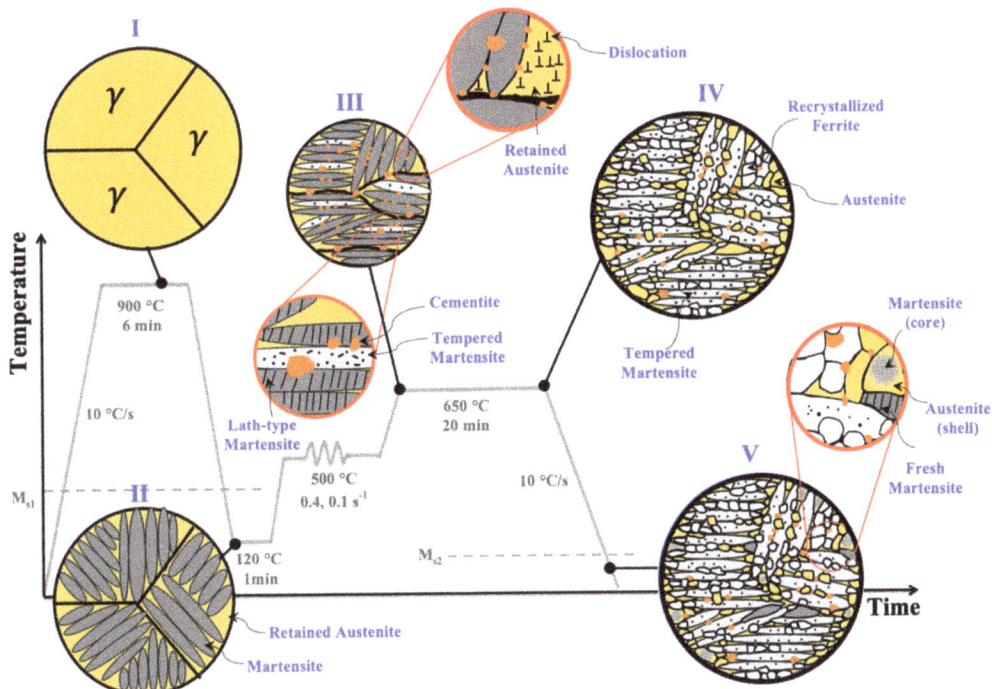

Figure 10. Schematic illustration of the microstructure evolution during the annealing (I), quenching to 120 °C (II), warm deformation at 500 °C (III), and IAT at 650 °C (IV) resulting in a heterogeneous multiphase microstructure (V).

4. Conclusions

A new processing route was employed to achieve a highly refined, multiphase, and heterogeneous microstructure in a medium Mn steel. Based on the detailed microstructural analysis, the main conclusions are summarized as follows:

1. The processing route comprising step quenching to 120 °C, warm deformation at 500 °C, and high-temperature partitioning at 650 °C for 20 min results in the occurrence of various desired microstructural evolutions such as martensite tempering and its partial recrystallization, carbide precipitation, partitioning of elements and austenite reversion.
2. The desired multiphase and heterogeneous microstructure of the studied medium Mn steel was composed of several phases including lath-shape tempered martensite, recrystallized ferrite, pearlite, 29% of ultrafine retained austenite (in both lath-type and equiaxed morphologies), some fresh martensite and undissolved carbides.
3. A core–shell structure of Mn-lean fresh martensite surrounded by Mn-rich austenite film was detected in the final microstructure of the samples. Despite warm deformation, during the intercritical annealing at 650 °C for 20 min, Mn can diffuse only to a short distance, and in the case of austenite grain size larger than the maximum diffusion distance, an Mn-rich layer around an Mn-lean core was created. The peripheral area of Mn enrichment was stabilized, while the core area tended to transform into fresh martensite during final cooling.
4. Employing a warm deformation strain of 40% at 500 °C provided a microstructure exhibiting a wide ferrite grain size range, in addition to a moderate level of recrystallization (i.e., 26%) compared to the microstructural features noticed earlier in the case of samples either deformed at 250 °C or without any deformation. In contrast,

the austenite grains displayed a high fraction (i.e., 73%) of dislocation-free grains in a narrow size range. This was attributed to the formation of new dislocation-free austenite grains, though the partial recrystallization of pre-existed austenite cannot be ruled out.

5. The calculation of M_s and SFE indicated that the RA in the sample deformed at 500 °C had lower stability in comparison to that in the sample deformed at 250 °C and also, the non-deformed sample. This indicates that by varying the deformation temperature, it is possible to finetune the microstructure stability and heterogeneity in terms of size, distribution, phase fraction, and morphology, known to be beneficial for mechanical properties, particularly tensile ductility.

The current results provide a better understanding of the various microstructural evolutions during the processing of MMnS and shed light on advanced microstructure design. It is shown that a judicious combination of several austenite stabilizing methods including warm deformation, pre-existed austenite, and high-temperature partitioning will be advantageous in creating a multiphase and heterogeneous microstructure, to be able to control the strengthening and deformation mechanisms, thereby enhancing the mechanical properties.

Author Contributions: Conceptualization, S.S., M.S. and P.K.; Formal analysis, S.S.; Funding acquisition, M.S. and J.K.; Investigation, S.S.; Methodology, S.S.; Project administration, J.K.; Supervision, J.K. and P.K.; Visualization, S.S. and V.J.; Writing—original draft, S.S.; Writing—review & editing, V.J., M.S. and P.K. All authors have read and agreed to the published version of the manuscript.

Funding: This research was funded by Academy of Finland, grant number 311934.

Institutional Review Board Statement: Not applicable.

Informed Consent Statement: Not applicable.

Conflicts of Interest: The authors declare no conflict of interest.

References

1. De Cooman, B.C.; Lee, S.J.; Shin, S.; Seo, E.J.; Speer, J.G. Combined intercritical annealing and Q&P processing of medium Mn steel. *Metall. Mater. Trans. A Phys. Metall. Mater. Sci.* **2017**, *48*, 39–45. [CrossRef]
2. Lee, Y.K.; Han, J. Current opinion in medium manganese steel. *Mater. Sci. Technol.* **2015**, *31*, 843–856. [CrossRef]
3. Lee, S.; Lee, S.J.; de Cooman, B.C. Austenite stability of ultrafine-grained transformation-induced plasticity steel with Mn partitioning. *Scr. Mater.* **2011**, *65*, 225–228. [CrossRef]
4. Xiong, X.C.; Chen, B.; Huang, M.X.; Wang, J.F.; Wang, L. The effect of morphology on the stability of retained austenite in a quenched and partitioned steel. *Scr. Mater.* **2013**, *68*, 321–324. [CrossRef]
5. Jimenez-Melero, E.; van Dijk, N.H.; Zhao, L.; Sietsma, J.; Offerman, S.E.; Wright, J.P.; van der Zwaag, S. Characterization of individual retained austenite grains and their stability in low-alloyed TRIP steels. *Acta Mater.* **2007**, *55*, 6713–6723. [CrossRef]
6. Yang, H.S.; Bhadeshia, H.K.D.H. Austenite grain size and the martensite-start temperature. *Scr. Mater.* **2009**, *60*, 493–495. [CrossRef]
7. Chen, S.L.; Cao, Z.X.; Wang, C.; Huang, C.X.; Ponge, D.; Cao, W.Q. Effect of volume fraction and mechanical stability of austenite on ductility of medium Mn steel. *J. Iron Steel Res. Int.* **2019**, *26*, 1209–1218. [CrossRef]
8. Han, J.; Lee, S.-J.; Jung, J.-G.; Lee, Y.-K. The effects of the initial martensite microstructure on the microstructure and tensile properties of intercritically annealed Fe–9Mn–0.05C steel. *Acta Mater.* **2014**, *78*, 369–377. [CrossRef]
9. Lee, S.; de Cooman, B.C. Effect of the intercritical annealing temperature on the mechanical properties of 10 Pct Mn multi-phase steel. *Metall. Mater. Trans. A* **2014**, *45*, 5009–5016. [CrossRef]
10. Cao, W.Q.; Wang, C.; Shi, J.; Wang, M.Q.; Hui, W.J.; Dong, H. Microstructure and mechanical properties of Fe–0.2C–5Mn steel processed by ART-annealing. *Mater. Sci. Eng. A* **2011**, *528*, 6661–6666. [CrossRef]
11. Xu, H.F.; Zhao, J.; Cao, W.Q.; Shi, J.; Wang, C.Y.; Wang, C.; Li, J.; Dong, H. Heat treatment effects on the microstructure and mechanical properties of a medium manganese steel (0.2C–5Mn). *Mater. Sci. Eng. A* **2012**, *532*, 435–442. [CrossRef]
12. Han, Q.; Zhang, Y.; Wang, L. Effect of Annealing Time on Microstructural Evolution and Deformation Characteristics in 10Mn1.5Al TRIP Steel. *Metall. Mater. Trans. A* **2015**, *46*, 1917–1926. [CrossRef]
13. Benzing, J.T.; Liu, Y.; Zhang, X.; Luecke, W.E.; Ponge, D.; Dutta, A.; Oskay, C.; Raabe, D.; Wittig, J.E. Experimental and numerical study of mechanical properties of multi-phase medium-Mn TWIP-TRIP steel: Influences of strain rate and phase constituents. *Acta Mater.* **2019**, *177*, 250–265. [CrossRef] [PubMed]

14. Speer, J.G.; Streicher, A.M.; Matlock, D.K.; Rizzo, F.; Krauss, G. Quenching and partitioning: A fundamentally new process to create high strength TRIP sheet microstructures. In Proceedings of the Symposium on the Thermodynamics, Kinetics, Characterization and Modeling of Austenite Formation and Decomposition, Chicago, IL, USA, 9–12 November 2003; pp. 505–522.
15. Speer, J.; Matlock, D.K.; de Cooman, B.C.; Schroth, J.G. Carbon partitioning into austenite after martensite transformation. *Acta Mater.* **2003**, *51*, 2611–2622. [CrossRef]
16. HajyAkbary, F.; Sietsma, J.; Miyamoto, G.; Furuhara, T.; Santofimia, M.J. Interaction of carbon partitioning, carbide precipitation and bainite formation during the Q&P process in a low C steel. *Acta Mater.* **2016**, *104*, 72–83. [CrossRef]
17. Seo, E.J.; Cho, L.; de Cooman, B.C. Application of Quenching and Partitioning Processing to Medium Mn Steel. *Metall. Mater. Trans. A Phys. Metall. Mater. Sci.* **2015**, *46*, 27–31. [CrossRef]
18. Seo, E.J.; Cho, L.; de Cooman, B.C. Kinetics of the partitioning of carbon and substitutional alloying elements during quenching and partitioning (Q&P) processing of medium Mn steel. *Acta Mater.* **2016**, *107*, 354–365. [CrossRef]
19. Ayenampudi, S.; Celada-Casero, C.; Sietsma, J.; Santofimia, M.J. Microstructure evolution during high-temperature partitioning of a medium-Mn quenching and partitioning steel. *Materialia* **2019**, *8*, 100492. [CrossRef]
20. Han, J.; da Silva, A.K.; Ponge, D.; Raabe, D.; Lee, S.M.; Lee, Y.K.; Lee, S.I.; Hwang, B. The effects of prior austenite grain boundaries and microstructural morphology on the impact toughness of intercritically annealed medium Mn steel. *Acta Mater.* **2017**, *122*, 199–206. [CrossRef]
21. Dai, Z.; Chen, H.; Ding, R.; Lu, Q.; Zhang, C.; Yang, Z.; van der Zwaag, S. Fundamentals and application of solid-state phase transformations for advanced high strength steels containing metastable retained austenite. *Mater. Sci. Eng. R Rep.* **2021**, *143*, 100590. [CrossRef]
22. Kim, J.-K.; Kim, J.H.; Suh, D.-W. Partially-recrystallized ferrite grains and multiple plasticity enhancing mechanisms in a medium Mn steel. *Mater. Charact.* **2019**, *155*, 109812. [CrossRef]
23. Bai, S.; Xiao, W.; Niu, W.; Li, D.; Liang, W. Microstructure and Mechanical Properties of a Medium-Mn Steel with 1.3 GPa-Strength and 40%-Ductility. *Materials* **2021**, *14*, 2233. [CrossRef] [PubMed]
24. Yan, S.; Liu, X.; Liang, T.; Zhao, Y. The effects of the initial microstructure on microstructural evolution, mechanical properties and reversed austenite stability of intercritically annealed Fe-6.1Mn-1.5Si-0.12C steel. *Mater. Sci. Eng. A* **2018**, *712*, 332–340. [CrossRef]
25. Hu, B.; Luo, H. A strong and ductile 7Mn steel manufactured by warm rolling and exhibiting both transformation and twinning induced plasticity. *J. Alloys Compd.* **2017**, *725*, 684–693. [CrossRef]
26. He, B.B.; Wang, M.; Huang, M.X. Resetting the austenite stability in a medium Mn steel via dislocation engineering. *Metall. Mater. Trans. A Phys. Metall. Mater. Sci.* **2019**, *50*, 2971–2977. [CrossRef]
27. Liu, L.; He, B.B.; Huang, M.X. Engineering heterogeneous multiphase microstructure by austenite reverted transformation coupled with ferrite transformation. *JOM* **2019**, *71*, 1322–1328. [CrossRef]
28. Chang, Y.; Wang, C.Y.; Zhao, K.M.; Dong, H.; Yan, J.W. An introduction to medium-Mn steel: Metallurgy, mechanical properties and warm stamping process. *Mater. Des.* **2016**, *94*, 424–432. [CrossRef]
29. Li, X.; Chang, Y.; Wang, C.; Hu, P.; Dong, H. Comparison of the hot-stamped boron-alloyed steel and the warm-stamped medium-Mn steel on microstructure and mechanical properties. *Mater. Sci. Eng. A* **2017**, *679*, 240–248. [CrossRef]
30. Tsuchiyama, T.; Inoue, T.; Tobata, J.; Akama, D.; Takaki, S. Microstructure and mechanical properties of a medium manganese steel treated with interrupted quenching and intercritical annealing. *Scr. Mater.* **2016**, *122*, 36–39. [CrossRef]
31. Ding, R.; Dai, Z.; Huang, M.; Yang, Z.; Zhang, C.; Chen, H. Effect of pre-existed austenite on austenite reversion and mechanical behavior of an Fe-0.2C-8Mn-2Al medium Mn steel. *Acta Mater.* **2018**, *147*, 59–69. [CrossRef]
32. He, B.B.; Hu, B.; Yen, H.W.; Cheng, G.J.; Wang, Z.K.; Luo, H.W.; Huang, M.X. High dislocation density-induced large ductility in deformed and partitioned steels. *Science* **2017**, *357*, 1029–1032. [CrossRef] [PubMed]
33. Lu, K. Making strong nanomaterials ductile with gradients. *Science* **2014**, *345*, 1455–1456. [CrossRef]
34. Wu, X.; Zhu, Y. Heterogeneous materials: A new class of materials with unprecedented mechanical properties. *Mater. Res. Lett.* **2017**, *5*, 527–532. [CrossRef]
35. Tasan, C.C.; Diehl, M.; Yan, D.; Zambaldi, C.; Shanthraj, P.; Roters, F.; Raabe, D. Integrated experimental–simulation analysis of stress and strain partitioning in multiphase alloys. *Acta Mater.* **2014**, *81*, 386–400. [CrossRef]
36. Wu, X.; Jiang, P.; Chen, L.; Yuan, F.; Zhu, Y.T. Extraordinary strain hardening by gradient structure. *Proc. Natl. Acad. Sci. USA* **2014**, *111*, 7197–7201. [CrossRef] [PubMed]
37. Wu, X.; Yang, M.; Yuan, F.; Wu, G.; Wei, Y.; Huang, X.; Zhu, Y. Heterogeneous lamella structure unites ultrafine-grain strength with coarse-grain ductility. *Proc. Natl. Acad. Sci. USA* **2015**, *112*, 14501–14505. [CrossRef] [PubMed]
38. Liu, L.; He, B.; Huang, M. The Role of Transformation-Induced Plasticity in the Development of Advanced High Strength Steels. *Adv. Eng. Mater.* **2018**, *20*, 1701083. [CrossRef]
39. Humphreys, J.; Rohrer, G.S.; Rollett, A. *Recrystallization and Related Annealing Phenomena*; Elsevier: Oxford, UK, 2017. [CrossRef]
40. Zhang, J.L.; Tasan, C.C.; Lai, M.J.; Yan, D.; Raabe, D. Partial recrystallization of gum metal to achieve enhanced strength and ductility. *Acta Mater.* **2017**, *135*, 400–410. [CrossRef]
41. Cai, Z.H.; Li, H.Y.; Jing, S.Y.; Li, Z.C.; Ding, H.; Tang, Z.Y.; Misra, R.D.K. Influence of annealing temperature on microstructure and tensile property of cold-rolled Fe-0.2C-11Mn-6Al steel. *Mater. Charact.* **2018**, *137*, 256–262. [CrossRef]
42. Li, X.; Song, R.; Zhou, N.; Li, J. An ultrahigh strength and enhanced ductility cold-rolled medium-Mn steel treated by intercritical annealing. *Scr. Mater.* **2018**, *154*, 30–33. [CrossRef]

43. Cai, M.; Huang, H.; Su, J.; Ding, H.; Hodgson, P.D. Enhanced tensile properties of a reversion annealed 6.5Mn-TRIP alloy via tailoring initial microstructure and cold rolling reduction. *J. Mater. Sci. Technol.* **2018**, *34*, 1428–1435. [CrossRef]
44. Sadeghpour, S.; Somani, M.C.; Kömi, J.; Karjalainen, L.P. A new combinatorial processing route to achieve an ultrafine-grained, multiphase microstructure in a medium Mn steel. *J. Mater. Res. Technol.* **2021**, *15*, 3426–3446. [CrossRef]
45. Field, D.P.; Bradford, L.T.; Nowell, M.M.; Lillo, T.M. The role of annealing twins during recrystallization of Cu. *Acta Mater.* **2007**, *55*, 4233–4241. [CrossRef]
46. Aashranth, B.; Davinci, M.A.; Samantaray, D.; Borah, U.; Albert, S.K. A new critical point on the stress-strain curve: Delineation of dynamic recrystallization from grain growth. *Mater. Des.* **2017**, *116*, 495–503. [CrossRef]
47. Hadadzadeh, A.; Mokdad, F.; Wells, M.A.; Chen, D.L. A new grain orientation spread approach to analyze the dynamic recrystallization behavior of a cast-homogenized Mg-Zn-Zr alloy using electron backscattered diffraction. *Mater. Sci. Eng. A* **2018**, *709*, 285–289. [CrossRef]
48. Padilha, A.F.; Plaut, R.L.; Rios, P.R. Annealing of cold-worked austenitic stainless steels. *ISIJ Int.* **2003**, *43*, 135–143. [CrossRef]
49. Reick, W.; Pohl, M.; Padilha, A.F. Recrystallization & phase Transformation Combined Reactions during Annealing of a Cold Rolled Ferritic/Austenitic Duplex Stainless Steel. *ISIJ Int.* **1998**, *38*, 567–571. [CrossRef]
50. Lee, S.J.; Lee, S.; de Cooman, B.C. Mn partitioning during the intercritical annealing of ultrafine-grained 6% Mn transformation-induced plasticity steel. *Scr. Mater.* **2011**, *64*, 649–652. [CrossRef]
51. Lis, J.; Morgiel, J.; Lis, A. The effect of Mn partitioning in Fe–Mn–Si alloy investigated with STEM-EDS techniques. *Mater. Chem. Phys.* **2003**, *81*, 466–468. [CrossRef]
52. Mohanty, R.R.; Girina, O.A.; Fonstein, N.M. Effect of heating rate on the austenite formation in low-carbon high-strength steels annealed in the intercritical region. *Metall. Mater. Trans. A* **2011**, *42*, 3680. [CrossRef]
53. Luo, H.W.; Qiu, C.H.; Dong, H.; Shi, J. Experimental and numerical analysis of influence of carbide on austenitisation kinetics in 5Mn TRIP steel. *Mater. Sci. Technol.* **2014**, *30*, 1367–1377. [CrossRef]
54. Tokizane, M.; Matsumura, N.; Tsuzaki, K.; Maki, T.; Tamura, I. Recrystallization and formation of austenite in deformed lath martensitic structure of low carbon steels. *Metall. Trans. A* **1982**, *13*, 1379–1388. [CrossRef]
55. Martin, D.; Ryde, L.; Eliasson, J.; Brask, J.; Hutchinson, B. Medium manganese steel and mechanism of austenite formation during reversion annealing. *Steel Res. Int.* **2021**, *92*, 2000381. [CrossRef]
56. Arribas, M.; Gutiérrez, T.; del Molino, E.; Arlazarov, A.; de Diego-Calderón, I.; Martin, D.; de Caro, D.; Ayenampudi, S.; Santofimia, M.J. Austenite reverse transformation in a Q&P route of Mn and Ni added steels. *Metals* **2020**, *10*, 862. [CrossRef]
57. Gramlich, A.; van der Linde, C.; Ackermann, M.; Bleck, W. Effect of molybdenum, aluminium and boron on the phase transformation in 4 wt.–% manganese steels. *Results Mater.* **2020**, *8*, 100147. [CrossRef]
58. Emami, M.; Askari-Paykani, M.; Farabi, E.; Beladi, H.; Shahverdi, H.R. Development of new third-generation medium manganese advanced high-strength steels elaborating hot-rolling and intercritical annealing. *Metall. Mater. Trans. A Phys. Metall. Mater. Sci.* **2019**, *50*, 4261–4274. [CrossRef]

Article

Casting and Characterization of A319 Aluminum Alloy Reinforced with Graphene Using Hybrid Semi-Solid Stirring and Ultrasonic Processing

Bernoulli Andilab *[ ], Payam Emadi and Comondore Ravindran

Centre for Near-Net-Shape Processing of Materials, Toronto Metropolitan University,
Toronto, ON M5B 2K3, Canada
* Correspondence: bandilab@ryerson.ca

Abstract: Advanced metallurgical processing techniques are required to produce aluminum matrix composites due to the tendency of the reinforcement particles to agglomerate. In this study, graphene nano-platelet reinforcement particles were effectively incorporated into an automotive A319 aluminum alloy matrix using a liquid metallurgical route. Due to its low density, it is a highly difficult task to produce an aluminum matrix composite reinforced with graphene. Hence, this study explored a novel approach to prevent particle floating to the melt surface and agglomeration. This was achieved via a hybrid semi-solid stirring of A319, followed by ultrasonic treatment of the liquid melt using a sonication probe. The microstructure and graphene particles were characterized using optical microscopy and scanning electron microscopy. Furthermore, the interfacial products produced with the incorporation of graphene in liquid aluminum were analyzed with X-ray diffraction. The tensile test results exhibited 10, 11 and 32% improvements in ultimate tensile strength, yield strength, and ductility of A319 reinforced with 0.05 wt.% addition of graphene. Analysis of strengthening models demonstrated primary contribution from Hall-Petch followed by CTE mismatch and load bearing mechanism. The results from this research enable the potential for using cost-effective, efficient and simple liquid metallurgy methods to produce aluminum reinforced graphene composites with improved mechanical properties.

Keywords: aluminum alloys; metal matrix composites; graphene; mechanical properties; metalcasting; strengthening mechanisms

Citation: Andilab, B.; Emadi, P.; Ravindran, C. Casting and Characterization of A319 Aluminum Alloy Reinforced with Graphene Using Hybrid Semi-Solid Stirring and Ultrasonic Processing. *Materials* **2022**, *15*, 7232. https://doi.org/10.3390/ma15207232

Academic Editor: Konstantin Borodianskiy

Received: 30 September 2022
Accepted: 13 October 2022
Published: 17 October 2022

Publisher's Note: MDPI stays neutral with regard to jurisdictional claims in published maps and institutional affiliations.

Copyright: © 2022 by the authors. Licensee MDPI, Basel, Switzerland. This article is an open access article distributed under the terms and conditions of the Creative Commons Attribution (CC BY) license (https://creativecommons.org/licenses/by/4.0/).

1. Introduction

High-strength aluminum (Al) alloys, such as the A319 alloy, are commonly used in the production of lightweight and high-performance automotive powertrain components. Continuous efforts are made by researchers to improve the properties of Al alloys including through the use of modification, grain refinement and heat treatment. An alternative strengthening method is through the reinforcement of metals with hard ceramic particles has gained impetus to produce metal matrix composites (MMCs). MMCs are manufactured by dispersing the reinforcements in the metal matrix. The reinforcements are typically applied to improve the properties of the base metal such as strength, ductility, and thermal conductivity. Aluminum and its alloys have attracted much attention as a base metal in MMCs [1,2]. Aluminum metal matrix composites (AMCs) are widely used as structural components in aircraft, aerospace, automobiles and various other fields [3]. The most commonly used reinforcements are silicon carbide (SiC), aluminum oxide (Al_2O_3) and titanium boride (TiB_2). These reinforcements have been found to increase the tensile strength, hardness, density and wear resistance of Al and its alloys [3].

In recent years, graphene has emerged as a material with high potential as a reinforcement particle for AMCs. Graphene is an atomic-scale honeycomb lattice made of carbon atoms. It has unique properties such as high strength and thermal conductivity [4,5].

Graphene is comparatively a newer type of carbon-based nanoparticles. Graphene can be effectively used as reinforcement for structural composites due to its unique characteristics (2-D single-atom-sheet sp^2 hybridized carbon atoms) and significantly enhanced mechanical and thermal properties [6].

The particle distribution plays a very vital role in the properties of the AMCs, which is improved by intensive shearing of the liquid melt or rigorous mixing of metallic powders. As a result, appropriate processing of AMCs reinforced with graphene is key in developing a material with enhanced properties. Optimized dispersion of graphene reinforcement in the Al matrix plays a critical role in improving the properties of AMCs. This is difficult to achieve because of the strong van der Waals forces between graphene particles, high specific surface area (2630 m^2/g) and low density of graphene (~1.8–2.2 g/cm^3) compared to Al (~2.7 g/cm^3). As a result, various processing techniques have been explored to determine the optimal method for incorporating graphene into Al.

The most commonly used processing route for the production of graphene reinforced AMCs is powder metallurgy [7–9]. However, it should be noted that powder metallurgy processes are typically lengthy and involve several steps, including final processing such as rolling or extrusion. For example, Bhaudaria et al. [3] investigated the combined effect of a nanocrystalline matrix and reinforcement with 0.5 wt.% graphene nano-platelets (GNPs) on pure Al using ball milling and spark plasma sintering. It was found that the ultimate tensile strength improved by 85% compared to that of pure Al. However, the increase in strength was accompanied by a significant decrease in ductility [3]. The decrease in ductility was attributed to cracking at the Al-graphene interface. It is also well known that agglomeration is an issue during processing of AMCs reinforced with graphene, which leads to porosity and a reduction in mechanical properties [10,11]. The results demonstrate excellent potential due to the significant strengthening enhancement, however the process involves multiple methods which can increase time and cost. Therefore, while powder metallurgy can enable the production of AMCs with enhanced mechanical properties, it may not be practical for industrial applications due to time and cost constraints.

As an alternative, liquid metallurgy techniques such as casting have been explored as a processing route to produce graphene reinforced AMCs. However, stir casting is not typically used compared to powder metallurgy techniques when manufacturing AMCs reinforced with graphene. This is due to the density difference between graphene and Al, which results in the floating of graphene particles at or near the melt surface. This leads to severe losses of graphene particles during melt treatment and casting. Further, the poor wettability between graphene and Al can inhibit the interfacial interaction which leads to agglomeration, thereby deteriorating the mechanical properties of the composite. Nonetheless, some studies have shown success in reinforcing Al with graphene. Liquid metallurgy processing techniques such as casting are of interest to the industry because they are practical, while enabling mass production capability. This is especially important for the automotive industry, where powder metallurgy techniques are less feasible.

Alipour and Eslami-Farsani [12] reported an optimal increase in tensile strength from 263 to 372 MPa after 0.5 wt.% addition of GNPs in an AA7068 Al alloy through a stir casting process. The main mechanism for this improvement in tensile strength was due to the grain refinement, according to the authors. The authors also found that beyond 0.5 wt.% addition, severe agglomeration of GNPs occurred. Similarly, Srivistava et al. [13] used the stir casting method to reinforce AA1100 Al alloy with GNPs. The results showed an increase in tensile strength and yield strength with increasing addition of GNPs up to 1.2%. In contrast, the elongation was negatively affected resulting in a brittle failure mode. Of note, while these studies implemented the stir casting technique, preprocessing via ball milling was required to prepare Al-graphene master alloys in order to incorporate graphene into the liquid Al melt. In the current study, the authors present a method which enables the incorporation of graphene without the requirement of preprocessing.

An alternative liquid metallurgy technique that has been successfully implemented to reinforce Al with graphene is via the pressure infiltration method [14,15]. It was found that the tensile strength of an Al-20Si alloy can be improved by 230% using the pressure infiltration method to reinforce 1.5 wt.% of graphene into the matrix [14]. This is also an effective method, however certain components such as automotive engine blocks and cylinder heads are not produced using pressure infiltration due to their complex geometry. Hence, there is a need for further exploration of methods that enable the production of high-strength AMCs reinforced with graphene.

Conventional casting and pressure infiltration have been implemented individually to reinforce graphene into an Al matrix. To that end, there is also potential to use multiple methods, such as semi-solid stirring and ultrasonic stirring, to achieve graphene reinforcement with excellent properties [16]. For example, Boostani et al. [17–19] implemented the combination of semi-solid stirring and non-contact ultrasonic processing to reinforce SiC encapsulated with graphene to enhance the mechanical properties of A356 Al alloy. However, the effect of the processing parameters can be further explored with varying graphene addition, as well as the effect of graphene on alternative aluminum alloys such as A319 Al alloy. As such, the processing of graphene in liquid metals is still in development, especially for varying alloys. Moreover, the current work by the authors implements the utilization of direct contact isothermal ultrasonic stirring with a sonication probe, which may be more suitable for industry applications. Furthermore, it is difficult to isolate the individual strengthening contribution of graphene due to the effect of SiC. Hence, there is potential to explore the individual effects of graphene on the mechanical properties of aluminum alloys.

In the present work, hybrid semi-solid stirring and ultrasonic processing is successfully implemented to reinforce graphene nano-platelets (GNPs) in A319 Al alloy for automotive cylinder heads and engine blocks. To the authors' knowledge, reinforcement of automotive A319 Al alloy with graphene has not been comprehensively investigated. In contrast to past works, this study demonstrates the isolated effects of semi-solid stirring and direct-contact ultrasonic stirring, as well as the combined effects. Furthermore, the results reveal the individual strengthening contribution of graphene, without the aid of a secondary reinforcement particle. The use of pre-processing techniques, such as ball-milling, were also not required to incorporate graphene in the liquid melt. Most studies involve the use of preprocessing methods such as ball-milling to first manufacture a master-alloy composite prior to casting. This has been found to be successful, although they typically increase the manufacturing time and cost. While there has been significant progress in research involving the incorporation of graphene as a reinforcement particle for Al using liquid metallurgy processes, there is still potential for improvement especially when considering automotive applications as the industry moves towards sustainable electric vehicles.

In this study, a simple and innovative, hybrid semi-solid stirring and ultrasonic processing is implemented to produce an A319/GNP reinforced composite with improved mechanical properties without the need for complex preprocessing methods. A comprehensive analysis on the individual effects of semi-solid stirring and ultrasonic processing on A319 was investigated. Subsequently, the combined effects were then explored and compared to the individual processes. This was achieved using optical microscopy, scanning electron microscopy, X-ray diffraction and tensile tests. The strengthening mechanisms were also studied using existing mathematical models and validated experimentally. The results from this research enable the potential for using cost effective, efficient and simple liquid metallurgy methods to produce aluminum reinforced graphene composites with improved mechanical properties.

2. Materials and Methods

Castings of A319-graphene MMCs were produced at the Centre for Near-net-shape Processing of Materials, Toronto, ON, Canada. Graphene particles (graphene nano-platelets) were obtained from XGSciences, Lansing, MI, USA for the production of high-strength GNP reinforced MMCs. The GNPs had an average diameter of 1 µm and thickness of 15 nm, according to the supplier data sheets. In this study, five experimental casting conditions were used and they involved semi-solid processing (SS) or ultrasonic processing (UST). The material designations for the experimental conditions are summarized in Table 1.

Table 1. Summary of casting conditions and designation.

Condition	GNP Addition (wt.%)	Semi-Solid (SS)	Ultrasonic Processing (UST)
A319	-	-	-
A319 + UST	-	-	*
A319 + 0.05 GNP + SS + UST	0.05	*	*
A319 + 0.05 GNP + SS	0.05	*	-
A319 + 0.1 GNP + SS	0.1	*	-

* indicates that the process was applied.

For each casting, approximately 1.5 kg of the alloy was melted at 750 °C in a SiC crucible using an electric resistance furnace. Subsequently, a ASTM B108-19 permanent tensile mold was preheated to 300 °C. A combination of semi-solid stirring and ultrasonic stirring was used to incorporate GNPs in the A319 melt. Once the alloy was molten, the furnace was cooled to 590 °C and held at this temperature to allow the A319 alloy to reach semi-solid state. This corresponds to a 0.2 solid fraction for the A319 alloy [20]. Subsequently, the GNPs were preheated to approximately 200 °C to aid in wetting of the particles with liquid Al and remove any moisture [3]. Following this, the GNPs were submerged into the semi-solid alloy and stirred mechanically with an impeller at 1000 RPM for approximately 10 min. Afterwards, the furnace temperature was raised back to 750 °C to allow the alloy to reach molten state. Upon reaching this temperature, ultrasonic treatment was applied directly to the melt using a Sonic Systems (Ilminster, UK) ultrasonic vibration device (model L500) with a frequency of 20 ± 1 kHz. For these experiments the vibrational amplitude was fixed at 24 µm, which was applied to the melt using a titanium alloy sonotrode for approximately 3 min. The ultrasonic treatment was carried out at approximately 750 °C and the sonotrode was immersed in the melt at least 10 mm below the melt surface. Following this, the melt was poured at a temperature of 720 °C into the preheated tensile mold. The chemical composition was determined from the cast samples using an Oxford Instruments Foundry-Master Pro (obtained from NDT Products Ltd., St. Catharines, ON, Canada) optical emission spectrometer and is given in Table 2.

Table 2. Average chemical composition of A319 alloy used in this study.

Element	Si	Cu	Mg	Ti	Fe	Mn	Zn	Ni	Al
wt.%	6.21	2.87	0.04	0.11	0.58	0.34	0.61	0.18	Bal.

After casting, tensile testing was performed on samples using a Instron United Universal Testing Machine Model STM-50kN (Instron, Massachusetts, United States) according to ASTM B 557M-15. Tensile tests on the dog bone shaped samples (12.95 mm diameter and 50.8 mm gauge length) were carried out with a cross head speed of 12 mm/min at ambient temperature. An extensometer was used to measure strain. After tensile testing, the samples were sectioned for fractography and microstructural analysis. Prior to metallographic preparation, the density of each of the casting conditions was measured using the Archimedes principle on the sectioned samples.

A Nikon Eclipse MA200 (obtained from Buehler, Lake Bluff, IL, United States) metallurgical optical microscope was used to analyze the as-cast grain structure. The samples were prepared for metallography by successive grinding steps using varying levels of SiC papers and subsequently polished using 5 µm alumina, 3 µm diamond suspension, 1 µm diamond suspension and 0.02 µm colloidal silica. The secondary dendrite arm spacing (SDAS) was measured using image analysis with Clemex Vision Pro software (Version Vision PE, Clemex Technologies Inc., Guimond, Longueuil, Quebec, Canada) and the linear intercept method. Furthermore, the GNP particles were examined using a JEOL JSM-6380LV scanning electron microscope (SEM) at 20 keV. Phase microanalysis was carried out using energy dispersive X-ray spectroscopy (EDX). The secondary phases and interfacial reaction products were identified by X-ray diffraction (X'Pert Pro PANalytical X-ray diffractometer). The samples were scanned at 0.05° for 2 s using monochromated Cu K_α radiation at 40 mA and 45 kV.

3. Results
3.1. Microstructural Characterization
3.1.1. As-Cast Microstructure

Optical microscopy and scanning electron microscopy were used to examine the as-cast grain structure and secondary phases present in the matrix. The grain structure of the castings was found to be dendritic. Furthermore, the as-cast microstructure of the A319 Al alloy consists of α-Al, eutectic Si, θ-Al_2Cu, and $Al_{15}(Fe,Mn)_3Si_2$ phases which were distributed within the interdendritic regions. The optical micrographs of the base A319 alloy can be seen in Figure 1. The phases were also confirmed with XRD as shown in Figure 5. In this study, it was found that the addition of GNPs demonstrated minor changes on the grain structure or the morphology of any secondary phases. The eutectic Si particles remained as coarse acicular flakes and the Al_2Cu phases were observed in both blocky and lamellar morphologies.

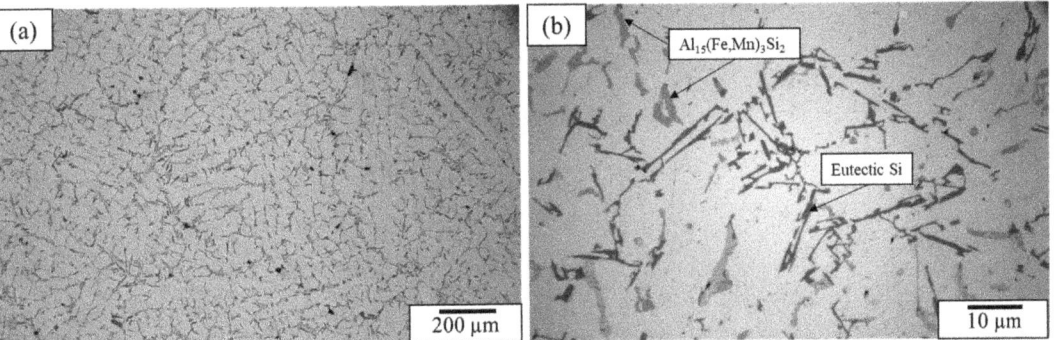

Figure 1. Optical micrographs of the A319 alloy captured at (**a**) 200× and (**b**) 1000× magnification.

The SDAS decreased slightly with 0.05 wt.% GNP addition (A319 + 0.05 GNP + SS + UST) to 16.0 ± 0.8 μm compared to 19.8 ± 2.4 μm with the base A319 casting (Table 3). This reduction in SDAS demonstrates successful incorporation of GNPs, since the addition of these particles promote grain refinement. Compared to the other conditions, the SDAS remained relatively the same with the highest SDAS reaching 22.2 ± 0.5 μm for the A319 + 0.05 GNP + SS composite. The experiments involving semi-solid stirring alone, resulted in severe particle agglomeration, hence the GNPs were not effectively incorporated in the melt. Similarly, the SDAS was not refined for the A319 + 0.05 GNP + SS and A319 + 0.1 GNP + SS composites demonstrating values of 22.2 ± 0.5 and 20.8 ± 0.6 μm, respectively.

Table 3. Summary of average SDAS measurements for casting conditions.

Sample	SDAS (μm)
A319	19.8 ± 2.4
A319 + UST	21.8 ± 2.0
A319 + 0.05 GNP + SS + UST	16.0 ± 0.8
A319 + 0.05 GNP + SS	22.2 ± 0.5
A319 + 0.1 GNP + SS	20.8 ± 0.6

3.1.2. Observation of Graphene Nano-Platelets

Casting experiments were performed to ensure that GNPs were embedded into A319 Al alloy for the production of A319 reinforced GNP composites. For this purpose, SEM analysis was carried out on the samples containing GNPs. Figure 2 demonstrates the general microstructure of the A319 + 0.05 GNP + SS + UST composite. The primary secondary phases observed include Al_2Cu and Fe-bearing phases. Eutectic Si was also present, as previously shown in Figure 1. However, under the SEM, eutectic Si becomes difficult to observe due to low atomic number. The results show that the Al_2Cu phases were homogeneously distributed, however they appear to be slightly fragmented as micro-spherical forms of Al_2Cu are present near the larger blocky phases. This is may be a result of the sonication effect, which is known to refine secondary phases due to cavitation induced by high frequency vibration.

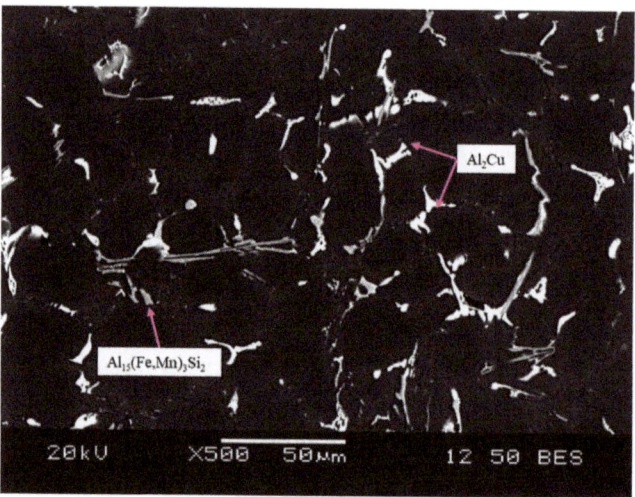

Figure 2. Backscattered SEM image of A319 + 0.05 GNP + SS + UST microstructure at 500×.

Additionally, GNP particles were observed in the matrix of the A319 Al alloy using SEM as seen in Figure 3. The GNPs were embedded longitudinal to the matrix examined under the SEM. Although, a number of particles appeared to be embedded at an angle to the surface (Figure 3b), potentially indicating a different plane. Furthermore, the analysis revealed that the GNP particles were typically observed near a eutectic Si phase as seen in Figures 4 and 5. Since, A319 predominantly consists of eutectic Si this may be related to the solidification mechanism whereby the GNPs are pushed into the interdendritic regions. Energy dispersive X-ray spectroscopy (EDX) was also performed to confirm the presence of GNP particles. This can be seen in Figure 4, where a different region showing the presence of GNPs is presented with a corresponding EDX linescan of a GNP particle (Figure 4b). The EDX linescan initially demonstrates an increase in energy corresponding to a eutectic Si particle with increasing distance, which is then followed by a peak in C energy that follows the contour of the GNP particle (platelet morphology).

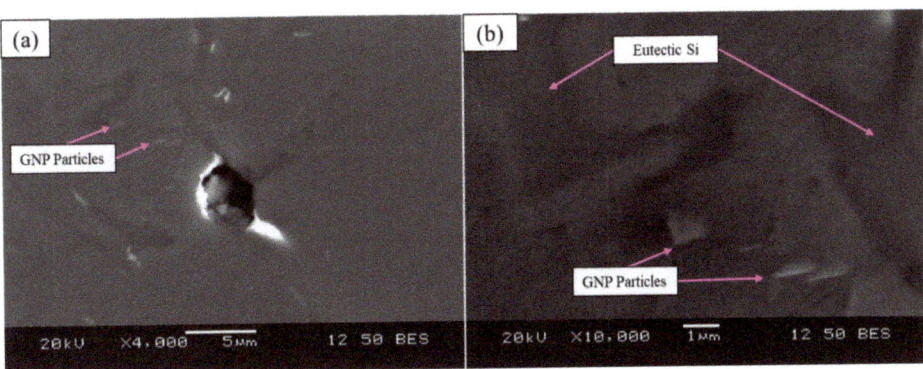

Figure 3. Backscattered SEM images of A319 + 0.05 GNP + SS + UST composite showing graphene nano-platelets embedded in Al matrix near eutectic Si at (**a**) 4000× and (**b**) 10,000× magnification.

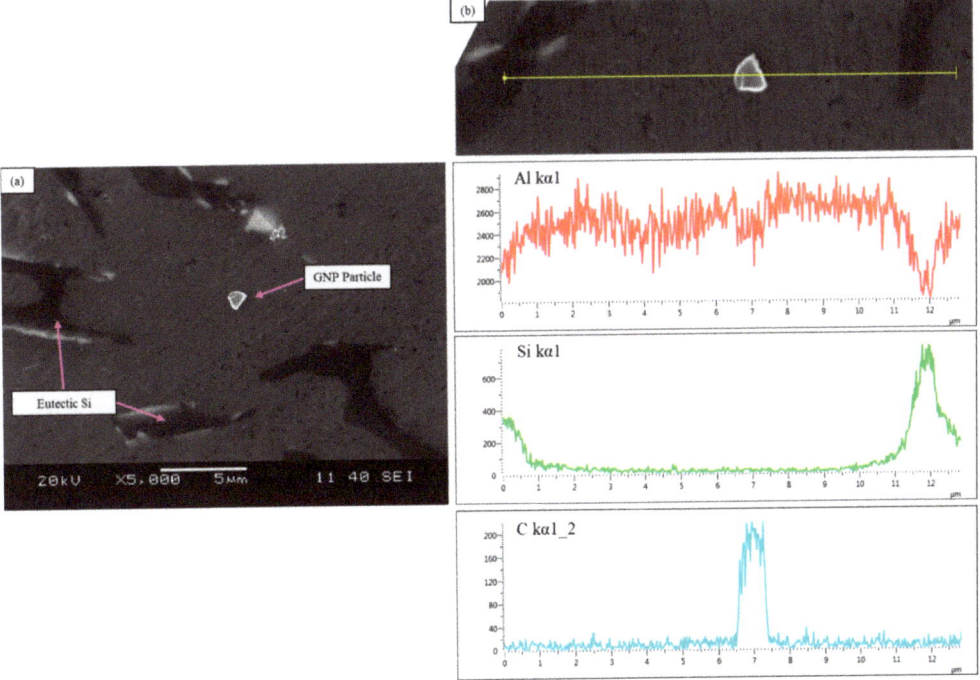

Figure 4. (**a**) Secondary electron image of GNP particle and (**b**) corresponding EDX linescan for A319 + 0.05 GNP + SS + UST composite.

The addition of GNPs can also promote the formation of interfacial products such as Al_4C_3. The high temperature processing, coupled with a relatively long processing time promotes breaking down of C atoms from graphene. Hence, it is thermodynamically favorable that a reaction can occur between C and available Al in the melt to produce Al_4C_3. X-ray diffraction was implemented to confirm the presence of Al_4C_3 with the addition of GNPs (Figure 5), since it is known that Al_4C_3 can deteriorate the mechanical properties of aluminum matrix composites. This was observed in a sample from the A319 + 0.05G + SS + UST condition, where a peak was identified at 43° (Figure 5b). Compared to the base

A319 casting, this peak does not occur at the same angle as seen in Figure 5a. Previous research has reported Al_4C_3 formation with diffraction angles of 31, 32, 33, 40 and 43° [21]. Similarly, this also correlates well to the findings by Alipour et al. [12] where the authors observed a peak at 43° corresponding to GNP addition, although it was not explicitly identified as Al_4C_3. However, identification of Al_4C_3 is often difficult even with the use of XRD. In this case, Al_4C_3 was occasionally observed in limited quantity according to the XRD results. This was likely due to the very low addition of GNP particles. Additionally, localized inhomogeneity could be a potential cause of the observation of Al_4C_3 in very limited quantity. Of note, it should be mentioned that the corresponding Al_4C_3 peak (43°) closely overlaps with the $Al_{15}(Fe,Mn)_3Si_2$ secondary phase, while it is possible for the Al_4C_3 interfacial product to reduce the mechanical properties, this was not observed in this study.

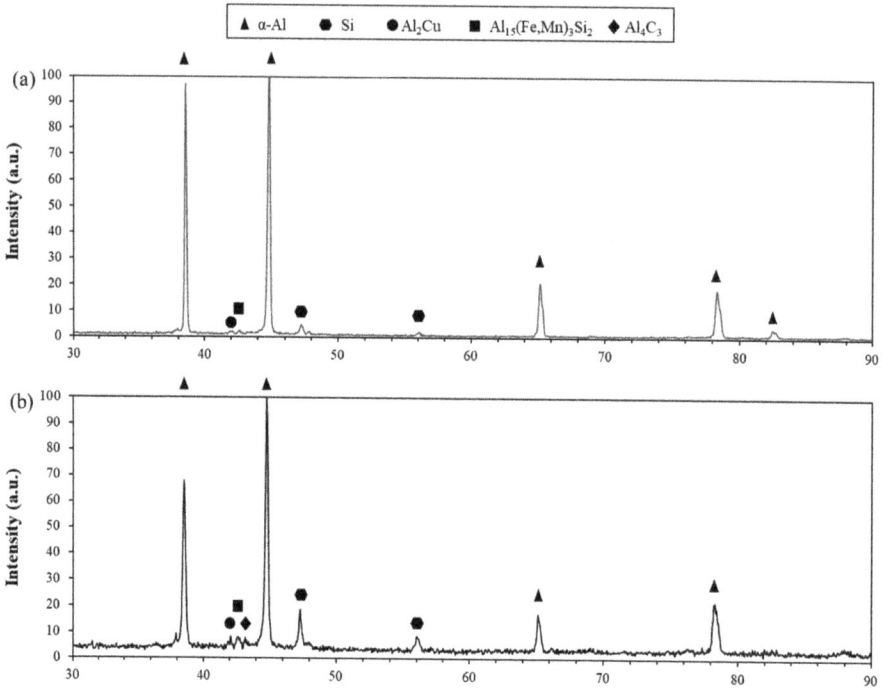

Figure 5. XRD analysis of (**a**) A319 base alloy and (**b**) A319 + 0.05 GNP + SS + UST composite.

3.2. Mechanical Properties

In this study, tensile testing was carried out to determine the mechanical properties of the composite castings. The mechanical properties of the casting conditions are shown in Figure 6. The base A319 condition exhibited a UTS of approximately 190 MPa. A319 with ultrasonic processing was also carried out to isolate the effects of ultrasonic processing on the microstructure and mechanical properties. It was found that there was a minor increase in UTS to 195 MPa for the A319 + UST samples. This can be attributed to the degassing effect and inclusion removal of ultrasonic processing, thereby promoting a cleaner melt through the removal of hydrogen gas and oxides. Experiments were also carried out to demonstrate the effect of semi-solid processing on incorporating GNPs in the melt. This was done for additions of 0.05 wt.% and 0.1 wt.% GNPs. The main purpose of using semi-solid processing was to prevent GNP particles from floating to the melt surface. This was found to be an effective method in preventing particle floating, however the results demonstrated a reduction in UTS to 180 and 176 MPa for the A319 + 0.05 GNP + SS and A319 + 0.1 GNP +

SS samples, respectively. Conversely, combining both ultrasonic processing and semi-solid processing resulted in an increase in UTS for an addition of 0.05 wt.% GNPs. The A319 + 0.05 GNP + SS + UST samples resulted in a UTS of 209 MPa. This was attributed to the combined effects of semi-solid processing in preventing particle floating and the ability of ultrasonic processing to de-agglomerate and distribute GNPs in the melt.

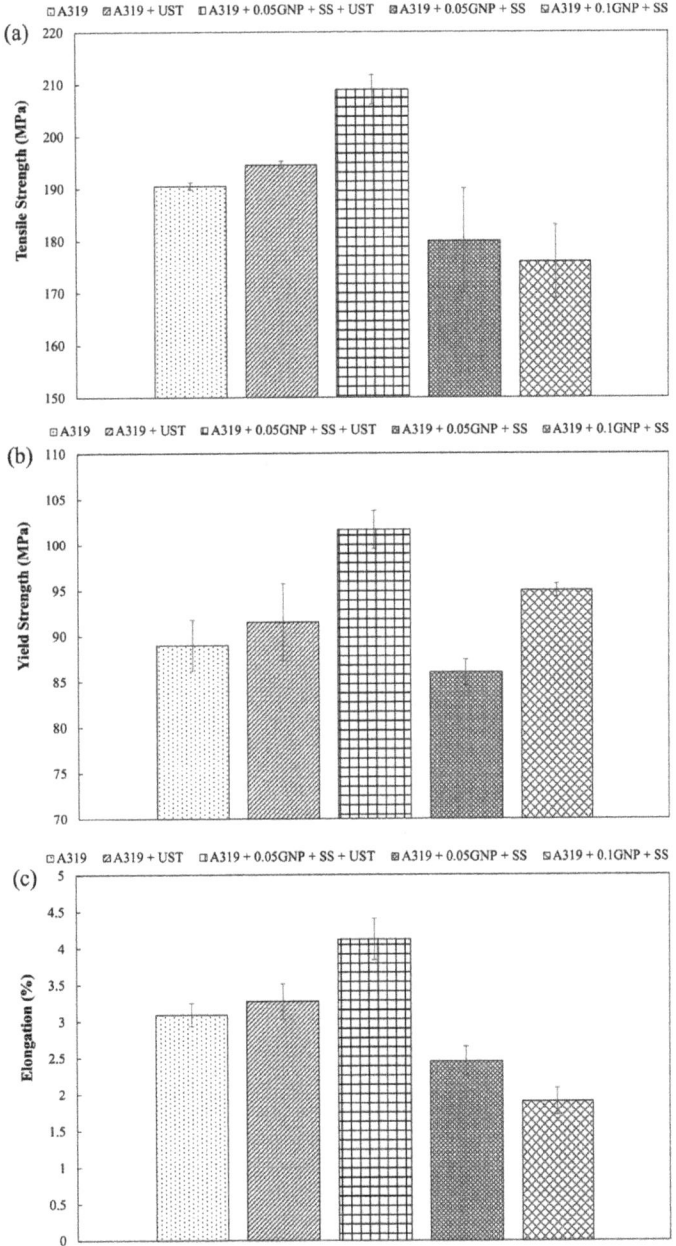

Figure 6. (a) Ultimate tensile strength, (b) yield strength and (c) elongation of cast samples.

In addition to the findings on the UTS, the yield strength was also the highest for the A319 + 0.05 GNP + SS + UST samples. The YS for these samples was measured at approximately 119 MPa compared to the base A319 cast samples, which demonstrated a YS of 107 MPa. As shown in Figure 6b, the other conditions resulted in similar YS values to the base condition and did not show improvements. The ductility of the conditions also followed a similar trend to the UTS results, displayed in Figure 6c. The ductility of the base A319 condition was measured at 3.1%. Additionally, the A319 + UST samples exhibited a slight increase in elongation at 3.3%, likely due to its degassing effect. While degassing is likely taking place during melt treatment, ultrasonic processing was not found to refine the grains as previously discussed. However, for the conditions involving semi-solid processing individually produced samples with reduced ductility compared to the base A319 casting. With semi-solid processing the A319-0.05 GNP + SS and A319-0.1 GNP + SS, the ductility reduced to approximately 2.4 and 1.9%, respectively, indicating a brittle fracture behavior. In contrast, the results showed an increase in ductility to 4.1% for the A319 + 0.05 GNP + SS + UST composite using the combination of semi-solid and ultrasonic processing.

The results suggest that while semi-solid processing can prevent floating of GNP particles to the melt surface, the process is not effective in de-agglomerating the GNP particles. Hence, ultrasonic processing is required de-agglomerate and distribute GNPs in the melt. As demonstrated by the UTS and YS results, the combination of ultrasonic processing and semi-solid processing demonstrates potential for producing AMCs reinforced with graphene with improved mechanical properties. In summary, the tensile test results demonstrated 10, 11 and 32% improvements in UTS, YS, and ductility of A319 reinforced with 0.05 wt.% addition of graphene through using hybrid semi-solid and ultrasonic processing.

3.3. Strengthening Models

In this study, theoretical models were used to determine the mechanisms for enhancing YS of the A319 + 0.05 GNP + SS + UST composite. A modified shear-lag and enhanced dislocation density model, developed by Fadavi Boostani et al. [18] was used to accurately predict the YS and the subsequent strengthening contributions as seen in Equation (1). It should be noted that due to the size of the particles used in the study, that the Orowan strengthening component was neglected.

$$\sigma_Y^{Gr} = \sigma_Y^{Al}(1 + \omega_L)(1 + \omega_T) + \sigma_Y^{Hall-Petch} \tag{1}$$

Equation (2) is composed of the strengthening effects in the form of load bearing (ω_L), thermally induced dislocations (ω_T), and Hall-Petch mechanisms on the yield strength of the A319 alloy used in this study to approximate the yield strength of A319 + 0.05 GNP + SS + UST. The Hall-Petch strengthening was approximated using Equation (2).

$$\sigma_Y^{Hall-Petch} = \sigma_i + kD^{-\frac{1}{2}} \tag{2}$$

where σ_i and k are the intrinsic stress (σ_i = 15.7 MPa) and material constant (k = 0.068 MPa/M$^{0.5}$) [22]. Using the measured SDAS in Table 3 for A319 + 0.05 GNP + SS + UST (D = 16 µm), the strengthening contribution from the Hall-Petch relationship exhibits a 16.2 MPa enhancement in YS. Of note, SDAS was measured for the samples in this study which has been found to accurately follow the Hall-Petch relationship in Al-Si cast alloys according to Ghasemalli et al. [23].

The load transfer from the Al matrix to graphene particles is known to contribute to the strengthening of the matrix material, this phenomenon can be expressed by Equation (3). This model is modified to account for the platelet-like morphology of graphene sheets or graphene nano-platelets. Hence, the load bearing parameter is defined as:

$$\omega_L = \sqrt{\frac{2G_m L^2}{E_f \lambda_{eff} t}} \tag{3}$$

where G_m, E_f, L, and t are the matrix shear modulus, Young's modulus, particle length and particle thickness, respectively. For the load bearing phenomenon, it is also important to take into account the inter-particle spacing, which is defined here as λ_{eff}:

$$\lambda_{eff} = 0.931\sqrt{\frac{0.306\pi dt}{V_{Gr}} - \frac{\pi d}{8}} - 1.061t \quad (4)$$

where d, t, and V_{Gr} are the particle length, thickness, and volume fraction, respectively. Using the values $d = 1$ µm, $t = 15$ nm, and $V_{Gr} = 0.00063$ results in an inter-particle spacing $\lambda_{eff} = 4.59$ µm. Then, using the calculated theoretical inter-particle spacing (λ_{eff}) value and inserting into Equation (3), $\omega_L = 0.03$.

There is a substantial difference in the coefficient of thermal expansion (CTE) of graphene and aluminum. This leads to an increase in dislocation density as a result of a plastic strain developed from the large ΔCTE. This can be expressed by Equation (5):

$$\omega_T = \frac{1.25 G_m b}{\sigma_Y^{Al}}\sqrt{\frac{12(T_{Fabrication} - T_{Test})(\alpha_m - \alpha_{Gr})V_{Gr}}{bd(1 - V_{Gr})}} \quad (5)$$

where G_m, b, d and σ_Y^{Al} are the matrix shear-modulus, the burgers vector, the diameter of graphene particles and the yield strength of the unreinforced A319 alloy, respectively. $T_{Fabrication}$ is the processing temperature (750 °C), while T_{Test} is the test temperature (300 °C). Using the values from Table 4, $\omega_T = 0.045$.

Table 4. Parameters used for calculating strengthening contributions [4,22,24].

Parameter	Unit	Al	Graphene
Burgers vector (b)	nm	0.25	-
Thermal expansion coefficient (α)	×10⁻⁶/°C	21.4	>−6
Shear modulus (G)	MPa	25	250
Young's modulus	GPa	70	1000
Intrinsic stress (σ_i)	MPa	15.7	-
Material constant (k)	MPa/M$^{0.5}$	0.068	-

Inserting the values $\omega_L = 0.03$, $\omega_T = 0.045$ and the YS enhancement from the Hall-Petch relationship into Equation (1) results in a total theoretical YS of 130 MPa. This is indeed close to the experimental YS value of 119 MPa for the A319 + 0.05 GNP + SS + UST composite, as previously discussed. This is approximately an 8.5% deviation from the theoretical YS determined from the model discussed. Hence, the mathematical models can accurately predict YS of AMCs reinforced with graphene.

3.4. Fracture Behavior

Fractography using SEM was carried out after tensile testing the cast samples. Figure 7a,b present the fractographic images of the A319-0.05 GNP + SS + UST and the A319-0.1 GNP + SS samples, identified as the best and worst conditions in this study. The A319-0.05 GNP + SS + UST samples exhibited a UTS, YS, and ductility of 209 MPa, 119 MPa, and 4.1%, respectively, as previously discussed. As a result, the fracture surface of these samples exhibited a ductile behavior compared to the other conditions as demonstrated by the dimples across surface (Figure 7a). This suggests that the GNPs displayed good interfacial bonding at the Al interface. Hence, this also contributed to the ductile fracture behavior as de-lamination of GNPs or micro-cracks were not observed. Fine cleavage planes can also be observed across the sample fracture surface. The A319-0.1 GNP + SS samples displayed UTS, YS, and ductility values of 176 MPa, 111 MPa, and 1.9%, respectively. The results show that the fracture surface exhibits a brittle fracture, presented in Figure 7b. This was due to agglomeration of GNP particles which resulted in a large void, which was a site for crack initiation and propagation. Additionally, tear ridges can be observed on the fracture surface which also show the direction of the crack propagation, whereby the fracture propagates away from the void region where the GNPs agglomerated. The results suggest that semi-solid processing should be accompanied by subsequent ultrasonic processing when adding the GNP particles. The semi-solid processing inhibits floating of particles to the melt surface, however during this process the particles have a tendency to agglomerate. Hence, following up this process with ultrasonic processing enables de-agglomeration of the GNP particles as shown in Figure 7.

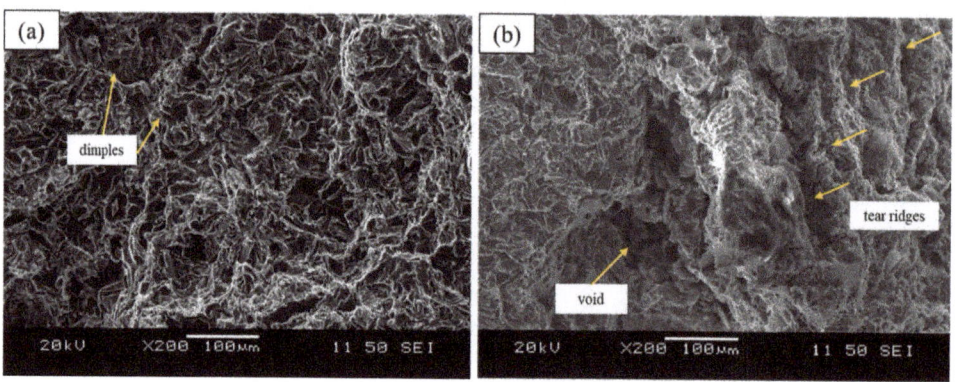

Figure 7. Fracture surface of (**a**) A319-0.05 GNP + SS + UST and (**b**) A319-0.1 GNP + SS samples.

4. Discussion

This study explores a processing route using a liquid metallurgy method to produce a A319 graphene reinforced MMC. The purpose of this study is to determine a process to effectively insert GNPs in automotive A319 Al alloy through liquid metallurgy, specifically casting. The results demonstrate that ultrasonic processing and semi-solid processing individually cannot effectively distribute GNPs in the liquid melt. Hence, a combination of both ultrasonic processing and semi-solid processing was performed to achieve effective addition of GNP particles. In order to demonstrate the effectiveness of combining both methods, casting conditions that isolate each process and the base condition were carried out. As a result, an A319-0.05 wt.% GNP composite (A319 + 0.05 GNP + SS + UST) was successfully produced with improved mechanical properties. This was achieved via a simple and innovative hybrid semi-solid and ultrasonic processing method. In order to demonstrate the synergistic effect of the hybrid method, castings of base A319, A319 with UST, and A319 with SS processing were also produced. Hence, these conditions demonstrate the isolated effects of UST and SS while also comparing to the A319 alloy without additions.

4.1. Effect of Processing and GNP Addition on A319 As-Cast Microstructure

The SDAS measurements of A319, A319 + UST, A319 + 0.05 GNP + SS, and A319 + 0.1 GNP + SS were measured as 19.8 ± 2.4, 21.8 ± 2.0, 22.2 ± 0.5, and 20.8 ± 0.6 µm, respectively (Table 3). Between these conditions the SDAS remained relatively unchanged. This indicated that UST processing alone did not refine the A319 alloy. Furthermore, the castings involving SS processing and GNP additions also did not show refinement in SDAS. While it is expected that GNP additions promote refinement of the alloy microstructure, this may be inhibited by agglomeration of GNPs. Agglomeration was indeed present in the A319 + 0.05 GNP + SS and A319 + 0.1 GNP + SS as demonstrated by the reduced mechanical properties (Figure 6) and the fracture surface (Figure 7b). In contrast, hybrid SS and UST processing enabled effective and homogeneous addition of GNPs into the melt. The SDAS of A319 + 0.05 GNP + SS + UST was measured as 16.0 ± 0.8 µm, which exhibit a slight reduction compared to the base A319 alloy. It has been previously reported that graphene, due to its high thermal conductivity, can change the solidification mechanism. As a result, it is possible for the particles to affect the temperature gradient ahead of the solidification front potentially leading to reduced grain size [18,25,26], while the SDAS refinement is minor, it still demonstrates slight refinement compared to the other conditions and it should be noted that only 0.05 wt.% of GNPs was added.

SEM analysis was carried out for observation of GNPs and analysis of the GNP interface with the Al matrix. The results showed that GNP particles were embedded in the Al matrix, typically observed near eutectic Si particles. This may indicate that during solidification graphene particles were pushed towards the interdendritic regions, as opposed to pinning at the boundaries of the dendrites. As previously discussed, this suggests that it is related to solidification mechanisms which are affected by the high thermal conductivity of graphene. Additionally, SEM and EDX confirmed the platelet morphology of the observed particle as well as the chemistry, demonstrated in Figure 4b. Figures 3b and 4a also demonstrates that the GNP particles have a strong and cohesive interface with the Al matrix. As a result, there was no observation of delamination between at the Al-GNP interface. This also correlates well to the fracture behavior of A319 + 0.05 GNP + SS + UST, which exhibited a relatively ductile fracture without the observation of micro-cracks, shown in Figure 7a.

4.2. Analysis of Interfacial Reactions

With the addition of reinforcing particles such as GNPs, it is also important to consider the interfacial interaction between the reinforcement particle and the matrix. The addition of reinforcement particles at high processing temperatures may produce reaction products, as the thermodynamic stability becomes weaker with elevated temperatures. Namely, Al_4C_3 can form if the GNP particles break down during the processing. This also is more likely to occur with increased processing times. Because preparation of the A319 + 0.05 GNP + SS + UST composite involves a multi-stage melt treatment, the processing time was longer compared to a conventional casting process. Figure 5 shows the XRD results of the A319 + 0.05 GNP + SS + UST composite, showing the peaks and corresponding reaction products. The peaks identified are standard for α-Al, Al_2Cu, Si and $Al_{15}(Fe,Mn)_3Si_2$. However, the results demonstrated that at 43° there was a peak, likely corresponding to Al_4C_3. Despite this, it was difficult to identify using XRD and the corresponding database likely due to the low addition of GNPs. The results correlate well to a study by Lee et al. [21] on the formation of Al_4C_3 in Al-Si alloys, where the authors determined that the transition temperature for the formation of Al_4C_3 is approximately 620 °C. In this study, the processing temperature reached a maximum of 750 °C, hence it is thermodynamically favorable for Al_4C_3 to form during the high temperature processing. Of note, it is likely that the amount of Al_4C_3 is minimal due to the low addition of 0.05 wt.% GNPs. Moreover, localized inhomogeneity and variation in composition, likely resulted in the interesting observation of Al_4C_3 in limited quantity. The direct observation of Al_4C_3 will be carried out using high-resolution transmission electron microscopy (HR-TEM) as part of a separate on-going study by the authors. Lastly, the limited presence of Al_4C_3 was not found to deteriorate the mechanical properties of the A319 + 0.05 GNP + SS + UST composite.

4.3. Enhancement of Mechanical Properties

In this study, the mechanical properties were evaluated for each of the castings (Figure 6). The results showed enhanced mechanical properties for the samples undergoing hybrid semi-solid and ultrasonic processing (A319 + 0.05 GNP + SS + UST). The UTS and YS of A319 + 0.05 GNP + SS + UST displayed a 10 and 11% increase, respectively, compared to the base A319 casting. A considerable increase in strength for a low addition of GNPs. Additionally, the ductility also increased from 3.1% to 4.1% compared to base A319, which is a 32% increase. The increase in mechanical properties demonstrate the effectiveness of the GNP particles as a reinforcement material. The results suggest that the

GNPs were effectively incorporated in A319 Al alloy using the hybrid method. This was shown in the SEM results whereby the GNP particles were observed and embedded in the Al matrix. The GNPs were typically oriented longitudinal to the analyzed surface; however, GNPs were also observed oriented at an angle to the surface (Figure 3b). Additionally, the GNPs displayed good interfacial bonding at the Al interface which correlates well to the mechanical properties. The interface between the GNP particle and the Al matrix was well observed in Figure 4. This also contributed to the ductile fracture behavior as de-lamination of GNPs or micro-cracks were not observed in this study. Additionally, clustering of GNPs was not observed under the SEM, which indicates that ultrasonic processing was effective in de-agglomerating clusters of GNPs into individual particles. The efficiency of the strengthening by GNP reinforcement likely depends on which planes the GNPs will inhabit as the properties are more dominant when the load is applied in the transverse direction. Modeling and prediction of the orientation of GNPs and solidification mechanisms should be further explored as this helps to understand how to fully optimize the mechanical properties.

As previously discussed, GNP particles were observed under the SEM embedded in the Al matrix. While GNPs are a contributor to the strength enhancement of A319, it should be noted that the sonication process also degasses the melt. Hence, there is an additional benefit of ultrasonic processing when producing a composite through casting. Interestingly, ultrasonic processing alone was found to exhibit slight improvements in the mechanical properties of A319, where the UTS and YS slightly increased from 190 to 195 MPa and 107 to 109 MPa, respectively. As previously mentioned, the slight improvement in mechanical properties is due to degassing of the melt. Although, it was found that UST processing did not refine the A319 alloy, which has been reported to occur in other alloys such as Mg [27,28].

In order to isolate the effects of semi-solid processing, castings involving the addition of GNPs were carried out only using semi-solid processing as the melt treatment. For both the A319 + 0.05 GNP + SS and A319 + 0.1 GNP + SS composites, the UTS decreased by 5.3 and 7.4%, respectively. Similarly, the fracture behavior was brittle in both castings with the worst condition (A319 + 0.1 GNP + SS) having a ductility of 1.9%. This was a result of severe agglomeration of the GNP particles, which can be seen in Figure 7b. In this study, semi-solid processing was used to prevent floating of GNP particles due to their much lower density compared to Al. While this process was effective in doing so, it is evident that mechanical mixing during semi-solid processing is not effective in preventing agglomeration of GNP particles. Hence, it is important to follow this process with ultrasonic processing to de-agglomerate and distribute the particles.

The strengthening contributions were also analyzed using theoretical mathematical models. The main strengthening contributors to YS are Hall-Petch, load bearing mechanism, and thermally induced dislocations as a result of CTE mismatch between graphene particles and Al. The theoretical YS enhancement from each mechanism was calculated to be 16.2, 3.0 and 4.8 MPa, respectively. Hence, the highest contribution to strength was due to the dendritic grain refinement through the Hall-Petch relationship. This is followed by CTE mismatch between graphene and the Al matrix, then by load bearing mechanism. As a result, the total contribution to YS enhancement is 24 MPa resulting in a theoretical YS of approximately 130 MPa. Although, compared to the experimental result there is a deviation by approximately 8.5%. Hence, it can be suggested that the model can be considered to predict the YS of AMCs reinforced with GNPs.

The discrepancy in the YS between the theoretical calculation and the experimental result can be attributed to some factors. The total calculated volume fraction does not fully represent that of the experiment due to losses throughout the casting process. The orientation of GNPs is not accounted for in the model, for example the strength properties are more dominant in the transverse loading direction compared to parallel with the corresponding plane. Furthermore, it is not fully known which planes the GNPs will inhabit during solidification, although it is expected that to achieve optimal strengthening enhancement this should be the (111) aluminum slip plane. The formation of interfacial products, such as Al_4C_3, is also not accounted for in the theoretical models. It can be expected that such compounds can act as either a strengthening phase or deteriorate the YS.

The results of this study demonstrated the effectiveness of reinforcing A319 with GNPs through a simple liquid metallurgy technique. To the authors' knowledge, reinforcement of automotive A319 Al alloy with graphene has not been comprehensively investigated. As a result, the findings in this study further contributes to light weight and development of high-strength materials for automotive powertrain components. The mechanical properties of A319 were enhanced with cost effective and low additions of GNPs using a hybrid semi-solid and ultrasonic melt treatment during the casting process. As a result, the formation of Al_4C_3 was minimal with the addition of 0.05 wt.% GNPs. This was an important finding due to the negative effect of Al_4C_3 on the mechanical properties. Lastly, it was found that theoretical strengthening models resulted in values close to the experimental results, although adjustments should be made to account for the orientation of GNPs in the metal matrix.

This study is significant since it clearly demonstrated that high cost and time-consuming preprocessing methods such as ball milling, spark plasma sintering and other methods are not required for efficient production of GNP reinforced Al composites. This is especially important for developing high strength materials for the automotive industry. Therefore, cost effective and efficient cast Al-GNP composites prepared through liquid metallurgy techniques, such as casting, can be used for next generation automotive vehicles leading to light weight and increased performance.

5. Conclusions

This research investigated a processing route using a casting process to produce a A319 graphene reinforced MMC. As a result, an A319-0.05 wt.% GNP composite (A319 + 0.05 GNP + SS + UST) was successfully cast with improved mechanical properties. This was achieved via a hybrid semi-solid and ultrasonic treatment method. In order to demonstrate the synergistic effect of the hybrid method, castings of base A319, A319 with UST, and A319 with SS processing were also produced. The alloy microstructure, analysis of graphene in the Al matrix, and the mechanical properties of the castings were examined. Subsequently, the strengthening contributions were also analyzed using theoretical mathematical models, which were then validated with the experimental results. The results from this research enable further understanding of a simple liquid metal processing route to produce aluminum matrix composites reinforced with graphene for high performance next generation automotive vehicles. The following conclusions can be drawn from this study:

1. The combination of semi-solid processing and ultrasonic processing was an effective and efficient method for the incorporation of GNPs in A319 Al alloy. This was attributed to prevention of particle floating during semi-solid stirring and the ability of ultrasonic processing to effectively de-agglomerate and distribute the GNPs in the liquid melt.
2. Semi-solid processing of A319 Al enabled the prevention of floating of GNPs to the melt surface, however it was ineffective in de-agglomerating GNP particles. This resulted in excessive agglomeration of GNPs promoting a brittle fracture and a reduction in mechanical properties.
3. Addition of 0.05 wt.% GNPs to A319 Al using the hybrid semi-solid and ultrasonic processing method resulted in an enhancement in UTS, YS, and ductility by 10, 11 and 32%, respectively. The SEM results revealed good interfacial bonding between GNP and Al-matrix. This was an important finding due to the very low addition of GNPs, which suggests a cost-effective incorporation of particles accompanied by the enhancement in mechanical properties.
4. X-ray diffraction results indicate the limited presence of Al_4C_3 with the addition of GNPs, which did not lead to any reduction in mechanical properties in the A319 + 0.05 GNP + SS + UST composite. This was attributed to localized inhomogeneity and

variation in composition, likely resulted in the interesting observation of Al_4C_3 in limited quantity.
5. The strengthening contributions for the A319 alloy reinforced with 0.05 wt.% GNPs (A319 + 0.05 GNP + SS + UST) were evaluated using mathematical models. The strength enhancements were primarily attributed to the Hall-Petch effect, followed by CTE mismatch and load bearing mechanism.
6. The mathematical models used to estimate the YS of the alloy corresponded well to the experimental results. The theoretical YS was calculated as 130 MPa, whereas the experimental YS was measured as 119 MPa. Thus, there is an 8.5% deviation and it can be suggested that the model is accurate in predicting YS of aluminum matrix composites reinforced with GNPs.

Author Contributions: Conceptualization, B.A. and C.R.; Methodology, B.A. and P.E.; Validation, B.A.; Formal Analysis, B.A.; Investigation, B.A.; Resources, C.R.; Data Curation, B.A.; Writing—Original Draft, B.A. and P.E.; Writing—Review and Editing, C.R.; Visualization, B.A. and P.E.; Supervision, C.R.; Project Administration, B.A., P.E. and C.R.; Funding Acquisition, C.R. All authors have read and agreed to the published version of the manuscript.

Funding: This research was funded by the Natural Sciences and Engineering Research Council of Canada (NSERC), RGPIN-2020-06096.

Institutional Review Board Statement: Not applicable.

Informed Consent Statement: Not applicable.

Data Availability Statement: Not applicable.

Acknowledgments: The authors are thankful to the Natural Sciences and Engineering Research Council of Canada (NSERC) for financial support of this project (RGPIN-2020-06096) and for the award of Canada Graduate Scholarship-Doctoral to Bernoulli Andilab (Grant number CGSD3-559982-2021) and Payam Emadi (CGSD3-535728-2019). The authors also thank Alan Machin and Qiang Li for their technical assistance. They are also grateful to Adam Belcastro, Mykola Sydorenko, Rohit Mishra and the members of the Centre for Near-net-shape Processing of Materials Lab for assistance with casting experiments and discussions.

Conflicts of Interest: The authors declare no conflict of interest.

References

1. Bartolucci, S.; Paras, J.; Rafiee, M.; Rafiee, J.; Lee, S.; Kapoor, D.; Koratkar, N. Graphene–aluminum nanocomposites. *Mater. Sci. Eng. A.* **2011**, *528*, 7933–7937. [CrossRef]
2. Chen, L.; Konishi, H.; Fehrenbacher, A.; Ma, C.; Xu, J.; Choi, H.; Xu, H.; Pfefferkorn, F.; Li, X. Novel nanoprocessing route for bulk graphene nanoplatelets reinforced metal matrix nanocomposites. *Scr. Mater.* **2012**, *67*, 29–32. [CrossRef]
3. Bhaudaria, A.; Singh, L.K.; Laha, T. Combined strengthening effect of nanocrystalline matrix and graphene. *Mater. Sci. Eng. A.* **2019**, *749*, 14–26. [CrossRef]
4. Lee, C.; Wei, X.; Kysar, J.; Hone, J. Measurment of elastic properties and intrinsic strength of monolayer graphene. *Science* **2008**, *321*, 385–388. [CrossRef]
5. Goyal, V.; Balandin, A. Thermal properties of the hybrid graphene-metal nanomicro-composites: Applications in thermal interface materials. *Appl. Phys. Lett.* **2012**, *100*, 073113. [CrossRef]
6. Snapp, P.; Kim, J.M.; Cho, C.; Leem, J.; Haque, M.F.; Nam, S. Interaction of 2D materials with liquids: Wettability, electrochemical properties, friction, and emerging directions. *NPG Asia Mater.* **2020**, *12*, 22. [CrossRef]
7. Ghasali, E.; Sangpour, P.; Jam, A.; Rajaei, H.; Shirvanimoghaddam, K.; Ebadzadeh, T. Microwave and spark plasma sintering of carbon nanotube and graphene reinforced aluminum matrix composite. *Arch. Civ. Mech. Eng.* **2018**, *18*, 1042–1054. [CrossRef]
8. Zhang, J.; Liu, Q.; Yang, S.; Chen, Z.; Liu, Q.; Jiang, Z. Microstructural evolution of hybrid aluminum matrix composites reinforced with SiC nanoparticles and graphene/graphite prepared by powder metallurgy. *Prog. Nat. Sci. Mater. Int.* **2020**, *30*, 192–199. [CrossRef]
9. Wu, Y.; Zhan, K.; Yang, Z.; Sun, W.; Zhao, B.; Yan, Y.; Yang, J. Graphene oxide/Al composites with enhanced mechanical properties fabricated by simple electrostatic interaction and powder metallurgy. *J. Alloys Compd.* **2019**, *775*, 233–240. [CrossRef]
10. Hu, Z.; Tong, G.; Lin, D.; Nian, Q.; Shao, J.; Hu, Y.; Saeib, M.; Jin, S.; Cheng, G.J. Laser Sintered Graphene Nickel Nanocomposites. *J. Mater. Process. Technol.* **2016**, *231*, 143–150. [CrossRef]

11. Rafiee, M.A.; Rafiee, J.; Srivastava, I.; Wang, Z.; Song, H.; Yu, Z.Z.; Koratkar, N. Fracture and Fatigue in Graphene Nanocomposites. *Small* **2010**, *6*, 179–183. [CrossRef]
12. Alipour, M.; Eslami-Farsani, R. Synthesis and characterization of graphene nanoplatelets reinforced AA7068 matrix nanocomposites produced by liquid metallurgy route. *Mater. Sci. Eng. A* **2017**, *706*, 71–82. [CrossRef]
13. Srivastava, A.K.; Sharma, B.; Saju, B.; Shukla, A.; Saxena, A.; Maurya, N. Effect of Graphene nanoparticles on microstructural and mechanical properties of aluminum based nanocomposites fabricated by stir casting. *World J. Eng.* **2020**, *17*, 859–866. [CrossRef]
14. Yang, W.; Chen, G.; Qiao, J.; Liu, S.; Xiao, R.; Dong, R.; Hussain, M.; Wu, G. Graphene nanoflakes reinforced Al-20Si matrix composites prepared by pressure infiltration method. *Mater. Sci. Eng. A* **2017**, *700*, 351–357. [CrossRef]
15. Yang, W.; Zhao, Q.; Xin, L.; Qiao, J.; Zhou, J.; Shao, P.; Yu, Z.; Zhang, Q.; Wu, G. Microstructure and mechanical properties of graphene nanoplates reinforced pure Al matrix composites prepared by pressure infiltration method. *J. Alloys Compd.* **2018**, *732*, 748–758. [CrossRef]
16. Du, X.; Du, W.; Wang, Z.; Liu, K.; Li, S. Ultra-high strengthening efficiency of graphene nanoplatelets reinforced magnesium matrix composites. *Mater. Sci. Eng. A* **2018**, *711*, 633–642. [CrossRef]
17. Boostani, A.; Mousavian, R.; Tahamtan, S.; Yazdani, S.; Khosroshahi, R.; Wei, D.; Xu, J.; Gong, D.; Zhang, X.; Jiang, Z. Graphene sheets encapsulating SiC nanoparticles: A roadmap towards enhancing tensile ductility of metal matrix composites. *Mater. Sci. Eng. A* **2015**, *648*, 92–103. [CrossRef]
18. Boostani, A.; Tahamtan, S.; Jiang, Z.; Wei, D.; Yazdani, S.; Khosroshahi, R.; Mousavian, R.; Xu, J.; Zhang, X.; Gong, D. Enhanced tensile properties of aluminium matrix composites reinforced with graphene encapsulated SiC nanoparticles. *Compos. Part A Appl. Sci. Manuf.* **2015**, *68*, 155–163. [CrossRef]
19. Boostani, A.; Yazdani, S.; Mousavian, R.; Tahamtan, S.; Khosroshahi, R.; Wei, D.; Brabazon, D.; Xu, J.; Zhang, X.; Jiang, Z. Strengthening mechanisms of graphene sheets in aluminium matrix nanocomposites. *Mater. Des.* **2015**, *88*, 983–989. [CrossRef]
20. Vandersluis, E.; Sediako, D.; Ravindran, C.; Elsayed, A.; Byczynski, G. Analysis of eutectic silicon modification during solidification of Al-6Si using in-situ neutron diffraction. *J. Alloys Compd.* **2018**, *736*, 172–180. [CrossRef]
21. Lee, J.C.; Byun, J.Y.; Park, S.B.; Lee, H.I. Prediction of Si contents to suppress the formation of Al_4C_3 in the SiCp/Al composite. *Acta Mater.* **1998**, *46*, 1771–1780. [CrossRef]
22. Frost, H.; Ashby, M. *Deformation-Mechanism Maps: The Plasticity and Creep of Metals and Ceramics*; Pergamon Press: Oxford, UK, 1982.
23. Ghassemali, E.; Riestra, M.; Bogdanoff, T.; Kumar, B.S.; Seifeddine, S. Hall-Petch equation in a hypoeutectic Al-Si cast alloy: Grain size vs. secondary dendrite arm spacing. *Procedia Eng.* **2017**, *207*, 19–24. [CrossRef]
24. Bao, W.; Miao, F.; Chen, Z.; Zhang, H.; Jang, W.; Dames, C.; Lau, C.N. Controlled ripple texturing of suspended graphene and ultrathin graphite membranes. *Nat. Nanotechnol.* **2009**, *4*, 562–566. [CrossRef]
25. EAgaliotis, M.; Rosenberger, M.R.; Ares, A.E.; Schvezov, C.E. Influence of the shape of the particles in the solidification of composite materials. *Procedia Mater. Sci.* **2012**, *1*, 58–63. [CrossRef]
26. Khan, M.A.; Rohatgi, P.K. A numerical study of thermal interaction of solidification fronts with spherical particles during solidification of metal-matrix composite materials. *Compos. Eng.* **1993**, *3*, 995–1006. [CrossRef]
27. Emadi, P.; Andilab, B.; Ravindran, C. Preparation and characterization of AZ91E/Al2O3 composites using hybrid mechanical and ultrasonic particle dispersion. *Mater. Sci. Eng. A* **2021**, *819*, 141505. [CrossRef]
28. Emadi, P.; Andilab, B.; Ravindran, C. Processing and Properties of Magnesium-Based Composites Reinforced with Low Levels of Al_2O_3. *Int. J. Metalcast.* **2022**, *16*, 1680–1692. [CrossRef]

Article

Study on the Effect of Pre-Refinement and Heat Treatment on the Microstructure and Properties of Hypoeutectic Al-Si-Mg Alloy

Ling Lin [1], Lian Zhou [1], Yu Xie [2], Weimin Bai [1], Faguo Li [1,*], Ying Xie [1], Mingxin Lu [1] and Jue Wang [1]

1. School of Materials Science and Engineering, Xiangtan University, Xiangtan 411105, China
2. Central Research Institute, Baoshan Iron & Steel Co., Ltd., Shanghai 201900, China
* Correspondence: author: lifaguo@xtu.edu.cn

Citation: Lin, L.; Zhou, L.; Xie, Y.; Bai, W.; Li, F.; Xie, Y.; Lu, M.; Wang, J. Study on the Effect of Pre-Refinement and Heat Treatment on the Microstructure and Properties of Hypoeutectic Al-Si-Mg Alloy. *Materials* 2022, *15*, 6056. https://doi.org/10.3390/ma15176056

Academic Editor: Konstantin Borodianskiy

Received: 25 July 2022
Accepted: 30 August 2022
Published: 1 September 2022

Publisher's Note: MDPI stays neutral with regard to jurisdictional claims in published maps and institutional affiliations.

Copyright: © 2022 by the authors. Licensee MDPI, Basel, Switzerland. This article is an open access article distributed under the terms and conditions of the Creative Commons Attribution (CC BY) license (https://creativecommons.org/licenses/by/4.0/).

Abstract: Hypoeutectic Al-Si-Mg alloys with a silicon content of around 10 wt % are widely used in the aerospace and automotive fields due to their excellent casting properties. However, the occurrence of "silicon poisoning" weakens the refinement effect of a conventional refiner system such as Al-5Ti-1B. In this paper, we proposed the "pre-refinement" method to avoid the "Si poisoning" to recover the refinement effect of Al-5Ti-1B. The core concept was to adjust the order of adding the Si element to form the TiAl$_3$ before forming the Ti-Si intermetallic compound. To prove the effectiveness of the "pre-refinement" method, three alloys of "pre-refinement", "post-refinement", and "non-refinement" of an Al-10Si-0.48Mg alloy were prepared and characterized in as-cast and heat-treatment states. The results showed that the average grain diameter of the "pre-refinement" alloy was 60.19% smaller than that of the "post-refinement" one and 81.34% smaller than that of the "non-refinement" one, which demonstrated that the proposed method could effectively avoid the "silicon poisoning" effect. Based on a refined grain size, the "pre-refinement" Al-10Si-0.48Mg alloy showed the best optimization effect in mechanical properties after a solid-solution and subsequent aging heat treatments. The best mechanical properties were found in the "pre-refinement" alloy with 2 h of solid solution treatment and 10 h of aging treatment: a hardness of 92 HV, a tensile strength of 212 MPa, and an elongation of 20%.

Keywords: Al-Si-Mg alloy; pre-refinement; grain refinement; microstructure; mechanical properties

1. Introduction

Al-Si-Mg casting alloys have been widely used in the aerospace, automotive, and 3C industries due to their excellent casting properties and outstanding strength-to-weight ratio. With the continuous development of the new-energy automotive industry, the use of Al-Si-Mg casting alloys will be further increased. The excellent performance of Al-Si-Mg casting alloys depend largely on the microstructure, which is characterized by the grain size of the matrix phase α-Al and the shape and distribution of the eutectic phase comprising silicon and Mg$_2$Si [1,2]. Grain refinement can produce more equiaxed crystal structures and reduce casting defects, and is an effective method to simultaneously improve the strength and toughness of casting alloys [3].

As early as 1930, Rosenhain et al. [4] found that adding the Ti element during the smelting process of Al alloys can refine the grain size, and related research on grain refinement has been conducted since then. Jones and Pearson [5] believe that the best grain-refining capacity needs a period of time to emerge; however, if the standing time during the casting is set too long, the grain size become coarser again. The decline in the refinement effect is attributed to the aggregation and precipitation of TiB$_2$ and TiC particles or the loss of their activity due to the change in the surface structure [6]. Grain refinement by adding refiners such as Al-Ti-B and Al-Ti-C is the conventional method to obtain fine grains for Al-Cu and Al-Mg alloys. Some alloying elements (such as Fe, Si, Mg) can improve the refining effect of grain refiners. However, when the alloying element content exceeds a certain value, the refining effect of the refiner decreases with the increase in the alloying element content [7,8]. However, for Al-Si alloys, it has been found that the refinement effect of the commonly used refiners is weakened when the content of Si element exceeds 3.5 wt %, which is known as the "Si poisoning" effect [9,10]. In the past, it was widely believed that the alloying elements with a toxic effect on the intermediate alloy would concentrate around TiB_2 or TiC particles and react with the Ti element, changing the surface physical and chemical properties of TiB_2 or TiC particles, reducing the surface activity of TiB_2 or TiC particles, and making the interfacial compatibility with melted aluminum worse. Schumacher and McKay found that the formation of $TiSi_2$ on basal faces of TiB_2 reduced the nucleation area and number of active sites for Al [11]. The reduction in α-Al growth restriction and the formation of silicide phases before solidification of α-Al was due to a strong exothermic interaction between titanium and silicon, which is the main mechanism of Si poisoning [12]. In fact, the small degree of grain refinement by additions of eutectic-forming elements (Cu, Mg and Si) is mainly attributed to their segregating power; however, they cannot form nucleant particles [13]. The nucleation work of α-Al on its surface increases and the grain refinement effect decreases [14,15]. It should be noted that poisoning occurs whether or not it is grain-refined with Al-Ti-B master alloys [16]. Therefore, we must find a way to circumvent the influence of Si on the nucleation and growth of α-Al.

Upon the addition of Al-5Ti-B refiner, the bulk $TiAl_3$ dissolves into the liquid, leaving TiB_2 as the nucleation particle for α-Al [17], then the Si segregates on the surface of TiB_2 particles and dissolves into a $TiAl_3$ two-dimensional compound ($TiAl_3$ 2DC). The $TiAl_3$ 2DC plays a key role in nucleation, and then the strong interaction between the Si atoms and Ti atoms can disturb the crystal structure of $TiAl_3$ 2DC to a certain extent, which as a result weakens the nucleation efficiency, and the so-called "Si-poisoning" occurs. Alternatively, an Al-Nb-Ti-B intermediate alloy rich in $(Nb,Ti)B_2$ particles with a "sandwich" structure is expected to be an effective refiner for high-silicon aluminum alloys [18]. Unfortunately, the addition of Nb elements also has the disadvantages of increasing the cost and the difficulty of accurate preparation.

The key of TiB_2 as the nucleation core of α-Al is to promise the function of the Ti-terminated (0001) surface as the nucleation template [19,20]. The "Si poisoning" lies in that the preferential combining of Si atoms and Ti atoms, which forms Ti-Si intermetallic compounds, destroys the Ti-terminated (0001) surface. Naturally, the "Si poisoning" can probably be avoided by adjusting the order of the addition of alloying elements into the Al alloy to ensure the Al atoms combine with Ti atoms before Si atoms. Studies have found [21,22] that the liquid Al atoms can form an orderly layered structure in front of the solid Ti atoms around the melting point of pure aluminum, which can benefit the heterogeneous nucleation. In this paper, a concept of "pre-refinement" is proposed to recover the refinement effect of the conventional Al-5Ti-B refiner for the hypoeutectic Al-Si-Mg alloy. The core concept was to adjust the order of adding the Si element to form the $TiAl_3$ before forming the Ti-Si intermetallic compound. The mechanical properties of the Al alloy were further improved through the subsequent heat-treatment process upon the grain-refining strengthening.

2. The Concept of the "Pre-Refinement" Design

The basic principle of "pre-refinement" was to ensure that the liquid Al atoms near the melting point of aluminum could form an orderly layered structure in front of solid Ti atoms. To satisfy this principle, the Al-5Ti-B refiner was firstly added to the Al-Mg alloy at a temperature close to the melting point of the alloy to ensure that Al atoms were orderly arranged on the Ti-terminated (0001) surface of the TiB_2 particles to form the $TiAl_3$ precursor. Si was then added, and as schemed, the Si atoms would not combine with Ti atoms to form a Ti-Si intermetallic compound due to the isolation of the Al atoms. During the subsequent cooling process, $TiAl_3$ formed around TiB_2 particles and α-Al grew up continuously on the TiAl₃ substrate, achieving heterogeneous nucleation to refine the grains. The schematic of the "pre-refinement" is shown in Figure 1.

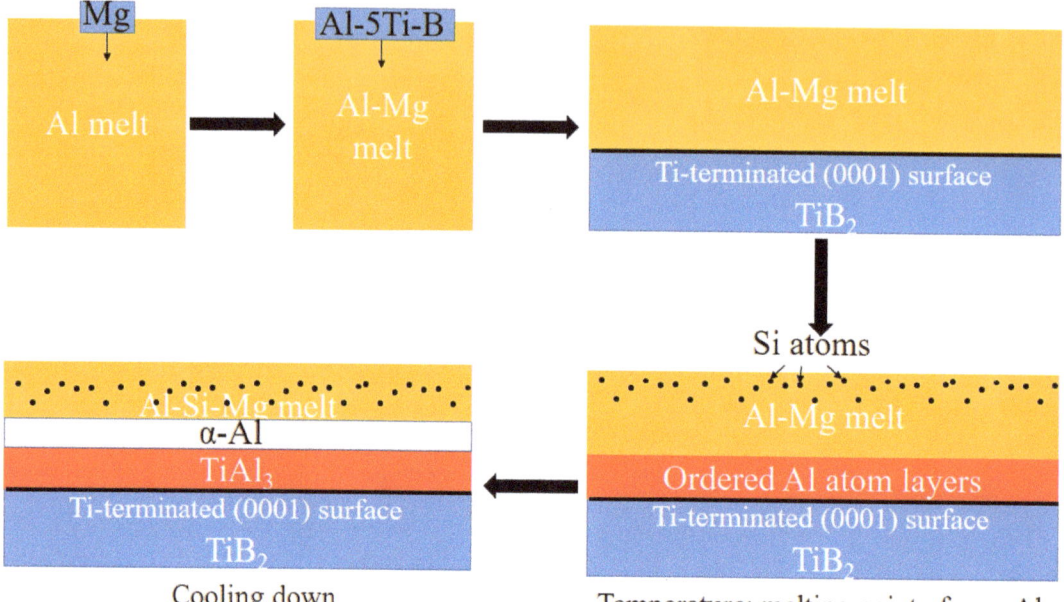

Figure 1. Schematic diagram of "pre-refinement" design concept.

Based on ZL 104 (ZAlSi9Mg, Chinese grade of aluminum alloy, 8.0–10.5 wt % Si, 0.2–0.5 wt % Mn, 0.17–0.35 wt % Mg), we expected to develop a new Al-10Si-0.48Mg alloy for automotive die casting integration. However, the Si content was more than 3.5 wt %, so the conventional method of adding an Al-Ti-B refiner could not produce the refining effect. Therefore, we attempted the "pre-refinement" method using the alloy Al-10Si-0.48Mg.

We used the results of thermodynamic calculation to help design the key parameters during the pre-refinement and the following heat-treatment process. The equilibrium solidification phase diagram of the Al-10Si-0.48Mg was calculated using the built-in database in the trial version of Thermo_Calc (Figure 2a; the database was restored according to reference [23]). The precipitation temperatures of the α-Al, Si, and Mg_2Si phases were 593.63, 574.57, and 506.94 °C, respectively. The liquid phase vanishing temperature was 565.13 °C. The percentage of each equilibrium phase at room temperature was as follows: 89.829% α-Al, 9.373% Si, and 0.802% Mg_2Si. The solidification paths of the Al-10Si-0.48Mg using the equilibrium and Scheil schemes (non-equilibrium solidification) were calculated using the trial version of Thermo_Calc (Figure 2b). The Mg_2Si precipitated at 559 °C of the non-equilibrium solidification. The final solidification temperature was 6.5 °C lower than that of the equilibrium solidification.

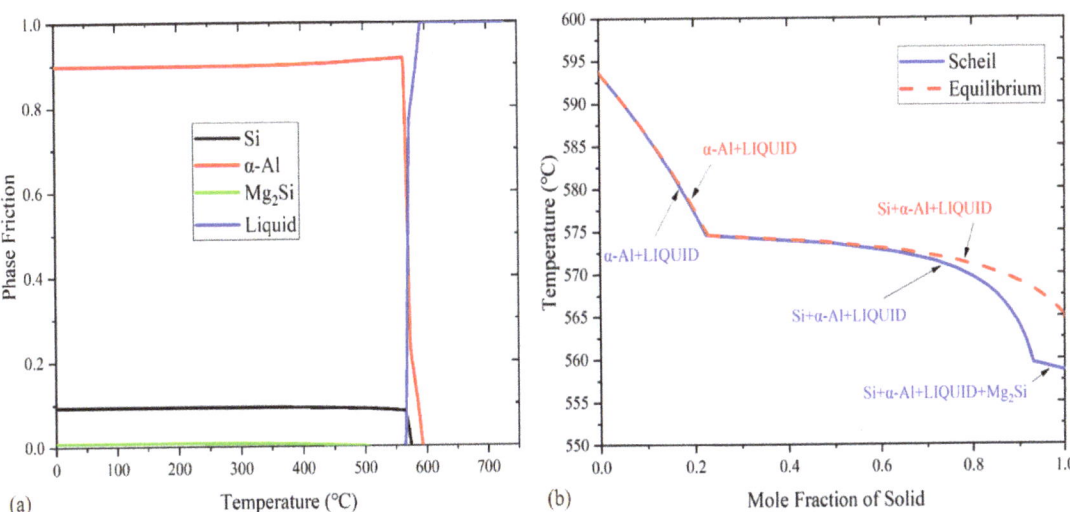

Figure 2. (**a**) Equilibrium solidification phase diagram for Al-10Si-0.48Mg; (**b**) calculated solidification paths for Al-10Si-0.48Mg according to the equilibrium and Scheil schemes.

3. Experimental Materials and Methods

The Al-10Si-0.48Mg alloy was casted in the experiment by using materials including high-purity Al strips (99.999 wt %), high-purity Mg particles (99.99 wt %), high-purity Al foil (99.999 wt %), high-purity Si blocks (99.999 wt %), an Al-5Ti-1B refiner, and a covering agent (50 wt % NaCl + 35 wt % KCl + 15 wt % NaF). The final smelted aluminum ingot size was $\Phi 40 \times 50$ mm. In the study, the macro and micro structure images were sampled in the middle of the ingot. The average grain size value was the average value of a total of 6 grain size values taken at the axial directions of 10, 25, and 40 mm and the radial directions of 5, 10, and 15 mm. To prove the effectiveness of "pre-refinement" design, the "pre-refinement", "post-refinement", and "non-refinement" Al-10Si-0.48Mg alloys were prepared. The preparation paths of the three types of alloys are shown in Figure 3. In the following, we will use the example of the pre-refinement preparation process to provide more details on the preparation process. During the preparation, the Al strips were put into the alumina ceramic crucible and melted at 700 °C, 33 °C higher than the melting point, in a well furnace (Xiangtan Samsung Instrument Co., Ltd., Xiangtan, China). Then, the furnace temperature was decreased to 675 °C and was held to add other materials. To prevent accidents such as fire caused by direct contact between Mg particles and the high-temperature liquid Al, the Mg particles were wrapped with aluminum foil and placed in the liquid. Then, the Al-5Ti-1B refiner was added into the Al-Mg liquid; it was necessary to gently stir with a graphite rod to ensure the refiner was evenly distributed in the alloy liquid to increase the refinement effect. After about 10 min, silicon blocks were pressed into the alloy liquid with a graphite rod so that the silicon blocks could be quickly and evenly dissolved in the alloy liquid. Finally, the liquid was casted into a graphite crucible that was quenched in air to room temperature.

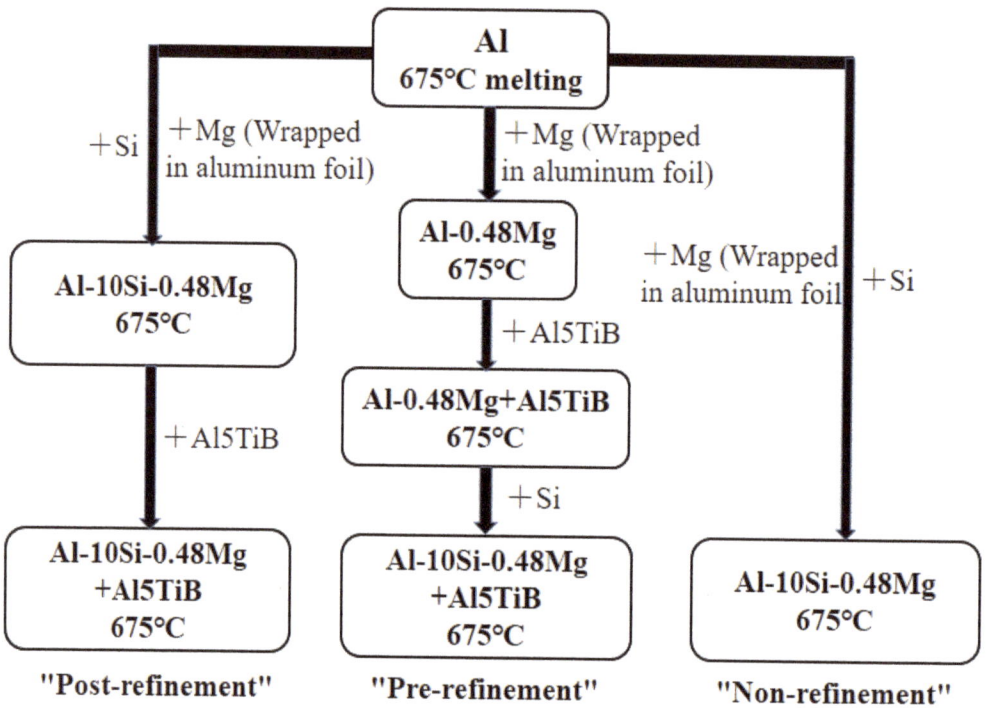

Figure 3. Schematic diagram of preparation paths of the three types of alloys: "post-refinement", "pre-refinement", and "non-refinement".

A solution heat treatment at 500–530 °C for 2–8 h after water quenching and artificial aging at 170–210 °C for 2–20 h are the commonly used heat treatments to improve the mechanical properties of alloys [24,25]. Likewise, the T6 heat-treatment process of a solid solution treatment followed by an aging treatment was carried out for the three types of as-cast alloys.

According to the calculation results shown in Figure 2, the temperatures of the solid solution treatment and aging treatment were 500 °C and 175 °C, respectively. The time of the solid solution treatment was 2 h or 8 h, and the time of the aging treatment was 10 h or 20 h. Thus, four groups of heat-treatment processes were used: 2 h solid solution treatment + 10 h aging treatment (denoted as 2 + 10), 2 h solid solution treatment + 20 h aging treatment (denoted as 2 + 20), 8 h solid solution treatment + 10 h aging treatment (denoted as 8 + 10), and 8 h solid solution treatment + 20 h aging treatment (denoted as 8 + 20). The roadmap for heat treatment is show in Figure 4.

The microstructure and composition distribution were characterized using OM (ZEISS, Zeiss, Jena, Germany), SEM with EDS (ZEISS EVO MA10, Zeiss, Jena, Germany), and XRD (Ultima IV, Rigaku Co., Tokyo, Japan). The microstructure hardness and tensile tests were carried out using a Vickers hardness tester (SHYCHVT-30, Laizhou Huayin hardness meter factory, Laizhou, China) and an electronic multifunctional stretching machine (WDW-100C, Jinan Fangyuan Instrument Co., Ltd., Jinan, China), respectively. The dimensions of the tensile sample are shown in Figure 5 (in mm).

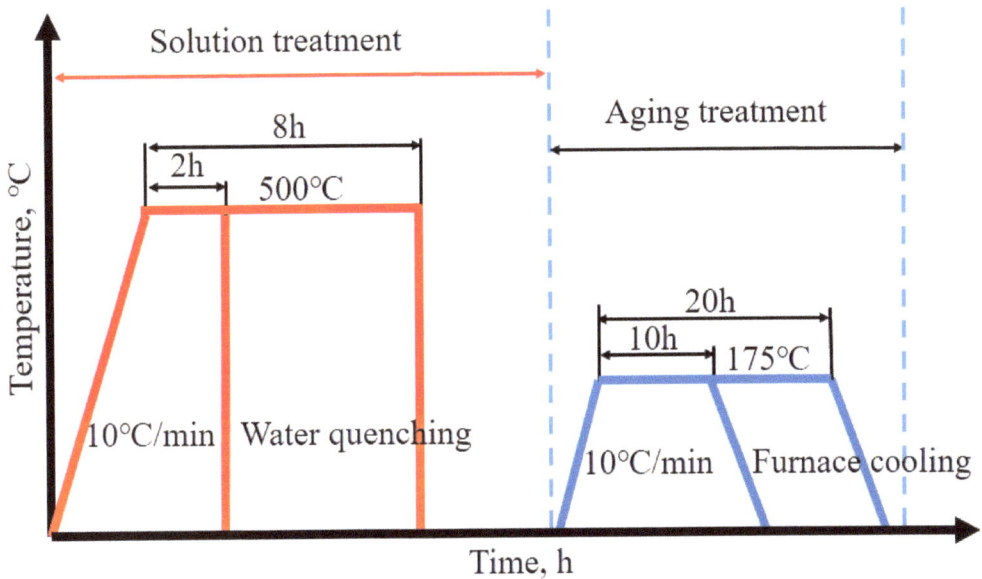

Figure 4. Roadmap for heat treatment.

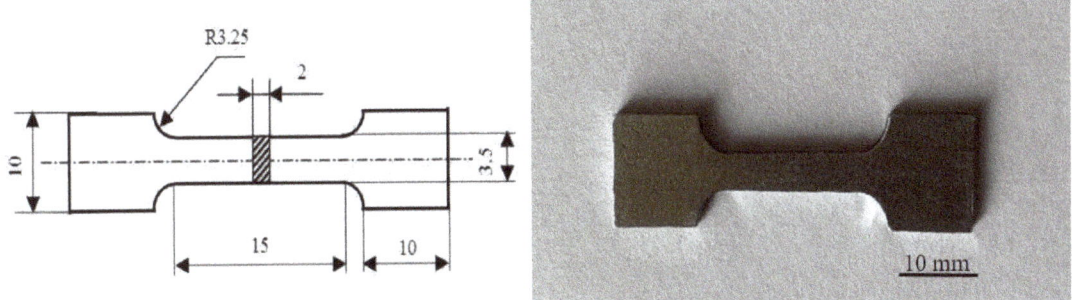

Figure 5. Dimensional drawing of tensile specimen and real specimen.

4. Results and Discussion

4.1. Microstructures of the As-Cast and Heat-Treated Alloys

Figure 6a shows the SEM image of the as-cast "pre-refinement" hypoeutectic Al-10Si-0.48Mg alloy. EDS data showed that Point A, Point B, and Point C were α-Al with the Al:Mg:Si equal to 90.5:9.5:0, eutectic phase with Al:Mg:Si equal to 39.86:18.63:41.51, and Si + (Al + Si) phase with Al:Mg:Si equal to 20.59:0:79.41. The phase constituents of the "pre-refinement" hypoeutectic Al-10Si-0.48Mg alloy after 8 h solution treatment + 20 h aging treatment remained α-Al, Mg_2Si, and Si phases, as shown in Figure 6b.

The metallographic structure of the "post-refinement" hypoeutectic Al-10Si-0.48Mg alloy is shown in Figure 6c, in which the phase in the gray region is the α-Al matrix, the black particle phase is the Mg_2Si intermetallic compound, and the needle block phase in the Al matrix is Si. It can be seen that the corner of the needle block was rounded. The α-Al phase was in the shape of large clusters with coarser grains compared with the pre-refinement one, indicating that the refinement effect of the Al-5Ti-1B refiner did not work well in the "post-refinement" preparation process. Due to the early addition of the

Si element in the "post-refinement" method, the Ti and Si combined first, so the "silicon poisoning" phenomena was thus obvious.

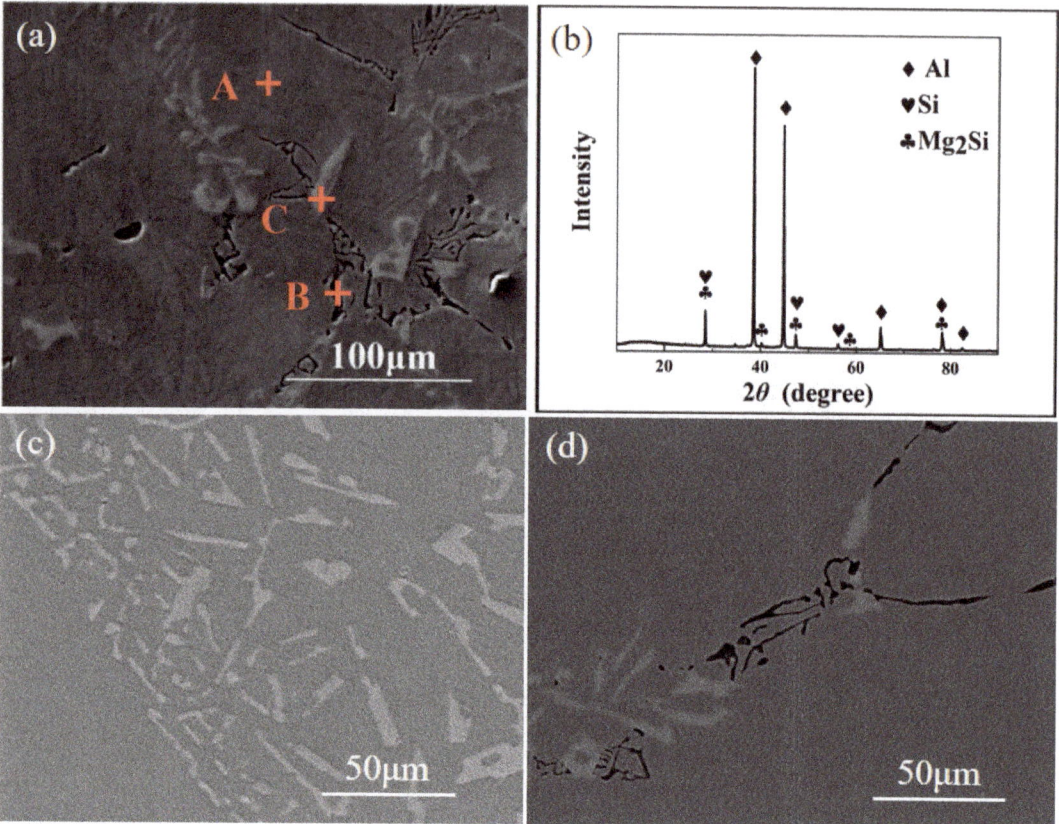

Figure 6. (**a**) SEM morphology of "pre-refinement" as-cast sample (A+, B+ and C+ are the positions of the spectral point scanning respectively); (**b**) XRD diffraction patterns of "pre-refinement" sample (8 + 20); (**c**) SEM morphology of "post-refinement" as-cast sample; (**d**) SEM morphology of "non-refinement" as-cast sample.

The metallographic structure of the "non-refinement" hypoeutectic Al-10Si-0.48Mg alloy is shown in Figure 6d, in which the grey region is the α-Al matrix, the black particle phase in the shape of Chinese character is the Mg_2Si intermetallic compound, and the black line phase is the Al-Si eutectic. As can be seen in Figure 6d, the matrix had much coarser dendrites than the above two types of alloys. The Chinese-character Mg_2Si phase had obvious angles appearing at the periphery, which can easily cause stress concentration in the preparation process and in service, and can damage the matrix and cause a large reduction in the mechanical properties of the alloy.

The macrostructures of the Al-10Si-0.48Mg alloy after different heat treatments are shown in Figure 7. The grain size distribution was uniform and the grain size of the pre-refinement sample was significantly smaller than that of the other two treatments. The grains of the pre-refinement samples showed fine equiaxed grains, the grains of the post-refinement samples showed coarse equiaxed grains, and the grains of the non-refinement samples showed coarse columnar grains.

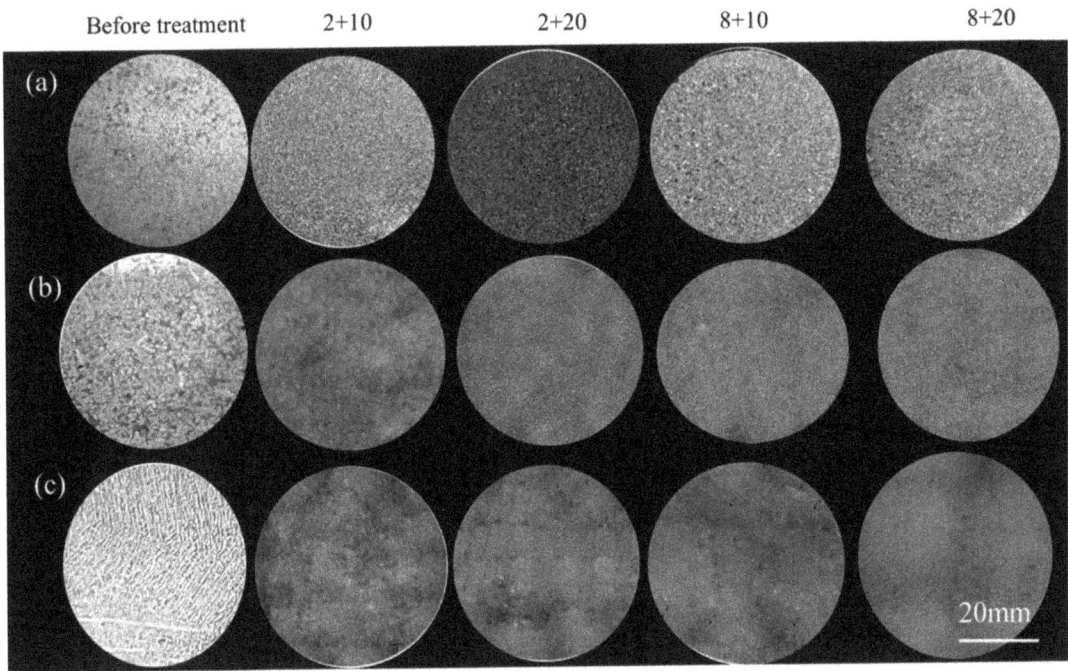

Figure 7. Macrostructures of Al-10Si-0.48Mg alloys after different heat treatments: (**a**) pre-refinement; (**b**) post-refinement; (**c**) non-refinement.

The grain size of an Al alloy can be approximately measured as the dendrite size of α-Al, which has polyhedral equiaxed grains. Under a polarizing microscope, different orientation grains have different colors. Because the grain color distribution has a certain randomness, it indicates that the grain orientation obtained by the casting process is randomly arranged. Figure 8 shows the polarized microstructure and the average size of α-Al dendrites after different preparation methods and heat treatments. As shown in Figure 8, the average α-Al dendrite sizes of the "pre-refinement", "post-refinement", and "non-refinement" Al-10Si-0.48Mg alloys were 284.91 μm, 715.61 μm, and 1526.82 μm, respectively. The average grain size of samples after multiple casting can reflect the refining ability of the three kinds of smelting processes. It can be seen the "pre-refinement" process had an obvious refinement effect on the Al-10Si-0.48Mg alloy; the grain size was only 18.66% of that of "non-refinement". Although the grain size of "post-refinement" was 46.87% of that of "non-refinement", the grain-refining capacity of the Al-5Ti-1B was obviously weakened by the "silicon poisoning".

Compared with the effect of the casting preparation on the grain size, the solution and aging treatments had little effect on the grain size. However, the morphology was affected by the heat treatment. As shown in Figure 8, the morphology of α-Al changed significantly in the "pre-refinement" Al-10Si-0.48Mg alloy during the heat treatments. As the solid solution treatment proceeded, the size of eutectic silicon decreased and the morphology changed from needle to spherical. With further solid solution treatment from 2 h to 8 h, the size of eutectic silicon increased. During the solid solution treatment, the eutectic silicon underwent two processes with time, including fragmentation or dissolution and the spherization of separated branches. If the dissolution time was too long, obvious pores would occur, leading to excessive burning. The coarsening of the microstructural composition and possible pores would have a negative impact on the mechanical properties. The coarsening of spheroidal eutectic silicon α-Al could be observed in either the "pre-refinement" or the "post-refinement" alloys, while only the process of breaking of eutectic silicon could be observed in the "non-refinement" alloys during the treatment.

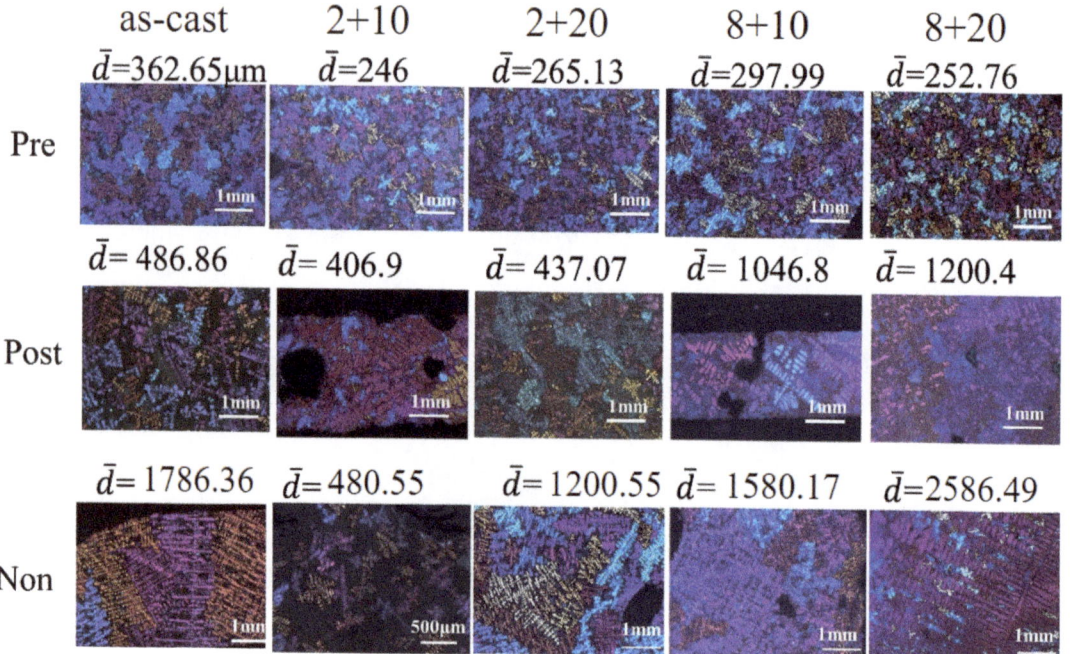

Figure 8. Dendrite structure and average grain diameter (μm) of the "pre-refinement", "post-refinement", and "non-refinement" as-cast and various heat-treated samples.

4.2. Mechanical Properties of the As-Cast and Heat-Treated Alloys

Al-Si-Mg cast alloys are usually heat treated to obtain the best combination of strength and ductility. Solution treatment and the following aging treatment are the commonly used heat treatment tools, and precipitation hardening is the main mechanism during heat treatment of Al-Si-Mg alloys. For the Al-Mg$_2$Si quasi-binary system, the Mg$_2$Si phase precipitates as follows: SSS → GP zone → β″ → β′ → β (Mg$_2$Si). These particles are invisible under the optical microscope, but this change can be indirectly observed with the change in alloy hardness and tensile strength due to the precipitates [26].

In this paper, the hardness of "pre-refinement", "post-refinement", and "non-refinement" as-cast and as-heated alloys of Al-10Si-0.48Mg alloy were tested. In the hardness test, six points on the surface of the sample were selected uniformly and randomly, and the hardness of the material was obtained by obtaining the average value of the six points. The measured hardness curve is shown in Figure 9. As can be seen in Figure 9, the effect of heat treatments was the most obvious in the "pre-refinement" alloy and the least in the "non-refinement" alloy. The highest hardness was achieved by the "pre-refinement" alloy with the 2 solid solution + 20 aging treatment.

Some studies showed that increasing the aging temperature led to the higher solid solubility of the Si element in α solid solution with almost no change in the Mg element. Thus, the Si content at higher aging temperature exceeds the content needed to form the reinforcing phase Mg$_2$Si, and the amount of precipitated Mg$_2$Si phase is dependent on the Mg content but independent of aging temperature [27].

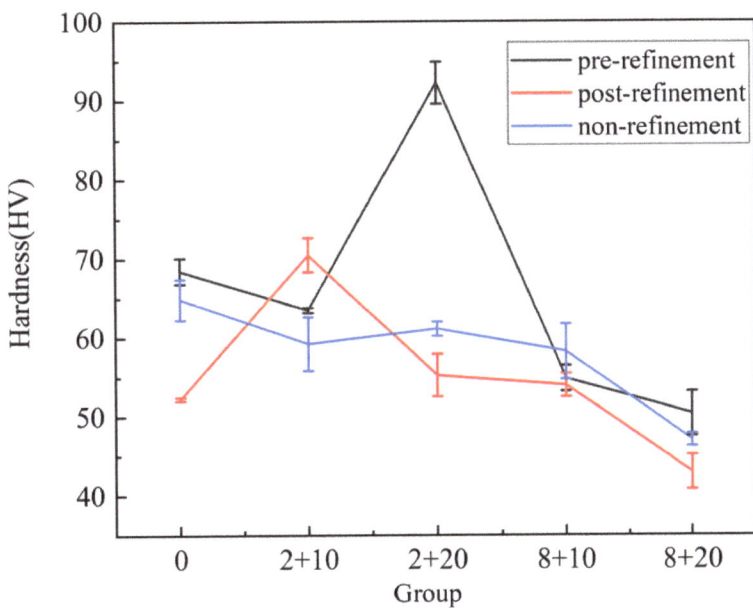

Figure 9. Hardness curves of "pre-refinement", "post-refinement", and "non-refinement" hypoeutectic Al-10Si-0.48Mg alloys with different heat-treatment methods.

The hardness decreased after the solid solution treatment because the dissolving of the secondary phases occurred during the casting process and then gradually increased during the aging process due to the precipitation of the secondary phase. As can be seen in Figure 9, the hardness of "pre-refinement" reached a peak value of about 92 HV with 2 h solid solution + 20 h aging. This was called the peak aged (PA) state and was related to the formation of a metastable intermetallic. With the aging time continually increasing, the hardness decreased, which corresponded to an over-aged state that coarsened the precipitates, lost compatibility with the matrix, and gradually reduced the hardness value [28]. Likewise, the hardness had a similar change with the varied heat-treatment condition for the "post-refinement" alloy and the "non-refinement", as shown by the curve in red and blue lines in Figure 9. However, it can be seen that the aging effect was small for the non-refinement alloys and the heat-treatment hardness was less than the as-cast hardness, indicating that few precipitates were produced during the aging.

Figure 10 shows the tensile curves of the three cast alloys under different heat treatment processes of "pre-refinement", "post-refinement", and "non-refinement". As can be seen in Figure 10a, the tensile strength of the hypoeutectic Al-10Si-0.48Mg alloy samples of each kind of alloy increased significantly after heat treatment and then gradually decreased the extension of aging time, which was in accord with the change in hardness and with the normal hardness–aging time relation. The elongation of the original hypoeutectic Al-10Si-0.48Mg alloy was not high for each kind of as-cast alloy due to the large secondary phases in the matrix, as shown in Figure 8. During the heat treatment, the large secondary phase dissolved into the matrix and then precipitated in a fine shape during aging, which benefited the elongation of the materials. It can be seen in Figure 10b that the elongation of the hypoeutectic Al-10Si-0.48Mg was increased by up to about 30%. The tensile strength of the alloy prepared using the "pre-refinement" method could reach about 212 MPa with 2 h solid solution + 10 h aging, which was a combination of a fine grain size and fine precipitates after the heat treatment.

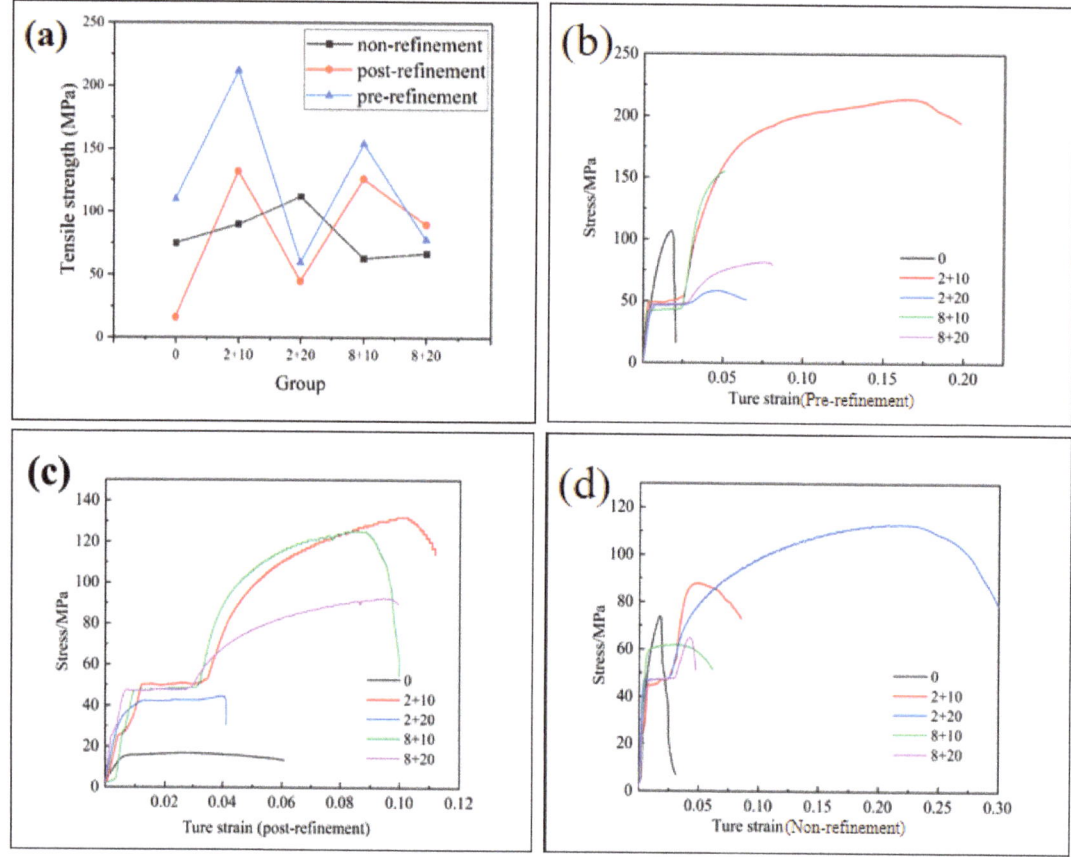

Figure 10. (a) Tensile strength of three casting alloys after "pre-refinement", "post-refinement", and "non-refinement" treatment; (b) tensile curves of "pre-refinement" casting alloys after different heat treatments; (c) tensile curves of "post-refinement" casting alloys after different heat treatments; (d) tensile curves of "non-refinement" casting alloys after different heat treatments.

Compared with the effect of heat treatment on the increase in mechanical properties for the as-cast alloy, it can be seen the same heat treatment had a greater effect on the mechanical properties of the alloys with a finer grain. Therefore, obtaining a desirable as-cast microstructure is the first step in obtaining more excellent mechanical properties of alloys whether followed by the heat treatment or not. For example, the "pre-refinement" concept used to refine the grain of as-cast alloy in this paper promises excellent comprehensive mechanical properties for Al-Mg-Si alloys after heat treatment.

5. Conclusions

In the paper, we proposed a concept of "pre-refinement" to avoid the "silicon poisoning" problem to refine hypoeutectic Al-Si alloys using a conventional refiner system such as Al-Ti-B. The core concept was to adjust the order of adding the Si element to form the $TiAl_3$ before forming the Ti-Si intermetallic compound. The feasibility of this method was demonstrated by the contrast experiments on the "pre-refinement", "post-refinement", and "non-refinement" Al-10Si-0.48Mg alloys. The grain size of the as-cast "pre-refinement" alloy was much smaller than either the "post-refinement" one or the "non-refinement" one, demonstrating that the "pre-refinement" method could effectively avoid the "silicon poisoning" effect and recover the refinement effect of the conventional refiner system. Based on a refined grain, the "pre-refinement" Al-10Si-0.48Mg alloy showed the best optimization effect in mechanical properties upon a solid solution and subsequent aging heat treatment. The tensile strength of the alloy prepared using the "pre-refinement" method could reach about 212 MPa with 2 h solid solution + 10 h aging with a good elongation of 20% compared with the "non-refinement" one with a tensile strength of 112 MPa and an elongation of about 30%.

Author Contributions: Data curation, Y.X. (Ying Xie) and M.L.; Investigation, M.L., L.L. and J.W.; Phase diagram calculation, W.B.; Project administration, F.L.; Writing—original draft preparation, L.L., L.Z. and J.W.; Writing—review and editing, F.L., Y.X. (Yu Xie) and W.B. All authors have read and agreed to the published version of the manuscript.

Funding: This research was funded by the Hunan Provincial Natural Science Foundation of China (No. 2021JJ30672) and the College Students' Innovation and Entrepreneurship Training Program of Xiangtan University.

Institutional Review Board Statement: Not applicable.

Informed Consent Statement: Not applicable.

Data Availability Statement: Not applicable.

Conflicts of Interest: The authors declare no conflict of interest.

References

1. Samuel, E.; Golbahar, B.; Samuel, A.M.; Doty, H.W.; Valtierra, S.; Samuel, F.H. Effect of grain refiner on the tensile and impact properties of Al–Si–Mg cast alloys. *Mater. Des.* **2014**, *56*, 468–479. [CrossRef]
2. Dong, X.X.; He, L.J.; Mi, G.B.; Li, P.J. Two directional microstructure and effects of nanoscale dispersed Si particles on microhardness and tensile properties of AlSi$_7$Mg meltspun alloy. *J. Alloys Compd.* **2015**, *618*, 609–614. [CrossRef]
3. Easton, M.A.; Qian, M.; Prasad, A.; StJohn, D.H. Recent advances in grain refinement of light metals and alloys. *Curr. Opin. Solid State Mater. Sci.* **2016**, *20*, 13–24. [CrossRef]
4. Rosenhain, W.; Grogan, J.D.; Schofield, T.H. Gas removal and grain refinement in aluminium alloys. *J. Inst. Meter.* **1930**, *44*, 305–318.
5. Jones, G.; Pearson, J. Factors affecting the grain-refinement of aluminum using titanium and boron additives. *Metall. Mater. Trans. B* **1976**, *7*, 223–234. [CrossRef]
6. Limmaneevichitr, L.; Eidhed, W. Fading mechanism of grain refine-ment of aluminum-silicon alloy with Al-Ti-B grain refiners. *Mater. Sci. Eng. A* **2003**, *349*, 197–206. [CrossRef]
7. Vabdyoussefi, M.; Worth, J.; Greer, A.L. Effect of instability of TiC particles on gain refinement of Al and Al-Mg alloys by addition of Al-Ti-C inoculations. *Mater. Sci. Technol.* **2000**, *16*, 1121–1128. [CrossRef]
8. Johnsson, M. Influence of Si and Fe on the grain refinement of aluminum. *Z. Met.* **1994**, *85*, 781–785.
9. Qiu, D.; Taylor, J.A.; Zhang, M.X.; Kelly, P.M. Mechanism for the poisoning effect of silicon on the grain refinement of Al-Si alloys. *Acta Mater.* **2007**, *55*, 1447–1456. [CrossRef]
10. Lee, Y.C.; Dahle, A.K.; Stjohn, D.H.; Hutt, J.E.C. The effect of grain refinement and silicon content on grain formation in hypoeutectic Al–Si alloys. *Mater. Sci. Eng. A* **1999**, *259*, 43–52. [CrossRef]
11. Schumacher, P.; Mckay, B.J. TEM investigation of heterogeneous nucleation mechanisms in Al-Si alloys. *J. Non-Cryst. Solids* **2003**, *317*, 123–128. [CrossRef]
12. Quested, T.E.A.; Dinsdale, T.; Greer, A.L. Thermodynamic evidence for a poisoning mechanism in the Al–Si–Ti system. *Mater. Sci. Technol.* **2006**, *22*, 1126–1134. [CrossRef]
13. Wang, F.; Liu, Z.L.; Qiu, D.; Taylor, J.A.; Easton, M.A.; Zhang, M.X. Revisiting the role of peritectics in grain refinement of Al alloys. *Acta Mater.* **2013**, *61*, 360–370. [CrossRef]
14. Schumaoher, P.S.; Greer, A.L. On the reproducibility of heterogeneous nucleation in amorphous Al85Ni10Ce5 alloys. *Mater. Sci. Eng. A* **1997**, *226–228*, 794–798. [CrossRef]
15. Schumaoher, P.S.; Greer, A.L. High-resolution transmission electron microscopy of grain-refining particles in amorphous aluminum alloys. In *Light Metals*; Hale, W., Ed.; Office of Scientific and Technical Information, U.S. Department of Energy: Warrendale, PA, USA, 1996; pp. 745–753.
16. Mark, A.E.; Arvind, P.; David, H.S. The Grain Refinement of Al-Si Alloys and the Cause of Si Poisoning: Insights Revealed by the Interdependence Model. *Mater. Sci. Forum* **2014**, *794–796*, 161–166.
17. Li, Y.; Hu, B.; Liu, B.; Nie, A.M.; Gu, Q.F.; Wang, J.F.; Li, Q. Insight into Si poisoning on grain refinement of Al-Si/Al-5Ti-B system. *Acta Mater.* **2020**, *187*, 51–65. [CrossRef]

18. Wu, D.Y.; Ma, S.D.; Jing, T.; Wang, Y.D.; Wang, L.S.; Kang, J.; Wang, Q.; Wang, W.; Li, T.; Su, R. Revealing the mechanism of grain refinement and anti Si-poisoning induced by (Nb, Ti)B$_2$ with a sandwich-like structure. *Acta Mater.* **2021**, *219*, 1359–6454. [CrossRef]
19. Zhang, H.L.; Han, Y.F.; Dai, Y.B.; Wang, J.; Sun, B.D. An ab initio molecular dynamics study: Liquid-Al/solid-TiB$_2$ interfacial structure during heterogeneous nucleation. *J. Phys. D* **2012**, *45*, 455307. [CrossRef]
20. Zhang, H.L.; Han, Y.F.; Dai, Y.B.; Lu, S.S.; Wang, J.; Zhang, J.; Shu, D.; Sun, B.D. An ab initio study on the electronic structures of the liquid-solid interface between TiB$_2$ (0001) surface and Al melts. *J. Alloys. Compd.* **2014**, *615*, 863–867. [CrossRef]
21. Zhang, H.L.; Han, Y.F.; Zhou, W.; Dai, Y.B.; Wang, J.; Sun, B.D. Atomic study on the ordered structure in Al melts induced by liquid/substrate interface with Ti solute. *Appl. Surf. Sci.* **2015**, *106*, 041606.
22. Fan, Z.; Wang, Y.; Zhang, Y.; Qin, T.; Zhou, X.R.; Thompson, G.E.; Pennycook, T.; Hashimoto, T. Grain refining mechanism in the Al/Al–Ti–B system. *Acta Mater.* **2015**, *84*, 292–304. [CrossRef]
23. Harald, F.; Tilo, G.; Hans, L.L.; Ferdinand, S. Investigation of the Al-Mg-Si system by experiments and thermodynamic calculations. *J. Alloys. Compd.* **1997**, *247*, 31–42.
24. China Mechanical Engineering Heat Treatment Society. *Heat Treatment Manual, the First Volume Process Basis*, 4th ed.; Mechanical Industry Press: South Norwalk, CT, USA, 2013; p. 486.
25. Said, B.; Zakaria, B.; Pascal, P.; Yann, B.P. Effects of heat treatment and addition of small amounts of Cu and Mg on the microstructure and mechanical properties of Al-Si-Cu and Al-Si-Mg cast alloys. *J. Alloys. Compd.* **2019**, *784*, 1026–1035.
26. Paray, F.; Gruzleski, G.E. Modification-a Parameter to Consider in the Heat Treatment of Al-Si Alloys. *Cast Metals* **1992**, *5*, 187–198. [CrossRef]
27. Siddiqui, R.A.; Abdullah, H.A.; Al-Belushi, K.R. Influence of aging parameters on the mechanical properties of 6063 aluminium alloy. *J. Mater. Processing Technol.* **2000**, *102*, 234–240. [CrossRef]
28. Scattergood, R.O.; Bacon, D.J. The Orowan mechanism in anisotropic crystals. *J. Exp. Theor. Phys.* **1975**, *31*, 179–198. [CrossRef]

Article

Development of a Thermomechanical Treatment Mode for Stainless-Steel Rings

Irina Volokitina [1], Ekaterina Siziakova [2,*], Roman Fediuk [3,4,*] and Alexandr Kolesnikov [5,*]

1. Department of Metallurgy and Mining, Rudny Industrial Institute, Rudny 111500, Kazakhstan; irinka.vav@mail.ru
2. Mineral Raw Material Processing Faculty, Saint Petersburg Mining University, 199106 St. Petersburg, Russia
3. Polytechnic Institute, Far Eastern Federal University, 690922 Vladivostok, Russia
4. Peter the Great St. Petersburg Polytechnic University, 195251 St. Petersburg, Russia
5. Department of "Life Safety and Environmental Protection" M. Auezov, South Kazakhstan University, Shymkent 160012, Kazakhstan
* Correspondence: sizyakova_ev@pers.spmi.ru (E.S.); fedyuk.rs@dvfu.ru (R.F.); kas164@yandex.kz (A.K.); Tel.: +7-7052566897 (A.K.)

Abstract: This article describes a technology for the thermomechanical treatment of stainless-steel piston rings. This technology makes it possible to obtain rings with an optimal combination of plastic and strength properties that is essential for piston rings. The following thermomechanical treatment is suggested for piston rings manufacturing: quenching at 1050 °C, holding for 30 min and cooling in water, then straining by the HPT method for eight cycles at cryogenic temperature and annealing at a temperature up to 600 °C. The resulting microstructure consisted of fine austenite grains sized 0.3 µm and evenly distributed carbide particles. Annealing above this temperature led to the formation of ferrite in the structure; however, preserving the maximum fraction of austenitic component is very important, since the reduction of austenite in the structure will cause a deterioration of corrosion resistance. The strength properties of steel after such treatment increased by almost two times compared with the initial ones: microhardness increased from 980 MPa to 2425 MPa, relative elongation increased by 20%. The proposed technology will improve the strength and performance characteristics of piston rings, as well as increase their service life, which will lead to significant savings in the cost of repair, replacement and downtime.

Keywords: severe plastic deformation; stainless steel; thermomechanical treatment; microstructure; mechanical properties

1. Introduction

The drive for the highest possible efficiency in manufacturing is reflected in the growing development of complex production processes and specialized materials [1–10]. In terms of manufacturing process development, this means a tendency to reduce the number of different production steps while using materials as efficiently as possible [11–15]. In terms of material performance, against an increasingly important, lightweight design, high strength with good ductility is of great importance [16,17]. Most manufacturing processes for processing metallic materials cause local or global changes in material properties. These production properties, meanwhile, can be used purposefully to increase component performance or reduce material usage in reverse order. Therefore, synergistic effects can be used to improve overall part production efficiency by adapting the production process to purposefully adjust certain local material properties [18,19].

The elastic properties of piston rings can be prolonged, and their failure reduced to a minimum by selecting the correct alloy grade and method of thermo-mechanical treatment. As a consequence, combustion engine manufacturers around the world are constantly searching for new technologies for piston rings production. Thus, improving the strength and performance characteristics of piston rings is an important technical task.

Conventional ultrafine-grained materials with grain sizes in the scale of several micrometers are usually manufactured using thermomechanical processes. An ultrafine-grained structure cannot be obtained by classical methods because of dynamic reduction and recrystallization processes, as well as limitations in formability [20–25]. Ultra-thin and nanocrystalline materials have been the subject of extensive research for a long time because their mechanical properties allow us to expect great potential as structural materials. Such materials can be made using two opposing approaches [26–29]. The first approach, often referred to as the "bottom-up process", is based on the aggregation of individual atoms or nanoscale particles to create a compact material. These include vapor deposition processes, electrolytic deposition processes and powder metallurgy methods [30–32]. These methods allow the production of nanocrystalline materials with very small grain sizes but with limited workpiece dimensions. They are not commonly used to make superfine cloth. The second approach, called the "top-down" process, is based on grinding the grains using severe plastic deformation (SPD). The SPD processes create ultra-thin cloths by applying extreme stretching at high hydrostatic pressures and low homologous temperatures [33–37]. The achievable minimum grain sizes depend on the material properties, but as a rule, grain sizes up to the nanocrystalline state are not achieved. The most used and investigated methods of severe plastic deformation are equal-channel angular pressing, multiple isothermal forging and high-pressure torsion.

Shear strain-based equal-channel angular pressing (ECAP) was first presented by Segal et al. [38] and is one of the most frequently used SPD methods. Over the past two decades, work related to ECAP of metallic materials has attracted considerable interest among scientists in both fields of metal physics and materials science. This interest arose because of the possibility to process large volumes [39–41]. One of the main purposes of such works is to grind metal grains to an ultrafine grained or nanostructured state. This ensures that such metals achieve a unique set of physical and mechanical properties [42–44]. The next objective of such studies is to investigate the mechanisms of formation of ultrafine grained structures in the SPD process, since ECAP can achieve very high degrees of strain. In this case, the shape and dimensions of the strained workpieces do not change. In recent works [45–47], ECAP was combined with cryogenic temperature, producing even better results.

In high-pressure torsion (HPT), shear strains are introduced into flat cylindrical specimens by torsion at high hydrostatic pressure stresses [48,49]. A number of different works [50,51] has found that SPD by the HPT method affects the structure of the material and increases the density of crystal lattice defects. Many scientists have succeeded in obtaining the smallest grain size microstructure in various materials using the HPT method rather than other SPD methods. As a result, it was possible to study the peculiarities of such a structure and evaluate its mechanical and physical properties. Material deformed by the HPT method is processed non-uniformly along the grains' radius. As a result, grains far from the center are more deformed than those in the center of the disk, so the microstructure becomes anisotropic. Since in our case the ring workpiece will be strained, this disadvantage can be avoided.

It should also be noted that when a certain accumulated degree of deformation is reached, the grinding process slows down and then stops altogether. This phenomenon still does not have an accepted explanation. However, one of the possible reasons is that there is some equilibrium between the strain grinding of the grains and their thermo-activated growth [52,53]. Therefore, grinding the structure to a nanocrystalline level by SPD only is not yet possible in most cases. As a result, new technologies are needed to continue the process of grinding the microstructure down to the nanocrystalline state.

There are several papers [54–56] that demonstrate the activation of new strain mechanisms by low temperatures. Such mechanisms suggest the possibility of continuing the process of grain refinement and reaching the nanoscale level. Lowering the temperature to cryogenic values provides improved mechanical properties and, as a result, increased wear resistance and hardness. The surface quality is also improved for polishing or finishing, which is necessary for piston rings (the presence of soft and ductile austenitic areas in the surface layer structure prevents the creation of a homogeneous mirror surface).

The main purpose of this article was to develop a new combined processing technology for piston rings used in internal combustion engines which will improve their performance and mechanical properties. This new technology combines high-pressure torsion with cryogenic temperature.

2. Materials and Methods

A special construction was developed to implement the HPT process on the existing equipment of the laboratory through the rectilinear movement of the upper die relative to the bed. The rectilinear motion of the upper die with the lower die attached to it transmits a torque due to the contact friction forcing directed at an oblique angle to the response part of the die. As a result, the rectilinear movement turns into a torsional movement.

Drawings were developed based on the analysis of scientific and technical literature and modeling in the Deform program package. The construction consists of several parts: an upper die which receives progressive motion from the press; a lower die which receives torque from the progressive motion of the upper die and the matrix itself, which holds the workpiece in the form of a ring (Figure 1).

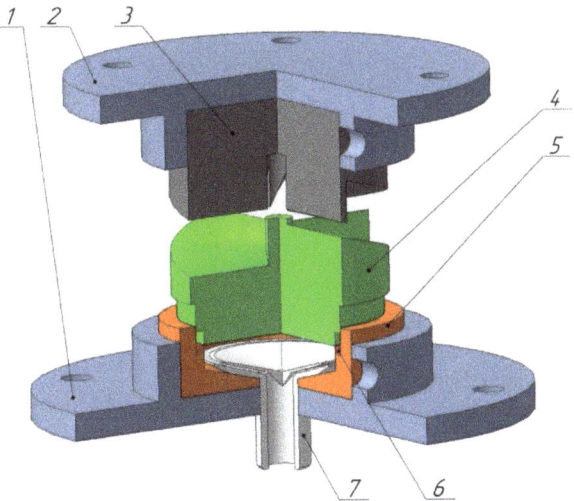

Figure 1. General view of the complete construction: 1—bottom carrier, 2—top carrier, 3—upper striker, 4—lower striker, 5—matrix, 6—piston ring, 7—nozzle.

Four spiral notches are present on the lower edge of the upper die. There is a cylindrical hole in the center of the upper die for the strain gauge rod and to ensure the alignment of the two dies.

The lower die has three steps. This structural solution is required for the straining of a ring workpiece (as in this case, rather than of a disk workpiece). The first pass (the second intermediate step) provides a contact with the side edge of the die where the workpiece is inserted. The second pass (the third lower step) provides contact with the workpiece along its inner radius, completely closing its cross section. According to this principle, the inner form of the matrix should also have a stepped form. The width of the step should correspond to the width of the circular workpiece to be machined.

A technological hole was made in the lower die to implement the high-pressure torsion process at cryogenic temperature. This hole was used to supply liquid nitrogen into the workpiece strain chamber. During the modeling, it was decided to use sprinkler-type nozzles, as, contrary to conventional nozzles, they allowed a uniform supply of nitrogen to the entire ring surface. These nozzles were made from polyurethane using a 3D printer.

The experiment itself and the assembly of the structure were carried out in the University laboratory on a single-curve hot-stamping press, model PB 6330-02, whose force was 1000 kN (Figure 2). Since there a martensitic transformation occurs in austenitic steels, the strain was applied out at cryogenic temperature and at room temperature for comparison. Therefore, depending on the temperature at which the workpieces were strained, the amount of martensite in the structure could vary greatly. The number of strain cycles was 8.

Figure 2. Construction fixed on the press.

The deformation blanks were ring-shaped, 76 mm in diameter, 3.5 mm wide and 3 mm in thickness. Since the piston rings do not work in aggressive media, AISI-304 austenitic stainless steel was chosen as the workpiece material. The initial structure before strain was obtained by quenching at 1050 °C, holding at this temperature for 30 min, and cooling in water. After such preheating, the γ-solid solution was fixed in chromium–nickel steel.

It is known that a significant disadvantage of strongly deformed materials is their very low ductility which limits the possibility of their practical application. The plastic properties of such material can be recovered by applying a final heat treatment. Therefore, a laboratory experiment was conducted to increase the ductility of the deformed samples obtained with the HPT method. The samples were cut into thin plates with a thickness of 5 mm after HPT and subjected to holding at temperatures of 300–650 °C for 15 min, followed by cooling in water.

The structure was studied using a JEM2100 transmission electron microscope (TEM) (Akishima, Japan) with a magnification range of 1000 to 50,000 times. Thin foil for the microstructure study was prepared by thinning using electrolytic polishing in an electrolyte consisting of 400 mL of H_3PO_4 and 60 g of CrO_3 at room temperature and voltage of 20 V, with current density of 2.5 A/cm². For a more objective interpretation of the structure and the analysis of transformations, an a EBSD analysis was carried out using Philips XL-30 SEM (Amsterdam, The Netherlands) with a field cathode at an accelerating voltage of 20 kV. The results were processed using the Tex SEM Lab software 4.2. The scans were performed on 50 μm × 50 μm sections at 0.2 μm increments. Given the experimental accuracy of the EBSD method, all low-angle boundaries with a disorientation of less than 2° were excluded from consideration.

The microhardness of the samples was measured by the Vickers method using a DM-8 automatic microhardness tester (Affri, Induno, Italy). The load was 1 N.

Mechanical uniaxial tension tests were performed at room temperature on an Instron 5882 machine at a deformation rate of 1.0 mm/min. The sample deformation was measured with an Instron strain gauge. Tensile tests were carried out on flat samples cut from the ring (working part dimensions: width of 3 mm, thickness of 3 mm and length of 6 mm) in accordance with GOST 1497-84 recommendations. Tensile tests of mechanical properties were carried out to determine strength and ductility characteristics: yield strength ($\sigma_{0.2}$), tensile strength (σ_B) and maximum elongation to failure (δ).

3. Results

The microstructure of AISI-304 stainless steel prior to HPT was coarse-grained with polyhedral grains, with an average size of 32 μm and the presence of twins. The structure contained ≈100% austenite (after a preliminary heat treatment—quenching). The microstructure obtained after applying the strain by the HPT method at room temperature and using cryogenic temperature is shown in Figure 3.

Figure 3. Microstructure of AISI-304 steel after straining by HPT: (**a**) at cryogenic temperature; (**b**) at room temperature.

EBSD analysis was performed to obtain additional information on grain size, texture and disorientation of the boundaries. Orientation maps of the microstructure of AISI-304 steel after eight cycles of high-pressure torsion deformation are shown in Figure 4.

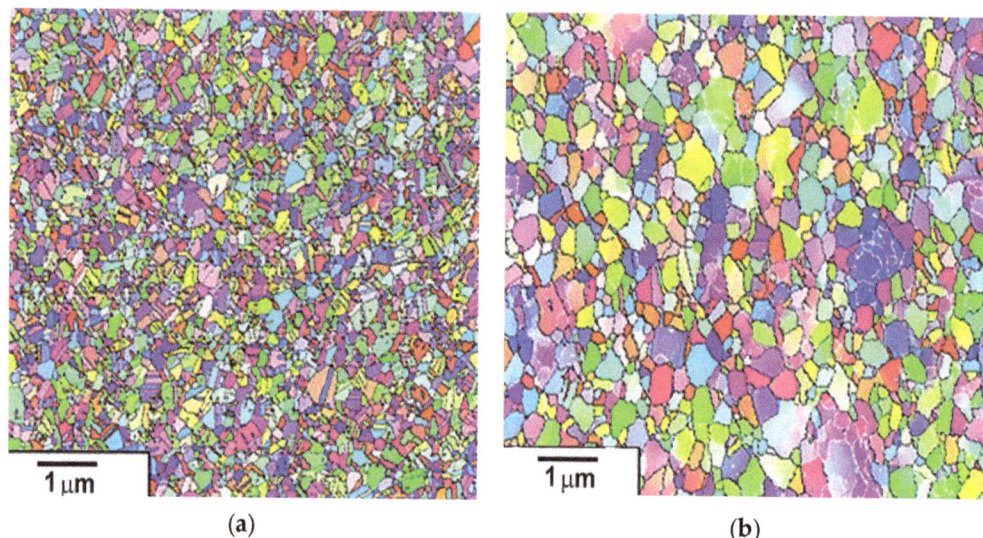

Figure 4. Microstructure orientation maps of AISI-304 steel after straining by HPT: (**a**) at cryogenic temperature; (**b**) at room temperature.

Tensile tests were carried out to determine the mechanical characteristics after metallographic studies. In addition, interrupted tensile tests were performed to record the progress of the start zone and the crack growth zone as fully as possible. Tests on the samples obtained after applying the strain by the HPT method at room temperature were carried out at room and at cryogenic temperature. Photographs of the samples' surface fractography shown in Figures 5 and 6 were analyzed using SEM. Each photo shows the average area of the fracture at 3000× magnification.

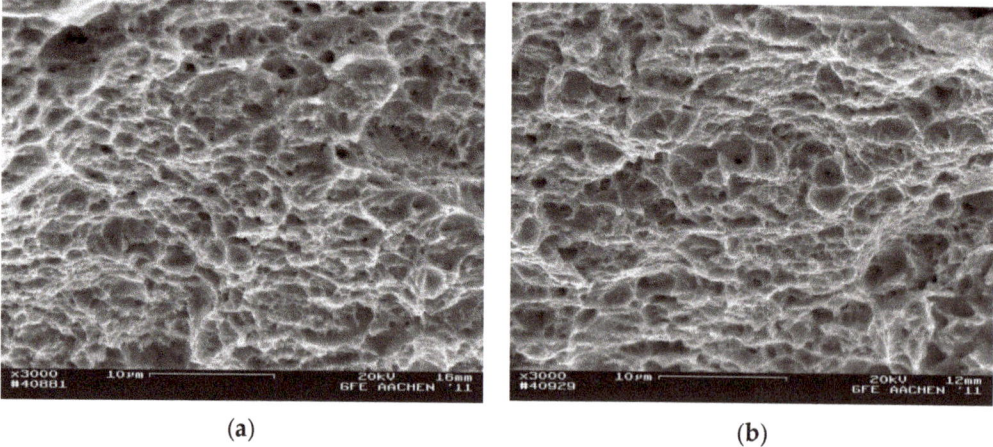

Figure 5. Fractography of the fracture surface in the crack start zone: (**a**) at cryogenic temperature; (**b**) at room temperature.

Samples after deformation through the HPT method were annealed at temperatures of 300–650 °C with an exposure time of 15 min to observe changes in the microstructure. This was done to check the possibility of preserving the mechanical properties and microstructure during heating. The evolution of the microstructure of the deformed samples during heating is shown in Figures 7 and 8.

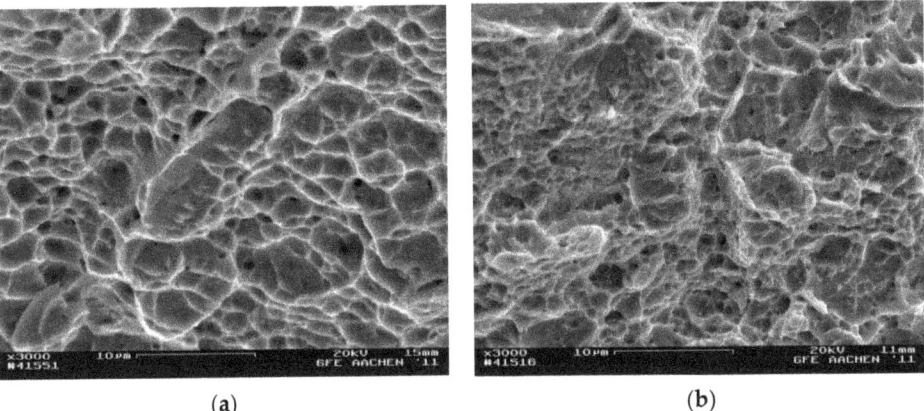

Figure 6. Fractography of the fracture surface in the crack growth zone: (**a**) at cryogenic temperature; (**b**) at room temperature.

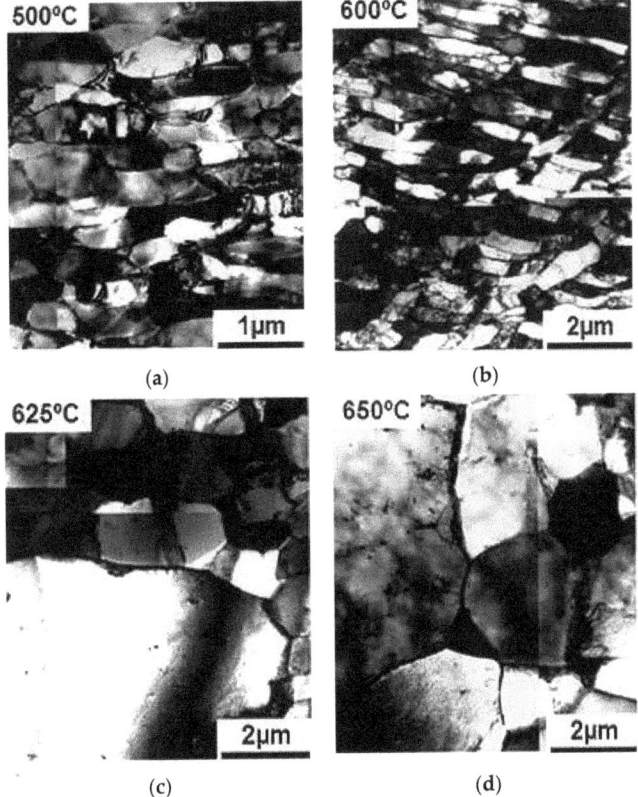

Figure 7. Microstructure of steel deformed with the HPT method at cryogenic temperature during heating: (**a**) 500 °C; (**b**) 600 °C; (**c**) 625 °C; (**d**) 650 °C.

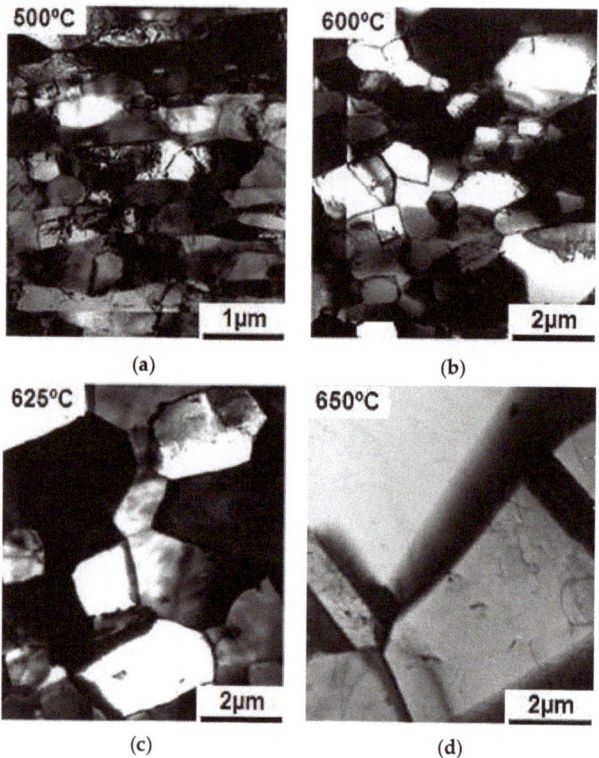

Figure 8. Microstructure of steel obtained with the HPT method at room temperature when heated at (**a**) 500 °C; (**b**) 600 °C; (**c**) 625 °C; (**d**) 650 °C.

After the strain and heating tests, mechanical tensile tests were carried out. The diagrams are shown in Figure 9.

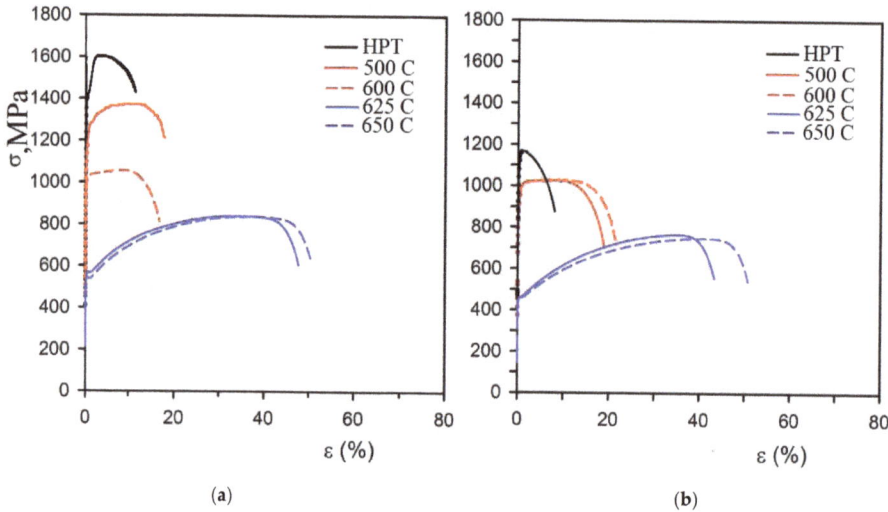

Figure 9. Tensile diagrams of samples obtained after HPT and heating during deformation: (**a**) at cryogenic temperature; (**b**) at room temperature.

4. Discussion

Analysis of the microstructure of the samples after deformation by the HPT method at room and cryogenic temperatures showed that eight strain cycles resulted in a homogeneous nanostructure in both cases (Figure 3). However, using cryogenic temperature resulted in a finer-grained structure. The cryogenic temperature resulted in a martensitic structure with a grain size of 0.2 μm (Figure 3a), whereas the room temperature resulted in a 0.5 μm microstructure consisting of a mixture of austenite and α-martensite (Figure 3b).

The EBSD analysis showed that the misorientation of the strain sub-boundaries increased during the deformation at both temperatures, i.e., the fraction of low-angle sub-boundaries decreased with increasing true strain. Strain twinning, which is characteristic at low strain rates, decreased after eight deformation cycles at room temperature, which was confirmed by the elimination of the peak with a misorientation angle of ~60. Intensive twinning was also observed at the eighth strain cycle when using cryogenic temperature. As a result, after eight deformation cycles with intensive cooling, the structure contained a large number of twins ~57%, while at room temperature, only 13% of twins were observed.

The distribution of boundaries on the angles of misorientation in both states was close. With conventional HPT, the fraction of low-angle boundaries was 18%, while with nitrogen treatment, it was 12%. The results showed that the total proportion of large-angle boundaries was at least 80%. This indicates the formation of a nanostructure with a predominance of large-angle boundaries in the workpieces.

The tensile tests showed that the obtained nanocrystalline structure had in both cases increased strength properties (Figure 9). In the initial state, we determined a yield strength of—275 MPa, a tensile strength of 515 MPa, and relative elongation of 40%. The formation of a nanocrystalline structure after eight cycles of strain by the HPT method at cryogenic temperature with a grain size of 45–50 nm led to an increase in the tensile strength to 1603 MPa compared to the initial state. The yield strength increased to 1282 MPa. The value of ductility decreased sharply up to 18% compared to the initial state. The samples deformed at room temperature showed the following mechanical properties: the tensile strength increased to 1198 MPa, and the yield strength increased to 1005 MPa. The value of ductility decreased up to 9% compared to the initial state.

The microhardness results correlated with the mechanical tensile test data and indicated that high-pressure torsion in the new die allowed obtaining a fairly homogeneous hardness across the entire cross section of the ring. After eight HPT cycles at cryogenic temperature, the microhardness increased almost three times compared to the initial state: from 980 MPa to 2715 MPa. Strain at room temperature resulted in an increase in microhardness to 2530 MPa. In this case, the main increase in hardness was in the first four passes—40%.

Low plastic properties were obtained based on the results of the study of mechanical properties using both deformation methods. This is a significant disadvantage of almost all samples obtained by severe plastic deformation methods [57–59]. Therefore, it is necessary to carry out additional research on crack growth during fracture.

The fracture surface of AISI 304 steel strained at cryogenic temperature in the crack start zone had a quasi-viscous nature. This was characterized by a ductile fracture mechanism. This was confirmed, as shown in Figure 5a, by the presence of spalling that alternated with dimpled fracture. The samples strained at room temperature had a brittle–ductile fracture in the crack start zone (Figure 5b).

The crack in the growth zone in the steel deformed at cryogenic temperature formed according to a microviscosity mechanism, and the pits were evenly spaced along the fracture surface (Figure 6a). The fracture surface in samples strained at room temperature was covered with spalls and pits indicating that the crack was spreading in a quasi-brittle way.

Based on the data obtained, cryogenic temperature allowed better plastic properties due to the martensite structure obtained in the samples, but these properties were still insufficient for further use. To increase the plastic properties, it is necessary to reduce the internal stress; this could be achieved by annealing. Annealing to 500 °C did not change the microstructure of samples processed at both cryogenic and room temperatures, and only a rearrangement of dislocations occurred (Figures 7a and 8a). Recrystallization began at 600 °C in the samples obtained by deformation at room temperature. This was shown by the separately occurring grains, while no annealing twins were observed yet (Figure 8c). There were still no changes in the microstructure in samples processed at cryogenic temperature (Figure 7c). When the samples obtained with room temperature deformation reached 625 °C, a completely recrystallized structure with an average grain size of 2 μm was observed, and annealing twins were observed (Figure 8c). The samples processed at cryogenic temperature started recrystallization at 625 °C which caused a strong growth of individual grains above 3 μm (Figure 7c). There were also annealing twins. Once the structure reached 650 °C, it became completely recrystallized with a grain size of 3 μm (Figure 7d). The effects of annealing at 650 °C on samples strained at room temperature did not practically differ from those on samples obtained at cryogenic temperature. The structure was fully recrystallized, with a grain size of 4 μm (Figure 8d).

The rate of grain growth in HPT-processed samples at cryogenic temperature after heating was several times faster than in samples HPT strained at room temperature. This can be explained by a more intense deformation occurring at cryogenic temperature and a correspondingly higher degree of recrystallization.

Another feature of the structural change during steel heating was the development of a reverse-phase transformation of strained martensite back to austenite or ferrite.

The thermal stability of steel after HPT was also studied by analyzing the dependence of its microhardness change on the annealing temperature. We observed a slight decrease in microhardness when heating specimens deformed at both cryogenic and room temperature. The decrease was greater as the is the heating temperature increased, but it remained at a level much higher than that of microhardness after hardening. Therefore, during annealing up to 500 °C, the microhardness of samples which had been deformed at cryogenic temperature decreased from 2715 to 2555 MPa, and that of samples deformed at room temperature decreased from 2530 to 2205 MPa.

At 600 °C, the samples obtained by deformation at room temperature showed a sharp decrease in microhardness, which reached 1365 MPa. This indicated the beginning of recrystallization processes. After annealing at 650 °C, microhardness decreased to 985 MPa.

The same was observed in the samples deformed at cryogenic temperature: when heating to 625 °C, we observed a sharp decrease in microhardness from 2555 to 1725 MPa; when annealing at 650 °C, microhardness was 1260 MPa.

5. Conclusions

The results showed that during high-pressure torsion at cryogenic and room temperatures during the first strain cycles, the difference in grain size and structure was not large, since the adiabatic effects were comparable. Cryogenic temperature became effective only after four cycles of deformation, when the defect density increased dramatically. The structure of samples strained at both temperatures was refined to a nanostructure level. Thus, straining AISI-316 steel with an initial grain size of 32 μm at room temperature led to an equiaxed homogeneous microstructure of 0.5 μm. The structure consisted of a mixture of austenite and α-martensite. Straining at cryogenic temperature resulted in an equiaxial homogeneous microstructure sized 0.2 μm, consisting of 90% α-martensite with a predominance of large-angle boundaries.

Austenitic steel after annealing at 600 °C showed an optimal combination of ductility and strength, which is essential for piston rings. Therefore, the following thermomechanical treatment is proposed for piston rings manufacturing: quenching at 1050 °C, holding for 30 min and cooling in water, then straining by the HPT method for eight cycles at cryogenic temperature and annealing at a temperature up to 600 °C. As a result of this treatment, the microstructure will consist of fine austenite grains sized 0.3 μm and evenly distributed carbide particles.

Author Contributions: Conceptualization, I.V., E.S. and R.F.; methodology, A.K.; investigation, I.V., A.K. and R.F.; data curation, A.K.; writing—original draft preparation, I.V. and A.K.; writing—review and editing, R.F.; supervision, E.S. and R.F.; project administration, I.V., A.K. and R.F.; funding acquisition, E.S. and R.F. All authors have read and agreed to the published version of the manuscript.

Funding: This research received no external funding.

Institutional Review Board Statement: Not applicable.

Informed Consent Statement: Not applicable.

Data Availability Statement: Data sharing is not applicable to this article.

Conflicts of Interest: The authors declare no conflict of interest.

References

1. Bazhin, V.Y.; Issa, B. Influence of heat treatment on the microstructure of steel coils of a heating tube furnace. *J. Min. Inst.* **2021**, *249*, 393. [CrossRef]
2. Pryakhin, E.I.; Sharapova, D.M. Understanding the structure and properties of the heat affected zone in welds and model specimens of high-strength low-alloy steels after simulated heat cycles. *CIS Iron Steel Rev.* **2020**, *19*, 60. [CrossRef]
3. Kolesnikov, A.; Fediuk, R.; Amran, M.; Klyuev, S.; Klyuev, A.; Volokitina, I.; Naukenova, A.; Shapalov, S.; Utelbayeva, A.; Kolesnikova, O.; et al. Modeling of Non-Ferrous Metallurgy Waste Disposal with the Production of Iron Silicides and Zinc Distillation. *Materials* **2022**, *15*, 2542. [CrossRef] [PubMed]
4. Aryshnskii, E.V.; Bazhin, V.Y.; Kawalla, R. Strategy of refining the structure of aluminum-magnesium alloys by complex microalloying with transition elements during casting and subsequent thermomechanical processing. *Non-Ferr. Met.* **2019**, *46*, 28. [CrossRef]
5. Lutskiy, D.S.; Ignativich, A.S. Study on hydrometallurgical recovery of copper and rhenium in processing of substandard copper concentrates. *J. Min. Inst.* **2021**, *251*, 723. [CrossRef]
6. Prokopchuk, N.R.; Globa, A.I.; Laptik, I.O. The properties of metal coatings enhanced with diamond nanoparticles. *Tsvetnye Met.* **2021**, *2021*, 50. [CrossRef]
7. Nadirov, K.S.; Zhantasov, M.K.; Bimbetova, G.Z.; Sadyrbayeva, A.S. Examination of optimal parameters of oxy-ethylation of fatty acids with a view to obtaining demulsifiers for deliquefaction in the system of skimming and treatment of oil: A method to obtain demulsifier from fatty acids. *Chem. Today* **2016**, *34*, 72–77.
8. Zhakipbaev, B.Y.; Zhanikulov, N.N.; Kolesnikova, O.G.; Akhmetova, K.; Kuraev, R.M.; Shal, A.L. Review of technogenic waste and methods of its processing for the purpose of complex utilization of tailings from the enrichment of non-ferrous metal ores as a component of the raw material mixture in the production of cement clinker. *Rasayan J. Chem.* **2021**, *14*, 997–1005. [CrossRef]
9. Kolesnikov, A.; Fediuk, R.; Kolesnikova, O.; Zhanikulov, N.; Zhakipbayev, B.; Kuraev, R.; Akhmetova, E.; Shal, A. Processing of Waste from Enrichment with the Production of Cement Clinker and the Extraction of Zinc. *Materials* **2022**, *15*, 324. [CrossRef]
10. Kenzhibaeva, G.S.; Botabaev, N.E.; Kutzhanova, A.N.; Iztleuov, G.M.; Suigenbaeva, A.Z.; Ashirbekov, K.A.; Kolesnikova, O.G. Thermodynamic Modeling of Chemical and Phase Transformations in a Waelz Process-Slag—Carbon System. *Refract. Ind. Ceram.* **2020**, *61*, 289–292. [CrossRef]
11. Kolesnikov, A.S. Kinetic investigations into the distillation of nonferrous metals during complex processing of waste of metallurgical industry. *Russ. J. Non-Ferr. Met.* **2015**, *56*, 1–5. [CrossRef]
12. Vasilyeva, N.; Fedorova, E. Big Data as a Tool for Building a Predictive Model of Mill Roll Wear. *Symmetry* **2021**, *13*, 859. [CrossRef]
13. Milyuts, V.G.; Tsukanov, V.V.; Pryakhin, E.I. Development of manufacturing technology for high-strength hull steel reducing production cycle and providing high-quality sheets. *J. Min. Inst.* **2019**, *239*, 536. [CrossRef]
14. Vasilyeva, N.V.; Ivanov, P.V. Development of a control subsystem to stabilize burden materials charging into a furnace. *J. Phys. Conf. Ser.* **2019**, *1210*, 12158. [CrossRef]
15. Maksarov, V.V.; Olt, J.; Keksin, A.I. The use of composite powders in the process of magnetic-abrasive finishing of taps to improve the quality of threads in articles made of corrosion-resistant steels. *Chernye Met.* **2022**, *49*, 49–55. [CrossRef]
16. Volokitina, I.; Kolesnikov, A.; Fediuk, R.; Klyuev, S.; Sabitov, L.; Volokitin, A.; Zhuniskaliyev, T.; Kelamanov, B.; Yessengaliev, D.; Yerzhanov, A.; et al. Study of the Properties of Antifriction Rings under Severe Plastic Deformation. *Materials* **2022**, *15*, 2584. [CrossRef] [PubMed]
17. Volokitina, I.; Vasilyeva, N.; Fediuk, R.; Kolesnikov, A. Hardening of Bimetallic Wires from Secondary Materials Used in the Construction of Power Lines. *Materials* **2022**, *15*, 3975. [CrossRef] [PubMed]
18. Zgonnik, P.V.; Kuzhaeva, A.A.; Berlinskiy, I.V. The Study of Metal Corrosion Resistance near Weld Joints When Erecting Building and Structures Composed of Precast Structures. *Appl. Sci.* **2022**, *12*, 2518. [CrossRef]
19. Vasilieva, N.V.; Fedorova, E.R. Process control quality analysis. *Tsvetnye Met.* **2020**, *10*, 70. [CrossRef]
20. Bolobov, V.I.; Popov, G.G. Methodology for testing pipeline steels for resistance to grooving corrosion. *J. Min. Inst.* **2021**, *252*, 854. [CrossRef]

21. Fediuk, R.S.; Yushin, A.M. The use of fly ash the thermal power plants in the construction. *IOP Conf. Ser. Mater. Sci. Eng.* **2015**, *93*, 012070. [CrossRef]
22. Maksarov, V.; Efimov, A.; Olt, J. Improving the quality of hole processing in welded products made of dissimilar materials with a new boring tool. *Int. J. Adv. Manuf. Technol.* **2022**, *118*, 1027. [CrossRef]
23. Volodchenko, A.A.; Lesovik, V.S.; Cherepanova, I.A.; Volodchenko, A.N.; Zagorodnjuk, L.H.; Elistratkin, M.Y. Peculiarities of non-autoclaved lime wall materials production using clays. *IOP Conf. Ser. Mater. Sci. Eng.* **2018**, *327*, 022021. [CrossRef]
24. Siziakova, E.V.; Ivanov, P.V. On the role of hydrated calcium carboaluminate in the improvement of the production technology of alumina from nephelines. *J. Phys. Conf. Ser.* **2020**, *1515*, 22048. [CrossRef]
25. Fediuk, R.S.; Smoliakov, A.K.; Timokhin, R.A.; Batarshin, V.O.; Yevdokimova, Y.G. Using thermal power plants waste for building materials. *IOP Conf. Ser. Earth Environ. Sci.* **2018**, *87*, 092010. [CrossRef]
26. Volokitina, I.; Kurapov, G. Effect of Initial Structural State on Formation of Structure and Mechanical Properties of Steels under ECAP. *Met. Sci. Heat Treat.* **2018**, *59*, 786–792. [CrossRef]
27. Kul'chitskii, A.A.; Kashin, D.A. The choice of a method for non-contact assessment of the composition of briquetted charge materials. *J. Phys. Conf. Ser.* **2019**, *1399*, 044108. [CrossRef]
28. Choi, I.-S.; Schwaiger, R.; Kurmanaeva, L.; Kraft, O. On the effect of Ag content on the deformation behavior of ultrafine-grained Pd–Ag alloys. *Scr. Mater.* **2009**, *61*, 64. [CrossRef]
29. Naizabekov, A.; Volokitina, I. Effect of the Initial Structural State of Cr–Mo High-Temperature Steel on Mechanical Properties after Equal-Channel Angular Pressing. *Phys. Met. Metallogr.* **2019**, *120*, 177–183. [CrossRef]
30. Fediuk, R.; Mosaberpanah, M.A.; Lesovik, V. Development of fiber reinforced self-compacting concrete (FRSCC): Towards an efficient utilization of quaternary composite binders and fibers. *Adv. Concr. Constr.* **2020**, *9*, 387.
31. Kolesnikov, A.S.; Sergeeva, I.V.; Botabaev, N.E.; Al'Zhanova, A.Z.; Ashirbaev, K.A. Thermodynamic simulation of chemical and phase transformations in the system of oxidized manganese ore–carbon. *Izv. Ferr. Metall.* **2017**, *60*, 759–765. [CrossRef]
32. Bolobov, V.I.; Latipov, I.U.; Popov, G.G. Estimation of the influence of compressed hydrogen on the mechanical properties of pipeline steels. *Energies* **2021**, *14*, 6085. [CrossRef]
33. Xu, J.; Li, J.; Shan, D.; Guo, B. Microstructural evolution and micro/meso-deformation behavior in pure copper processed by equal-channel angular pressing. *Mater. Sci. Eng. A* **2016**, *664*, 114. [CrossRef]
34. Muszka, K.; Zych, D.; Lisiecka-Graca, P.; Madej, L.; Majta, J. Experimental and Molecular Dynamic Study of Grain Refinement and Dislocation Substructure Evolution in HSLA and IF Steels after Severe Plastic Deformation. *Metals* **2020**, *10*, 1122. [CrossRef]
35. Volokitina, I.; Volokitin, A. Evolution of the Microstructure and Mechanical Properties of Copper during the Pressing—Drawing Process. *Phys. Met. Metallogr.* **2018**, *119*, 917–921. [CrossRef]
36. Fediuk, R.; Smoliakov, A.; Muraviov, A. Mechanical properties of fiber-reinforced concrete using composite binders. *Adv. Mater. Sci. Eng.* **2017**, *2017*, 2316347. [CrossRef]
37. Valiev, R.Z.; Estrin, Y.; Horita, Z.; Langdon, T.G.; Zehetbauer, M.J.; Zhu, Y.T. Producing Bulk Ultrafine-Grained Materials by Severe Plastic Deformation. *JOM* **2006**, *58*, 33–39. [CrossRef]
38. Segal, V.M. Materials processing by simple shear. *Mater. Sci. Eng. A* **1995**, *197*, 157–164. [CrossRef]
39. Lezhnev, S.N.; Volokitina, I.; Kuis, D.V. Evolution of Microstructure and Mechanical Properties of Compo-site Aluminum-Based Alloy during ECAP. *Phys. Met. Metallogr.* **2018**, *119*, 810–815. [CrossRef]
40. Murashkin, M.Y.; Sabirov, I.; Kazykhanov, V.U. Enhanced mechanical properties and electrical conductivity in ultrafine-grained Al alloy processed via ECAP-PC. *J. Mater. Sci.* **2013**, *48*, 4501–4509. [CrossRef]
41. Fediuk, R.; Yushin, A. Composite binders for concrete with reduced permeability. *IOP Conf. Ser.-Mater. Sci. Eng.* **2016**, *116*, 012021. [CrossRef]
42. Volokitina, I.; Naizabekov, A. CuZn36 brass microstructure and mechanical properties evolution at equal channel angular pressing. *J. Chem. Technol. Metall.* **2020**, *55*, 586–591.
43. Dao, M.; Lu, L.; Asaro, R.; De Hosson, J.T.M.; Ma, E. Toward a quantitative understanding of mechanical behavior of nanocrystalline metals. *Acta Mater.* **2007**, *55*, 4041–4065. [CrossRef]
44. Valiev, R.Z.; Islamgaliev, R.K.; Alexandrov, I.V. Bulk nanostructured materials from severe plastic deformation. *Prog. Mater. Sci.* **2000**, *45*, 103–189. [CrossRef]
45. Volokitina, I. Effect of Cryogenic Cooling after ECAP on Mechanical Properties of Aluminum Alloy D16. *Met. Sci. Heat Treat.* **2019**, *61*, 234–238. [CrossRef]
46. Volokitina, I. Evolution of the Microstructure and Mechanical Properties of Copper under ECAP with Intense Cooling. *Met. Sci. Heat Treat.* **2020**, *62*, 253–258. [CrossRef]
47. Nayan, N.; Narayana Murty, S.V.S.; Jha, A.K.; Pant, B.; Sharma, S.C.; George, K.M.; Sastry, G.V.S. Mechanical properties of aluminium-copper-lithium alloy AA2195 at cryogenic temperatures. *Mater. Des.* **2014**, *58*, 445–450. [CrossRef]
48. Zhilyaev, A.P.; Ringot, G.; Huang, Y.; Cabrera, J.M.; Langdon, T.G. Mechanical behavior and microstructure properties of titanium powder consolidated by high-pressure torsion. *Mater. Sci. Eng. A* **2017**, *688*, 498–504. [CrossRef]
49. Zhilyaev, A.P.; Langdon, T.G. Using high-pressure torsion for metal processing: Fundamentals and applications. *Prog. Mater. Sci.* **2008**, *53*, 893–979. [CrossRef]
50. Kawasaki, M.; Ahn, B.; Lee, H.; Zhilyaev, A.; Langdon, T.G. Using high-pressure torsion to process an aluminum–magnesium nanocomposite through diffusion bonding. *J. Mater. Res.* **2015**, *31*, 88–99. [CrossRef]

51. Volokitin, A.; Naizabekov, A.; Volokitina, I.; Lezhnev, S.; Panin, E. Thermomechanical treatment of steel using severe plastic deformation and cryogenic cooling. *Mater. Lett.* **2021**, *304*, 130598. [CrossRef]
52. Jabir, H.A.; Abid, S.R.; Murali, G.; Ali, S.H.; Klyuev, S.; Fediuk, R.; Vatin, N.; Promakhov, V.; Vasilev, Y. Experimental Tests and Reliability Analysis of the Cracking Impact Resistance of UHPFRC. *Fibers* **2020**, *8*, 74. [CrossRef]
53. Lezhnev, S.; Volokitina, I.; Koinov, T. Research of influence equal channel angular pressing on the microstructure of copper. *J. Chem. Technol. Metall.* **2014**, *49*, 621–630.
54. Kon'Kova, T.N.; Mironov, S.Y.; Korznikov, A.V. Severe cryogenic deformation of copper. *Phys. Met. Metallogr.* **2010**, *109*, 171–176. [CrossRef]
55. Li, Y.S.; Tao, N.R.; Lu, K. Microstructural evolution and nanostructure formation in copper during dynamic plastic deformation at cryogenic temperatures. *Acta Mater.* **2008**, *56*, 230–241. [CrossRef]
56. Nadig, D.S.; Ramakrishnan, V.; Sampathkumaran, P.; Prashanth, C.S. Effect of cryogenic treatment on thermal conductivity properties of copper. *AIP Conf. Proc.* **2012**, *1435*, 133–139. [CrossRef]
57. Xu, C.; Horita, Z.; Langdon, T.G. The evolution of homogeneity in processing by high-pressure torsion. *Acta Mater.* **2006**, *55*, 203–212. [CrossRef]
58. Verma, D.P.; Pandey, S.A.; Bansal, A.; Upadhyay, S.; Mukhopadhyay, N.K.; Sastry, G.V.S.; Manna, R. Bulk Ultrafine-Grained Interstitial-Free Steel Processed by Equal-Channel Angular Pressing Followed by Flash Annealing. *J. Mater. Eng. Perform.* **2016**, *25*, 5157–5166. [CrossRef]
59. Langdon, T.G. The characteristics of grain refinement in materials processed by severe plastic deformation. *Rev. Adv. Mater. Sci.* **2006**, *13*, 6–14.

Article

Precipitation Evolution in the Austenitic Heat-Resistant Steel HR3C upon Creep at 700 °C and 750 °C

Liming Xu [1], Yinsheng He [2], Yeonkwan Kang [2], Jine-sung Jung [2] and Keesam Shin [1,*]

[1] School of Materials Science and Engineering, Changwon National University, Changwon 51140, Korea; xulimings2016@gmail.com
[2] KEPCO Research Institute, Korea Electric Power Corporation, Daejeon 34056, Korea; yinshenghe@kepco.co.kr (Y.H.); yeonkwan.kang@kepco.co.kr (Y.K.); jinesung.jung@kepco.co.kr (J.-s.J.)
* Correspondence: keesamgg@gmail.com; Tel.: +82-55-2130-3696

Abstract: HR3C (25Cr-20Ni-Nb-N) is a key material used in heat exchangers in supercritical power plants. Its creep properties and microstructural evolution has been extensively studied at or below 650 °C. The precipitation evolution in HR3C steel after creep rupture at elevated temperatures of 700 °C and 750 °C with a stress range of 70~180 MPa is characterized in this paper. The threshold strength at 700 °C and 750 °C were determined by extrapolation method to be $\sigma_{10^5}^{700}$ = 57.1 MPa and $\sigma_{10^5}^{750}$ =37.5 MPa, respectively. A corresponding microstructure investigation indicated that the main precipitates precipitated during creep exposure are Z-phase (NbCrN), $M_{23}C_6$, and σ phase. The dense Z-phase precipitated dispersively in the austenite matrix along dislocation lines, and remained stable (both size and fraction) during long-term creep exposure. $M_{23}C_6$ preferentially precipitated at grain boundaries, and coarsening was observed in all creep specimens with some continuous precipitation of granular $M_{23}C_6$ in the matrix. The brittle σ phase formed during a relatively long-term creep, whose size and fraction increased significantly at high temperature. Moreover, the σ phases, grown and connected to form a large "island" at triple junctions of grain boundaries, appear to serve as nucleation sites for high stress concentration and creep cavities, weakening the grain boundary strength and increasing the sensitivity to intergranular fracture.

Keywords: HR3C steel; microstructure; creep rupture; precipitates

Citation: Xu, L.; He, Y.; Kang, Y.; Jung, J.-s.; Shin, K. Precipitation Evolution in the Austenitic Heat-Resistant Steel HR3C upon Creep at 700 °C and 750 °C. *Materials* 2022, 15, 4704. https://doi.org/10.3390/ma15134704

Academic Editor: Konstantin Borodianskiy

Received: 31 May 2022
Accepted: 29 June 2022
Published: 5 July 2022

Publisher's Note: MDPI stays neutral with regard to jurisdictional claims in published maps and institutional affiliations.

Copyright: © 2022 by the authors. Licensee MDPI, Basel, Switzerland. This article is an open access article distributed under the terms and conditions of the Creative Commons Attribution (CC BY) license (https:// creativecommons.org/licenses/by/ 4.0/).

1. Introduction

HR3C (25Cr-20Ni-Nb-N) is an advanced austenitic steel, used as a superheater and reheater in ultra-supercritical (USC) boilers for its high-temperature oxidation resistance and outstanding creep strength [1,2]. Nevertheless, the microstructural degradation, especially precipitation behavior, inevitably affects the mechanical properties during long-term creep or exposure to temperatures of 600 °C and above.

In austenitic heat-resistant steel, a large number of research works reveal that the types of precipitates during long-term service exposure or creep are mainly NbCrN (Z-phase), $M_{23}C_6$, and σ phase at 600 °C or 650 °C. The precipitation and growth of different precipitates are closely related to the microstructural stability and mechanical properties [3–5]. Z-phase is a typical precipitate with a tetragonal structure with a = 0.3073 nm and b = 0.7391 nm, which has a strong hardening effect on the performance of the HR3C steel [6]. The fine dispersion of the Z-phase can pin dislocations and increase the strength, as Golański et al. and Bin, W et al. reported [7,8]. $M_{23}C_6$ carbide, as a metastable phase, is preferentially precipitated at the grain boundaries in the early stage of service exposure [9]. Zhang et al. discovered the growth and coarsening of $M_{23}C_6$ at the grain boundary, increasing the tendency of intergranular cracking [10]. The FeCr-type σ phase has a tetragonal structure with a = 0.88 nm and b = 0.45 nm, which is a common precipitate in stainless steels such as AISI304, AISI321, AISI347, and other similar types [11–13]. The presence of the σ phase greatly decreases the plasticity, toughness strength, and corrosion resistance of heat-resistant steel after long-term service, as Cao et al. reported [14].

However, since the temperature of the superheater and reheater in practical applications can be overheated to above 650 °C, additional research on the relationship between microstructural evolution and mechanical properties at a more elevated temperature is essential. In this study, microscopic observation and phase analysis were used to investigate the effect of various precipitates on the creep behaviors at temperatures up to 700 °C and 750 °C under different stresses.

2. Materials and Methods

In this work, the as-received HR3C boiler tube steel was domestically manufactured with the following specifications: outer diameter of 57 mm and wall thickness of 4.5 mm. The chemical composition of HR3C steels in this study is listed in Table 1.

Table 1. Chemical composition of the HR3C steel (wt. %).

Element	Cr	Ni	Nb	C	N	Mn	Si	Fe
ASTM A213	24.0~26.0	19.0~22.0	0.20~0.60	0.04~0.10	0.15~0.35	2.00 max	1.50 max	Balance
* EDS result	24.4	18.4	0.4	5.9	0.3	1.2	0.4	49.0

* EDS is carried out with a large area signal acquisition mode.

The samples for creep testing were manufactured with a diameter of 4 mm and gauge length of 25 mm according to ASTM E8. In accordance with ASTM E 139-11, the creep tests were carried out at 700 °C and 750 °C under different stresses of 70~180 MPa with the creep tester (ATS arm ratio creep tester, Series 2320 Lever Arm). The temperature of the samples was monitored by thermocouple and controlled by an induction heating system.

The microstructure was analyzed by JSM-6510 scanning electron microscope (SEM) (JEOL Ltd., Tokyo, Japan, equipped with INCA EDS, operated at an accelerating voltage of 20 kV), JSM-7900F high-resolution scanning electron microscope (HRSEM) (JEOL Ltd., Tokyo, Japan, equipped with EDS), and Philips CM200 transmission electron microscope (TEM) (FEI Ltd., Hillsboro, OR, USA, operated at 200 kV). Specimens for SEM were prepared by mechanical grinding, polishing, and final etching with Kalling's #2 reagent. Specimens for ECCI (Electron Channeling Contrast Imaging) and TEM observation were prepared by mechanical grinding, polishing, and final electrolytic twin-jet polishing (Struers TenuPol-5) with a solution of 10% perchloric acid in ethanol at 15 V for 40 s. The microhardness was tested by a Vickers hardness tester (Future-Tech. JP/FM-7) under a load of 200 gf for more than 50 iterations for each specimen.

3. Results and Discussion

3.1. Creep Rupture Test

The specific conditions and results of the creep tests are displayed in Table 2.

Table 2. Creep test conditions and time till rupture.

State	Stress (MPa)	Time (h)
As-received	-	-
700 °C	180	398
	150	1463
	120	3945
	90	9411
750 °C	140	222
	110	925
	90	2310
	70	4680

The relationship between the applied stress and microhardness as a function of time till rupture is shown in Figure 1. The curve in Figure 1a shows that the long-term creep rupture strengths of HR3C can be expressed by linear regression:

$$\sigma = A \log t + B \tag{1}$$

where t is the creep rupture time, σ is the applied stress, and A and B are constants which are related to the material and test temperature. Therefore, the threshold strength of HR3C steel at 700 °C and 750 °C after 10^5 h creep test can be determined by the extrapolation method as follows: $\sigma^{700}_{10^5}$ = 57.1 MPa, $\sigma^{750}_{10^5}$ = 37.5 MPa. According to the safety requirement of ASTM standard (KA-SUS310J1 TB) [15], the threshold strength for 10^5 h is 55 MPa at 700 °C and 25 MPa at 750 °C.

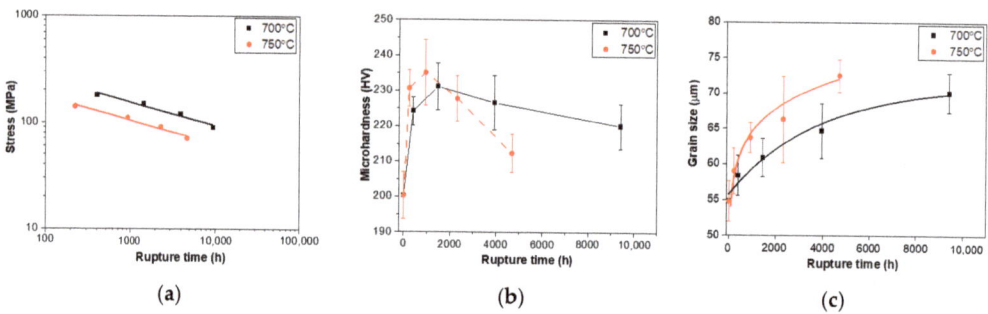

Figure 1. (a) The log-log plot of stress versus rupture time, (b) microhardness versus rupture time, and (c) grain size versus rupture time.

Figure 1b shows the variation of Vickers microhardness with rupture time. The microhardness increases rapidly at the early stages of the creep test and reaches the maximum value of 231 HV at 700 °C/1463 h and 235 HV at 750 °C/2310 h. Then, the microhardness decreases again with prolonged creep duration. The early increase in microhardness results from the continuous precipitation of $M_{23}C_6$ in the early stage of the creep.

The average grain size of HR3C steel was measured with low-magnification optical microscope images by the average grain intercept (AGI) method according to ASTM E112-13. The variation in grain size is shown in Figure 1c below. According to the ideal grain growth,

$$d^2 - d_0^2 = kt,$$

where d_0 is the initial grain size, d is the final grain size and k is a temperature-dependent constant given by an exponential law:

$$k = k_0 \exp(-Q/RT), \tag{2}$$

where k_0 is a constant, T is the absolute temperature and Q is the activation energy for boundary mobility.

The calculated value of k with the fitting curves is 0.162 and 0.372 at 700 °C and 750 °C, respectively, which means the grain growth rate at 750 °C is much faster than at 700 °C.

3.1.1. Fracture Morphology

SEM micrographs of the rupture surface at 700 °C and 750 °C are shown in Figures 2 and 3, respectively. Figures 2a–c and 3a,b show a totally intergranular brittle fracture feature with cleavage and a rock candy fracture surface. However, with decreasing applied stress and increasing time to rupture, a dimpled morphology appears on the grain facets at 700 °C/90 MPa/9411 h in Figure 2d and 750 °C/90 MPa/2310 h, 750 °C/70 MPa/4680 h in Figure 3c,d. Creep is a time-dependent deformation under a constant load or stress at elevated temperatures. Generally, the creep of a metal has three stages.

When a high stress is applied, there is no steady stage (secondary creep stage) of continuous microstructural changes under the service condition. The growth of $M_{23}C_6$ and σ phase at the grain boundaries and the continuous precipitation of granular $M_{23}C_6$ in the matrix occur at a low creep rate in the second creep stage. Coarse $M_{23}C_6$ and σ phase at the grain boundaries serve as high stress concentration and creep cavity nucleation sites, leading to the propagation of cracks on the grain boundaries and, eventually, intergranular fracture. Some partial dimple morphology formed on the fracture facets due to the cavity developed on the surface of $M_{23}C_6$ surrounded by the ductile matrix [16]. Figure 2e, the electropolished fracture surface, shows large fractured σ phases on the grain boundary, indicating the σ phases are insignificant to the strengthening.

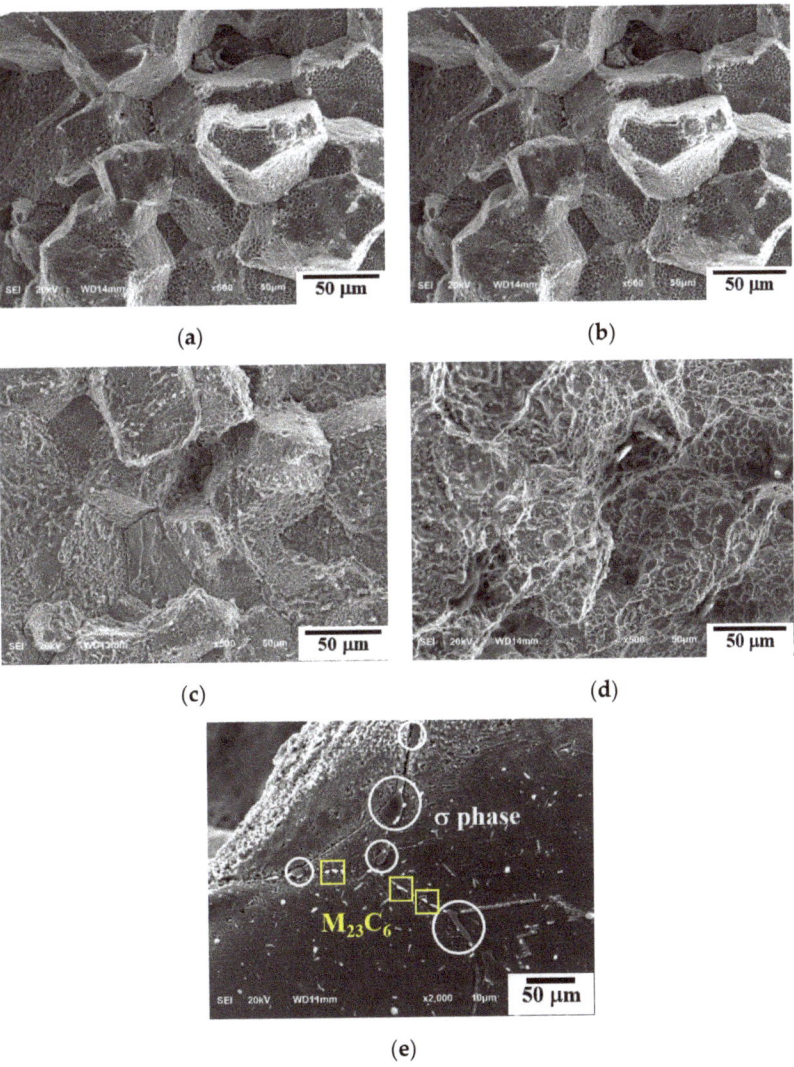

Figure 2. SEM micrographs of crept fractures: (**a**) 700 °C/180 MPa/398 h, (**b**) 700 °C/150 MPa/1463 h, (**c**) 700 °C/120 MPa/3945 h, (**d**) 700 °C/90 MPa/9411 h, and (**e**) electropolished fracture surface of (**d**).

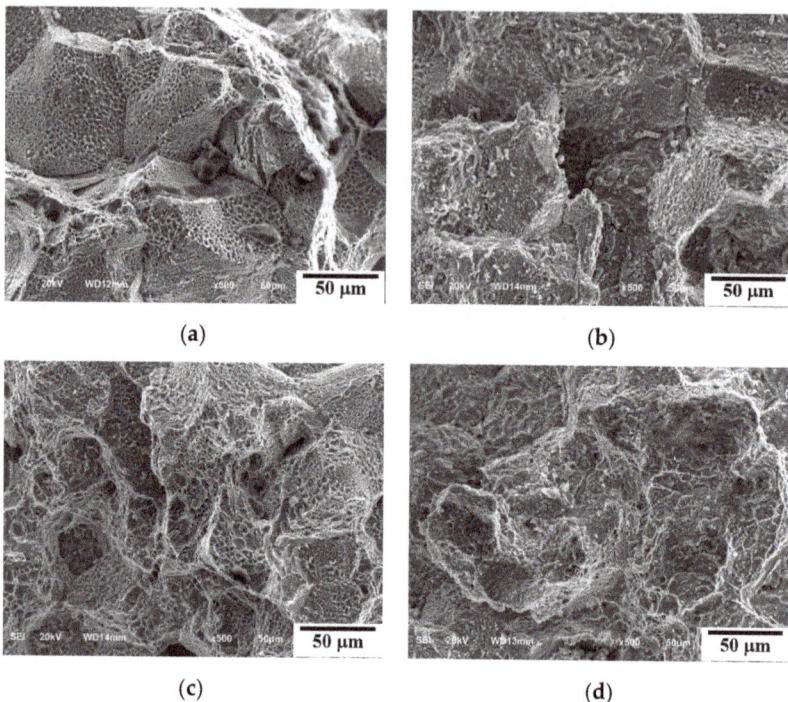

Figure 3. SEM micrographs of crept fractures: (**a**) 750 °C/150 MPa/222 h, (**b**) 750 °C/110 MPa/925 h, (**c**) 750 °C/90 MPa/2310 h, and (**d**) 750 °C/70 MPa/4680 h.

3.1.2. Precipitation Behavior

Typical precipitates of HR3C steel after the creep rupture test and the corresponding EDS results are shown in Figure 4. The EDS shows excessive precipitation at the grain boundaries and inside the grains. The large undissolved particles inside the grains are Cr- and Nb-rich nitrides at site 1 and site 2 (S1 and S2) in Figure 4a, and are the primary Z-phase. The continuous chain-like precipitates in the grain boundary are identified as Cr-rich $M_{23}C_6$ carbides at site 3 and site 4 (S3 and S4) in Figure 4a. In addition, the large blocky particles (~2 μm) are Fe- and Cr-rich σ phases at site 5 and site 6 (S5 and S6).

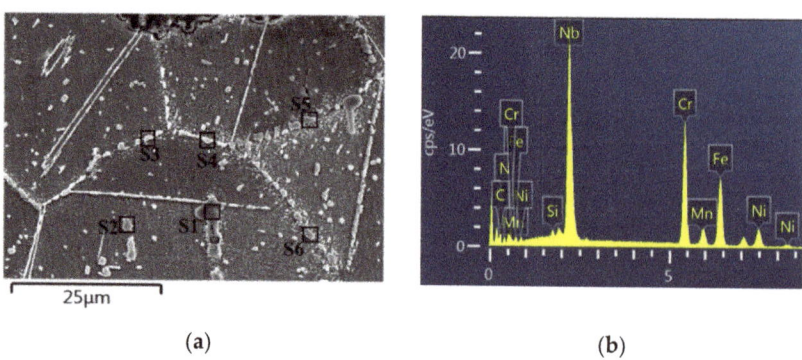

Figure 4. *Cont.*

Ele.	Wt. %					
	NbCrN		$M_{23}C_6$		σ phase	
	S1	S2	S3	S4	S5	S6
C	9.10	8.35	9.75	11.35	6.97	10.97
N	5.54	7.69	2.32	1.93	0.08	0.07
Si	0.44	0.19	3.07	3.53	0.91	0.38
Cr	23.78	29.30	45.31	40.03	33.69	32.10
Mn	0.46	0.38	0.61	0.39	1.13	1.04
Fe	19.29	10.03	19.60	21.64	45.00	39.83
Ni	7.08	2.91	18.62	20.79	12.09	15.34
Nb	34.32	41.14	0.71	0.35	0.14	0.26
Total	100.00	100.00	100.00	100.00	100.00	100.00

(c)

Figure 4. EDS analysis of precipitates in the 700 °C/90 MPa/9411 h crept specimen: (**a**) SEM micrographs in the near-fracture region, (**b**) the corresponding spectrum of site 1 (S1) and (**c**) EDS results of S1~S6 in (**a**) with red letters for major elements.

Figure 5 shows the Z-phase particles in the as-received and creep rupture specimens of 700 °C/90 MPa/9411 h. In the as-received specimen, the Z-phase precipitated in the austenite matrix uniformly, as shown in Figure 5. These coarse undissolved particles are considered as primary Z-phase with a size of ~1 μm. This was also reported by Zieliński, A [17]. In specimens with longer times to rupture, a fine Z-phase is observed, as shown in Figure 5c,d, also called the secondary Z-phase, with a size of ~50 nm. By interacting with the dislocations, present in high density, this fine dispersion of Z-phase enhances the strength of the matrix, as reported by Hu et al. [18].

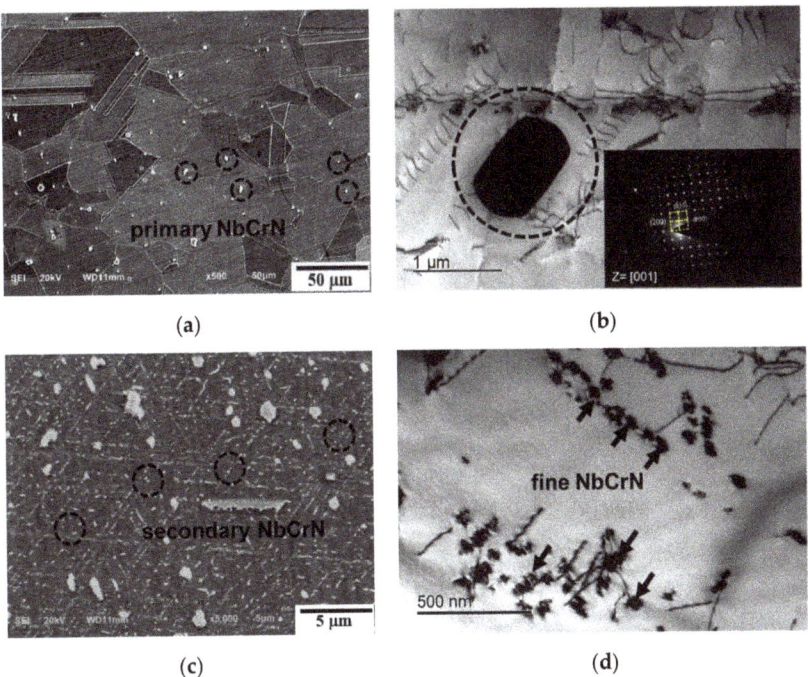

Figure 5. SEM and TEM micrographs of Z-phase in the specimen of: (**a**,**b**) as-received, (**c**,**d**) 700 °C/90 MPa/9411 h.

The SEM micrographs of a triple junction of the grain boundaries of creep-ruptured specimens at 700 °C and 750 °C are shown in Figures 6 and 7. The $M_{23}C_6$ precipitation is a diffusion-type phase transformation controlled by the driving force for nucleation and the diffusion of the C and Cr atoms in the austenite steel. Therefore, as shown in Figures 6a and 7a, the rod-like $M_{23}C_6$ particles (~200 nm) preferentially precipitated at the grain boundaries at the early stage of creep due to the higher interfacial energy of grain boundaries and its higher atom diffusion rate compared to those of the grain interiors [19,20]. With prolonged creep time, the $M_{23}C_6$ particles at the grain boundaries coarsened (up to ~600 nm) and gradually grew into chains, thereby decreasing the pinning efficiency. Meanwhile, the granular $M_{23}C_6$ particles (~200 nm) continuously precipitated in the matrix with the extension of the creep rupture time.

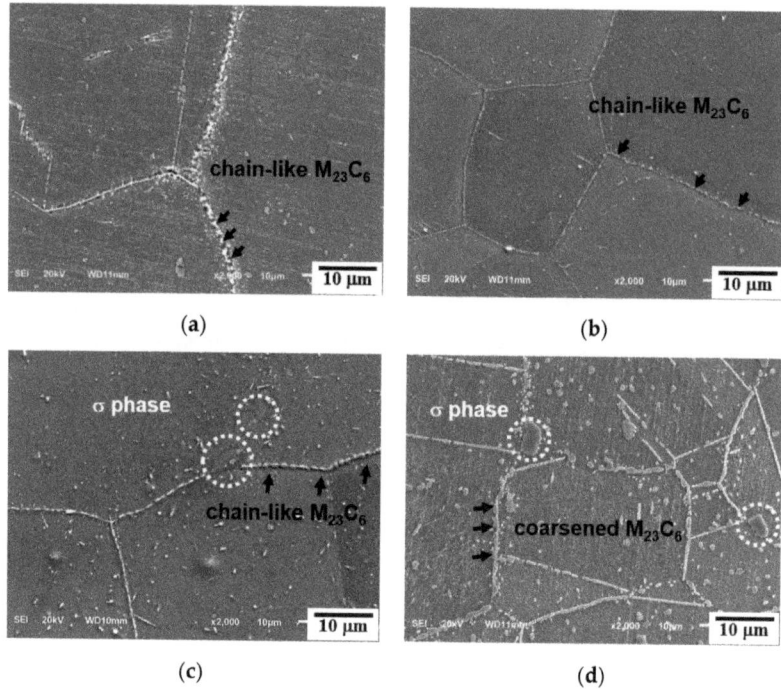

Figure 6. SEM micrographs from cross-sectional fractures of creep specimens: (**a**) 700 °C/180 MPa/398 h, (**b**) 700 °C/150 MPa/1463 h, (**c**) 700 °C/120 MPa/3945 h, and (**d**) 700 °C/90 MPa/9411 h.

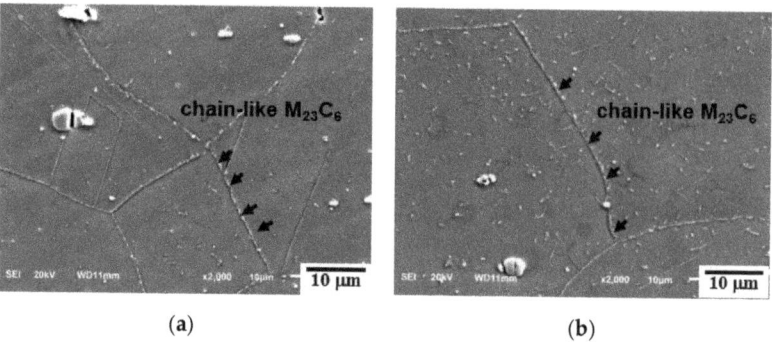

Figure 7. Cont.

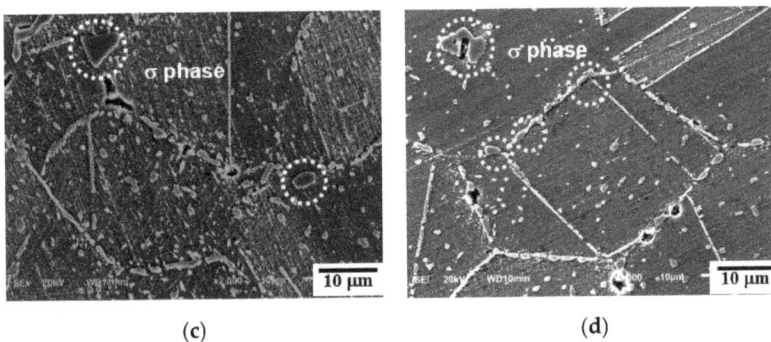

Figure 7. SEM micrographs from cross-sectional fractures of creep specimens: (**a**) 750 °C/150 MPa/222 h, (**b**) 750 °C/110 MPa/925 h, (**c**) 750 °C/90 MPa/2310 h, and (**d**) 750 °C/70 MPa/4680 h.

Figure 8 shows the distribution of $M_{23}C_6$ in the 9411 h crept specimen. The chain-like $M_{23}C_6$ carbides distributed along the grain boundaries and fine granular $M_{23}C_6$ carbides precipitated in the grain interior are shown in Figures 8b and 8c, respectively. Consistent with Figures 6a and 7a, the preferred precipitation of $M_{23}C_6$ carbides at the grain boundaries occurred due to the grain boundary having a higher interfacial energy and a faster diffusion rate for alloying atoms than in the grain interior [21]. The $M_{23}C_6$ carbides precipitated in the grain boundaries could provide good creep resistance due to the pinning effects of grain boundaries at the early stage of creep. However, $M_{23}C_6$ carbide is metastable, with low thermodynamic stability [22]. The coarsening of $M_{23}C_6$ carbides noticeably weakened the grain boundaries and increased the risk of embrittlement. In addition, the growth of Cr-rich $M_{23}C_6$ consumed Cr from the matrix and contributed to the formation of a Cr-depleted region near the grain boundaries, resulting in intergranular corrosion [23–25]. On the other hand, the fine $M_{23}C_6$ precipitated in the grain interior increased the strength of the matrix through precipitate hardening and preventing the motion of dislocations according to the Orowan law, as shown in Figure 8c.

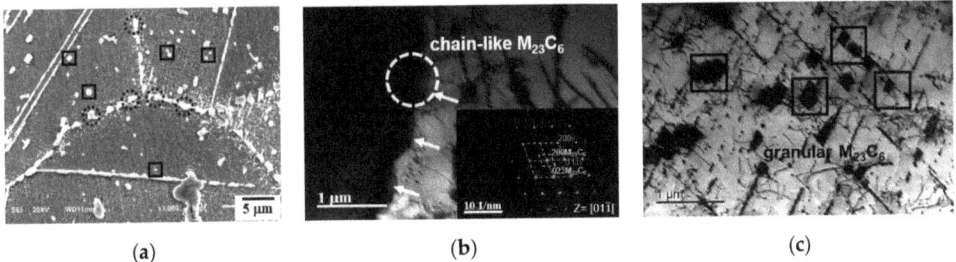

Figure 8. Distribution of $M_{23}C_6$ in the 9411 h crept specimen: (**a**) SEM micrographs of two kinds of $M_{23}C_6$, (**b**) TEM micrographs of coarsened $M_{23}C_6$ distributed along the grain boundary, and (**c**) TEM micrographs of fine $M_{23}C_6$ in the grain interior. Precipitates in dotted circles indicate chain-like $M_{23}C_6$ at the grain boundary and precipitates in solid rectangles indicate granular $M_{23}C_6$ in the grain interior.

The ECCI micrographs of precipitations in the grain boundaries are shown in Figure 9. It can be seen that only $M_{23}C_6$ carbides are observed in the grain boundaries at the beginning of the creep process, Figure 9a,c. At creep times of 3945 h at 700 °C and 2310 h at 750 °C, the blocky σ phase was observed at the grain boundaries. It can be seen that the blocky σ phase grew and connected to a larger σ phase "island", serving as nucleation sites for high stress concentration and creep cavities and leading to the propagation of cracks on the grain boundaries and an eventual intergranular fracture, as shown in Figure 9d.

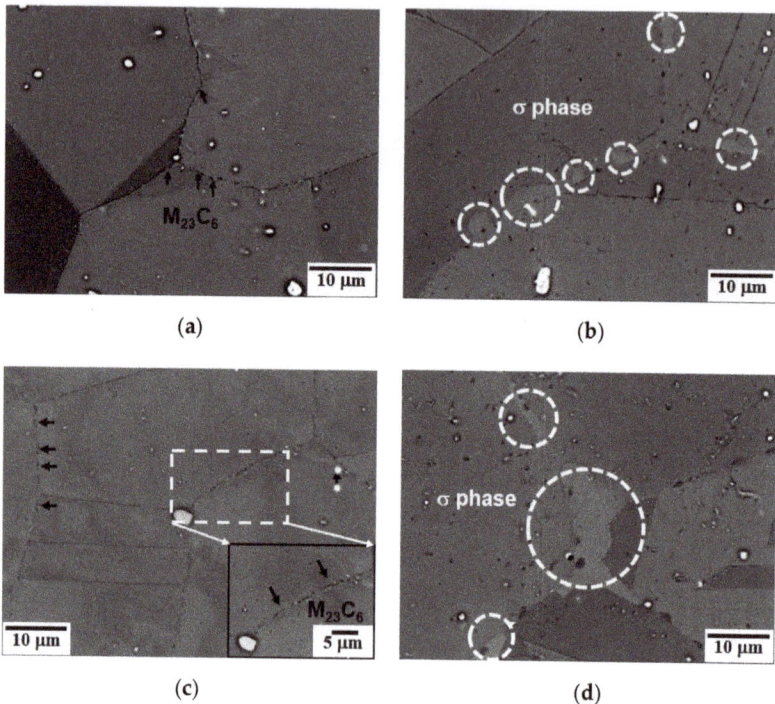

Figure 9. ECCI micrographs from cross-sectional fractures of creep specimens: (**a**) 700 °C/180 MPa/398 h, (**b**) 700 °C/90 MPa/9411 h, (**c**) 750 °C/150 MPa/222 h, and (**d**) 750 °C/70 MPa/4680 h.

The EBSD micrographs of 398 h and 9411 h crept specimens at 700 °C are shown in Figure 10. As shown in Figure 10a–c, there is no σ phase observed at 398 h. At 9411 h, the blocky σ phase formed along the high-angle grain boundaries, as presented in Figure 10d–f.

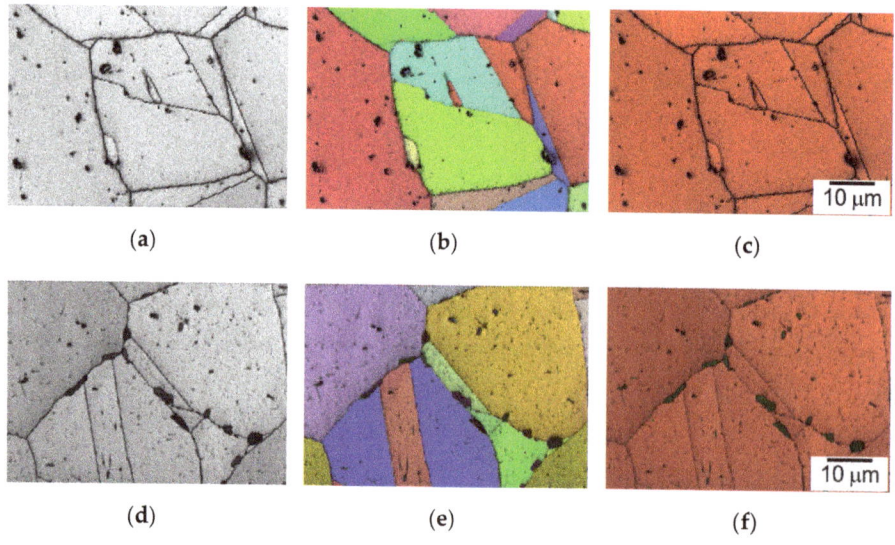

Figure 10. *Cont.*

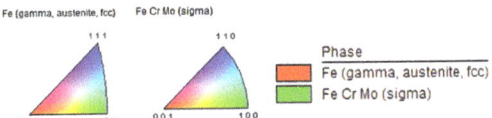

Figure 10. EBSD micrographs of 398 h crept specimen: (a) image quality, (b) inverse pole figure, and (c) phase map with grain boundary; and 9411 h crept specimen with σ phase: (d) image quality, (e) inverse pole figure, and (f) phase map with grain boundary.

Figure 11 shows the area fractions of Z-phase, $M_{23}C_6$, and σ phase at the two temperatures. Note that the fraction of Z-phase changes little with creep time and temperature compared to the $M_{23}C_6$ and σ phase due to its slower nucleation and growth rate [10]. In addition, the size of the fine Z-phase remains s at ~50 nm. In contrast, the fraction of $M_{23}C_6$ and σ phase grow linearly with creep time. The growth rate of $M_{23}C_6$ gradually decreases at both temperatures. This has two main reasons: (1) though the preferred $M_{23}C_6$ precipitate at the grain boundaries is easily coarsened with increasing creep time, the fine $M_{23}C_6$ carbides in the grain interior are stable; (2) the precipitation of σ phase consumes the Fe and Cr, presumably suppressing the formation of $M_{23}C_6$ carbides [11]. The early increase in microhardness results from the continuous precipitation of $M_{23}C_6$ in the early stage of the creep as shown in Figure 2b. The growth of the σ phase is linear at the grain boundaries (the first three data points are σ phase free). The growth rate of σ phase at 750 °C is almost twice that at 700 °C.

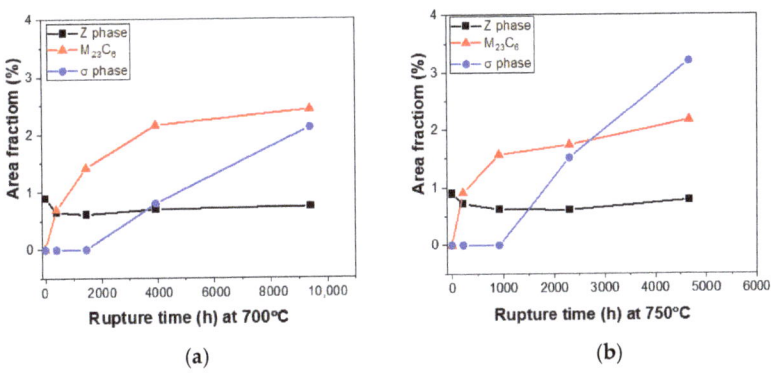

Figure 11. Precipitation evolution of Z-phase, $M_{23}C_6$, and σ phase at (a) 700 °C and (b) 750 °C.

In summary, the microstructural evolution of HR3C in this study can be characterized as follows: (1) before the creep test, the HR3C specimen had a Z-phase distributed both on the boundary and in the interior of the grains, whereas both the σ phase and $M_{23}C_6$ were not present; (2) the creep-ruptured specimens had high-density $M_{23}C_6$ precipitated mostly at the grain boundary from the early stage of the creep test, and grain interior $M_{23}C_6$ appeared at the later stage of the creep test; σ phases (up to 10 μm) were found mostly at the grain boundary, whereas the Z-phase appeared very stable and did not show much difference compared to the before test. The coarse σ phases were observed as fractured at the creep rupture surface.

4. Conclusions

Standard creep rupture strength tests were carried out for HR3C steel at 700 °C and 750 °C. The results show that the HR3C steel in this study has good creep performance. The major precipitates in this alloy, i.e., Z-phase, $M_{23}C_6$, and σ phase, are densely distributed on and along the grain boundaries, with some presence in the grain interior. The nucleation and growth of these phases have significant effects on the creep rupture behavior:

1. The dense Z-phase, including primary coarse Z-phase (~1 μm) and secondary fine Z-phase (~50 nm), dispersively precipitated in the matrix along the dislocation lines. Moreover, it showed high stability (both the size and the area fraction of ~0.8%) against coarsening with the extension of creep time;
2. The $M_{23}C_6$ preferentially precipitated at the grain boundaries and coarsened distinctly from ~200 nm to ~600 nm after a creep rupture of 9411 h. Meanwhile, granular $M_{23}C_6$ continuously precipitated in the matrix with the extension of creep rupture time and kept a relatively stable size of ~200–300 nm under long-term creep exposure;
3. The σ phase did not observe in the early stage of creep exposure till 700 °C/120 MPa/3945 h and 750 °C/120 MPa/2310 h. The fraction of the σ phase grew linearly with increasing time to rupture and the growth rate of σ phase at 750 °C was higher than at 700 °C;
4. All the crept HR3C specimens showed the intergranular brittle fracture under different stresses. As the time to creep rupture increased (low creep stress), partial dimple morphology formed on fracture facets by the void nucleation of $M_{23}C_6$. Coarse $M_{23}C_6$ and σ phase at the grain boundaries served as nucleation sites for high stress concentrations and creep cavities and led to the propagation of cracks on the grain boundaries and an eventual intergranular fracture;
5. Creep rupture mechanism and corrosion: the creep rupture specimens showed a typical intergranular fracture with small dimples caused by the decoupling of $M_{23}C_6$, indicating that the grain boundary was weakened due to the dense precipitation of this phase, even though the matrix was ductile enough to show dimples. The coarse and brittle σ phases do not play any significant role in strengthening. The depletion of Cr in the periphery of the grain boundary is expected to be a cause of corrosion.

Author Contributions: Conceptualization, Y.H. and L.X.; methodology, L.X. and Y.H.; validation, Y.H., J.-s.J. and K.S.; formal analysis, L.X.; investigation, L.X.; resources, Y.K. and Y.H.; data curation, L.X.; writing—original draft preparation, L.X.; writing—review and editing, K.S.; visualization, L.X.; supervision, J.-s.J. and K.S.; project administration, Y.K., Y.H. and J.-s.J.; funding acquisition, Y.K. and J.-s.J. All authors have read and agreed to the published version of the manuscript.

Funding: This research was funded by the Korea Institute of Energy Technology Evaluation and Planning (KETEP) and the Ministry of Trade, Industry, and Energy (MOTIE) of the Republic of Korea (no. 20217410100050).

Institutional Review Board Statement: Not applicable.

Informed Consent Statement: Not applicable.

Data Availability Statement: Not applicable.

Conflicts of Interest: The authors declare no conflict of interest.

References

1. Hu, Z.F. Heat-resistant steels, microstructure evolution and life assessment in power plants. *Therm. Power Plants* **2012**, *10*, 195–226.
2. Wang, J.Z.; Liu, Z.D.; Bao, H.S.; Cheng, S.C.; Bin, W. Effect of aging at 700 °C on microstructure and mechanical properties of S31042 heat resistant steel. *J. Iron Steel Res.* **2013**, *20*, 54–58. [CrossRef]
3. Zhou, Y.; Liu, Y.; Zhou, X.; Liu, C.; Yu, J.; Huang, Y.; Li, H.; Li, W. Precipitation and hot deformation behavior of austenitic heat-resistant steels: A review. *J. Mater. Sci. Technol.* **2017**, *33*, 1448–1456. [CrossRef]
4. Iseda, A.; Okada, H.; Semba, H.A.; Igarashi, M. Long term creep properties and microstructure of SUPER304H, TP347HFG, and HR3C for A-USC boilers. *Energy Mater.* **2007**, *2*, 199–206. [CrossRef]
5. Gharehbaghi, A. Precipitation Study in a High Temperature Austenitic Stainless Steel Using Low Voltage Energy dispersive X-ray Spectroscopy. Master's Thesis, Royal Institute of Technology (KTH), Stockholm, Sweden, 2012.
6. Jack, D.H.; Jack, K.H. Crystal Structure of Nb Cr N. *J. Iron Steel Inst.* **1972**, *209*, 790–792.
7. Golański, G.; Zieliński, A.; Sroka, M.; Słania, J. The effect of service on microstructure and mechanical properties of HR3C heat-resistant austenitic stainless steel. *Materials* **2020**, *13*, 1297. [CrossRef] [PubMed]
8. Bin, W.; Liu, Z.C.; Cheng, S.C.; Liu, C.M.; Wang, J.Z. Microstructure evolution and mechanical properties of HR3C steel during long-term aging at high temperature. *J. Iron Steel Res.* **2014**, *21*, 765–773.

9. Hong, H.U.; Nam, S.W. The occurrence of grain boundary serration and its effect on the M23C6 carbide characteristics in an AISI 316 stainless steel. *Mater. Sci. Eng. A* **2002**, *332*, 255–261. [CrossRef]
10. Zhang, Z.; Hu, Z.; Tu, H.; Schmauder, S.; Wu, G. Microstructure evolution in HR3C austenitic steel during long-term creep at 650 °C. *Mater. Sci. Eng. A* **2017**, *681*, 74–84. [CrossRef]
11. Lee, J.; Kim, I.; Kimura, A. Application of small punch test to evaluate sigma-phase embrittlement of pressure vessel cladding material. *J. Nucl. Sci. Technol.* **2003**, *40*, 664–671. [CrossRef]
12. Ji, Y.S.; Park, J.; Lee, S.Y.; Kim, J.W.; Lee, S.M.; Nam, J.; Shim, J.H. Long-term evolution of σ phase in 304H austenitic stainless steel: Experimental and computational investigation. *Mater. Charact.* **2017**, *128*, 23–29. [CrossRef]
13. Schwind, M.; Källqvist, J.; Nilsson, J.O.; Ågren, J.; Andrén, H.O. σ-phase precipitation in stabilized austenitic stainless steels. *Acta Mater.* **2000**, *48*, 2473–2481. [CrossRef]
14. Cao, T.S.; Cheng, C.Q.; Zhao, J.; Wang, H. Precipitation behavior of σ phase in ultra-supercritical boiler applied HR3C heat-resistant steel. *Acta Metall. Sin.* **2019**, *32*, 1355–1361. [CrossRef]
15. Sawada, K.; Kimura, K.; Abe, F. Data Sheets on the Elevated-Temperature Properties of 25Cr–20Ni–Nb–N Stainless Steel Tube for Power Boilers (KA-SUS310J1 TB). NIMS Creep Data Sheet No. 58: National Institute for Materials Science. 2011. Available online: https://mdr.nims.go.jp/concern/publications/v118rf25g (accessed on 28 June 2022).
16. Wang, R.; Duan, M.; Zhang, J.; Chen, G.; Miao, C.; Chen, X.; Li, J.; Tang, W. Microstructure Characteristics and Their Effects on the Mechanical Properties of As-Served HR3C Heat-Resistant Steel. *J. Mater. Eng. Perform.* **2021**, *30*, 4552–4561. [CrossRef]
17. Zieliński, A.; Sroka, M.; Hernas, A.; Kremzer, M. The effect of long-term impact of elevated temperature on changes in microstructure and mechanical properties of HR3C steel. *Arch. Metall. Mater.* **2016**, *61*, 761–765. [CrossRef]
18. Hu, Z.F.; Zhang, Z. Investigation the effect of precipitating characteristics on the creep behavior of HR3C austenitic steel at 650 °C. *Mater. Sci. Eng. A* **2019**, *742*, 451–463. [CrossRef]
19. Hong, H.U.; Rho, B.S.; Nam, S.W. Correlation of the $M_{23}C_6$ Precipitation Morphology with Grain Boundary Characteristics in Austenitic Stainless Steel. *Mater. Sci. Eng. A* **2001**, *318*, 285–292. [CrossRef]
20. Alsmadi, Z.Y.; Abouelella, H.; Alomari, A.S.; Murty, K.L. Stress-Controlled Creep–Fatigue of an Advanced Austenitic Stainless Steel at Elevated Temperatures. *Materials* **2022**, *15*, 3984. [CrossRef]
21. Zieliński, A.; Golański, G.; Sroka, M. Evolution of the microstructure and mechanical properties of HR3C austenitic stainless steel after ageing for up to 30,000 h at 650–750 °C. *Mater. Sci. Eng. A* **2020**, *796*, 139944. [CrossRef]
22. Vujic, S.; Standströ, R.; Sommitsch, C. Precipitation evolution and creep strength modeling of 25Cr20NiNbN austenitic steel. *Mater. High Temp.* **2015**, *32*, 607–618. [CrossRef]
23. Terada, M.; Escriba, D.M.; Costa, I.; Materna-Morris, E.; Padilha, A.F. Investigation on the intergranular corrosion resistance of the AISI 316L(N) stainless steel after long time creep testing. *Mater. Charact.* **2008**, *59*, 663–668. [CrossRef]
24. Kaneko, K.; Futunaga, T.; Yamada, K.; Nakada, N.; Kikuchi, M.; Saghi, Z.; Barnad, J.S.; Midgley, P.A. Formation of $M_{23}C_6$—Type precipitates and chromium–depleted zone in austenite stainless steel. *Scr. Mater.* **2011**, *65*, 509–512. [CrossRef]
25. Yan, J.; Gu, Y.; Sun, F.; Xu, Y.; Yuan, Y.; Lu, Y.; Yang, Z.; Dang, Y. Evolution of microstructure and mechanical properties of a 25Cr-20Ni heat resistant alloy after long-term. *Mater. Sci. Eng. A* **2016**, *675*, 289–298. [CrossRef]

Disclaimer/Publisher's Note: The statements, opinions and data contained in all publications are solely those of the individual author(s) and contributor(s) and not of MDPI and/or the editor(s). MDPI and/or the editor(s) disclaim responsibility for any injury to people or property resulting from any ideas, methods, instructions or products referred to in the content.

Materials Editorial Office
E-mail: materials@mdpi.com
www.mdpi.com/journal/materials

www.mdpi.com

MDPI
St. Alban-Anlage 66
4052 Basel
Switzerland

www.ingramcontent.com/pod-product-compliance
Lightning Source LLC
LaVergne TN
LVHW070723100526
838202LV00013B/1156